# Earth Systems Data and Models

Volume 6

The book series Earth Systems Data & Models publishes state-of-the-art research and technologies aimed at understanding processes and interactions in the earth system. A special emphasis is given to theory, methods, and tools used in earth, planetary and environmental sciences for: modeling, observation and analysis; data generation, assimilation and visualization; forecasting and simulation; and optimization. Topics in the series include but are not limited to: numerical, data-driven and agent-based modeling of the earth system; uncertainty analysis of models; geodynamic simulations, climate change, weather forecasting, hydroinformatics, and complex ecological models; model evaluation for decision-making processes and other earth science applications; and remote sensing and GIS technology. The series publishes monographs, edited volumes and selected conference proceedings addressing an interdisciplinary audience, which not only includes geologists, hydrologists, meteorologists, chemists, biologists and ecologists but also physicists, engineers and applied mathematicians, as well as policy makers who use model outputs as the basis of decision-making processes.

Feng Qian · Shengli Pan · Gulan Zhang

# Tensor Computation for Seismic Data Processing

## Linking Theory and Practice

Springer

Feng Qian
School of Information and Communication Engineering and Center for Information Geoscience
University of Electronic Science and Technology of China
Chengdu, China

Shengli Pan
School of Cyberspace Security
Beijing University of Posts and Telecommunications
Beijing, China

Gulan Zhang
State Key Laboratory of Oil and Gas Reservoir Geology and Exploitation and School of Geoscience and Technology
Southwest Petroleum University
Chengdu, China

ISSN 2364-5830 ISSN 2364-5849 (electronic)
Earth Systems Data and Models
ISBN 978-3-031-78899-4 ISBN 978-3-031-78900-7 (eBook)
https://doi.org/10.1007/978-3-031-78900-7



This Springer imprint is published by the registered company Springer Nature Switzerland AG
The registered company address is: Gewerbestrasse 11, 6330 Cham, Switzerland

*Dedicated to our family, friends, colleagues,*
*and to all who are interested in tensor*
*computation for seismic data processing.*

# Preface

The subject of seismic data processing invests a significant effort to offer mathematical formulas to quantify abstract physical concepts and understand behind the processing flows. High dimensionality is a ubiquitous feature of modern seismic data, both within post-stack and pre-stack seismic data. These well-constructed formulas are always presented in the form of vectors and matrices by classical seismic data analysis methods, where multidimensional seismic data has to be unfolded into vectors or matrices for quantification. However, such vectorization or matricization fails to exploit the multilinear geological structure in data and may contribute to suboptimal performance in seismic data processing. A tensor serves as a higher-order extension of vectors and matrices, often regarded as a multidimensional array. Tensors provide a natural way to represent multidimensional seismic data, and tensor computation can help prevent possible damage to the high-dimensional geological structure of the subsurface that occurred in classical seismic data analysis. The latest developments in tensor computation theory are prompting us to reinvent the way we move from classical matrix-based processing methods to tensor-based processing methods for seismic data.

This book provides a bridge between the theory of tensor computation and the practice of seismic data processing by presenting carefully detailed tensor solutions to selected problems in the data processing sequences and showing the prominent potential of the tensor approach. Includes well-known tensor decomposition devoted entirely to dealing with all practical processing problems, such as tensor singular value decomposition, tensor deep learning, tensor dictionary learning, etc. These tensor methods have been proven to be superior to the traditional matrix-based processing methods for high-dimensional seismic data, using synthetic and real field example tests. For each method, we provide a detailed theoretical mathematical deduction by a step-by-step analysis and described in detail the pros and cons, which run throughout all chapters of the book.

This book is mainly intended to be a reference for researchers and graduate students, as well as industrial engineers, who are interested in tensor computation applied to high-dimensional seismic data processing. This book has introductory parts on both seismic data processing and tensor computation and advanced parts on tensor-based seismic data processing methods and practical applications. New researchers should find it easy to stay up to date with the latest advancements in these specific techniques through the advanced sections, while the introductory sections will help them begin their own seismic data processing research.

We extend our heartfelt thanks to both current and former members of our team, as well as our wonderful collaborators, for making this publication possible. We also thank our editors, Qiao Shu, Hemavathy Manivannan, and Christian Witschel, for their advice and support. This work was partially funded by the National Natural Science Foundation of China under grant numbers 41874155 and 42130812.

Chengdu, China — Feng Qian
Beijing, China — Shengli Pan
Chengdu, China — Gulan Zhang
December 2024

# Contents

# Acronyms

| | |
|---|---|
| 3D-DnCNN | 3D denoising convolutional neural network |
| ADMM | Alternating direction method of multipliers |
| AE | Autoencoder |
| AltMin | Alternating minimization |
| AVO | Amplitude variation with offset |
| BP | Backpropagation |
| CNNs | Convolutional neural networks |
| CP | CANDECOMP/PARAFAC |
| CPD | Canonical polyadic decomposition |
| DA | Denoising autoencoder |
| DCAE | Deep convolutional auto-encoder |
| DcL | Dictionary learning |
| DCT | Discrete cosine transform |
| DL | Deep learning |
| DMAE | Deep matrix autoencoder |
| DMSSA | Damped-multichannel singular spectrum analysis |
| DnCNN | Denoising convolutional neural network |
| DNNs | Deep neural networks |
| DTAE | Deep tensor autoencoder |
| DTV | Directional total variation |
| FFT | Fast Fourier transform |
| FNN | Fully connected neural network |
| FPTNN | Framelet-based order-tensor neural network |
| FR-Net | Footprint removal network |
| Fxy | Fxy filtering |
| GANs | Generative adversarial networks |
| HOSVD | Higher-order singular value decomposition |
| HRSC | Hybrid rank-sparsity constraint |
| HT | Hierarchical tucker |
| JTDL | Joint tensor dictionary learning |
| LRA | Low-rank approximation |

| | |
|---|---|
| LRTA | Low-rank tensor approximation |
| LS | Least squares |
| MC | Matrix completion |
| M-D | Multidimensional |
| MDL | Matrix deep learning |
| MSE | Mean squared error |
| MSSA | Multichannel singular spectrum analysis |
| PMF | Parallel matrix factorization |
| ReLU | Rectified linear unit |
| RSE | Relative square error |
| RTDL | Robust tensor deep learning |
| RTPCA | Robust tensor principal component analysis |
| SGD | Stochastic gradient descent |
| SGK | Sequential generalized K-means |
| SGRDL | Statistics-guided residual dictionary learning |
| SNR | Signal-to-noise ratio |
| SOTA | State of the art |
| SR | Sparse representation |
| SSA | Singular spectrum analysis |
| SSDA | Stacked sparse DA |
| SSTF | SVD-free structured tensor factorization |
| STC | Structured tensor completion |
| STF | Structured tensor factorization |
| STWNN | Structured tensor weighted nuclear norm |
| SURE | Stein's unbiased risk estimate |
| SVD | Singular value decomposition |
| SWT | Stationary wavelet transform |
| TBP | Tensor backpropagation |
| TCNN | Tensor convolutional neural network |
| TDcL | Tensor dictionary learning |
| TDL | Tensor deep learning |
| TKD | Tucker decomposition |
| t-NN | Tensor neural network |
| TNN | Tensor nuclear norm |
| t-OSSVD | Tensor optimal shrinkage of singular value decomposition |
| t-product | Tensor-tensor product |
| t-SVD | Tensor singular value decomposition |
| TV | Total variation |
| UTV | Unidirectional total variation |
| VSP | Vertical seismic profile |

# Chapter 1
# Introduction

The main advantage of tensor computation is the ability to avoid within-data structure loss in flat-view matrix computation-based data analysis when a tensor is used as a rational representation for high-dimensional data. Tensor computation has witnessed tremendous applications in various science and engineering disciplines, from signal processing [1, 2] to computer vision [3, 4], data mining [5, 6], and seismic data processing covered in this book [7, 8], to mention a few. Despite active research on tensor computation over the past four decades, there has been no publication of a systematic review focused this work for seismic data processing in book form. Therefore, we decided to write this book to draw more attention of geophysical readers.

In the realm of geophysics, where tensor computation meets high dimensional seismic data, tensors serve as an advanced tool to represent the information hidden within data to capture higher-order similarities among vector-valued variables. By extending matrix-based methods, tensor computation methods are developed for estimating parameters of latent variable models such as low-rank approximation, e.g., [9, 10], and dictionary learning, e.g., [11]. Another application of tensor computation is constructing multilinear maps (i.e., functions) forwards and backward in deep learning called tensor deep learning, constituting higher-order generalizations of linear mappings represented by matrices. For details on these tensor computation models, please refer to Chap. 2 of this book.

In summary, the pursuit of high-dimensional structure preservation has motivated the development of tensor-based seismic data processing methods, resulting in more optimized tensor solutions compared to conventional matrix-based processing methods. In the following section, we begin by discussing how the high-dimensional traits of seismic data lead to limitations in traditional matrix-based processing methods, thus prompting the evolution of tensor-based processing methods that are applied in typical real-world processing scenarios.

 
F. Qian et al., *Tensor Computation for Seismic Data Processing*,
Earth Systems Data and Models 6, https://doi.org/10.1007/978-3-031-78900-7_1

## 1.1 Motivation

### 1.1.1 High Dimensional Modern Seismic Data

Over the past century, the primary goal of seismic reflection exploration has been to identify subsurface geological formations with the potential to contain oil and gas reserves [12]. Before the mid-1980s, most seismic surveys were conducted using single lines, which were interpreted in two dimensions (2D). These 2D survey lines were typically spaced several kilometers apart, making it challenging to accurately interpolate subsurface stratigraphic features. With advancements in computational power and seismic technologies, however, geophysicists have been able to gather seismic data along more closely spaced survey lines and integrate this data to create three-dimensional (3D) models of the subsurface geology [13, 14]. As a result, 3D seismic exploration has become the standard practice in modern industry. The raw data acquired through these methods can be organized using various parameters, leading to high-dimensional datasets, such as 3D post-stack, 4D pre-stack, and 5D wide-azimuth pre-stack data.

High-dimensional seismic data comprises both the desired signals, such as reflections and refractions, and unwanted noise, including footprint artifacts, erratic noise, and random noise. The primary reflections in raw seismic data are frequently contaminated by waves resulting from various noise sources, such as footprint artifacts, ground roll, air waves, and unidentifiable random noise. Therefore, a key challenge in geophysical signal processing is the effective separation of desired signals from these noise components. The accuracy of seismic data interpretation is heavily dependent on the quality of processed data, particularly with regard to its signal-to-noise ratio (SNR).

### 1.1.2 Conventional Seismic Data Processing

In conventional matrix-based processing approaches, high-dimensional seismic data is usually represented as seismic images (i.e., matrix-form data), where the relationships between pixels are analyzed using techniques such as spatial-domain filtering, sparse representation (SR), low-rank approximation (LRA), and deep learning (DL).

*The spatial-domain filtering*: This method, known as spatial-domain filtering, works by applying a filter mask to the image, where each pixel is modified by computing the sum of the products between the filter coefficients and the corresponding pixels covered by the mask. Examples of spatial-domain filtering techniques include structure-oriented filtering [15, 16], median filtering [17], mean filtering [18], and morphological filtering [19].

*The sparse representation methods*: Sparse representation (SR) has emerged as a key approach for denoising seismic images, which can be divided into two main categories: analytical sparsifying transforms and data-driven sparse coding (SC). In analytical sparsifying transforms, the assumption is that clean seismic images can be represented sparsely in a predefined transform domain, while the noise cannot. Commonly employed transforms include the Hilbert–Huang transform [20], wavelet transform [21, 22], curvelet transform [23, 24], shearlet transform [25, 26], framelet transform [27–29], seislet transform [30], and Radon transform [31]. However, relying solely on a single analytical sparsifying transform often restricts denoising efficiency, as no single transform can adequately capture all the image features. On the other hand, sparse coding (SC) methods allow for learning a sparse representation of seismic data by combining basis elements known as a dictionary, which is better adapted to seismic images. Examples of such techniques include fast [32, 33], statistics-based [34–36], constrained [37, 38], adaptive [39], multitask [40, 41], and double-sparsity [42] dictionary learning (DcL) approaches.

*The low-rank approximation methods*: In a strict sense, sparse representation models act as efficient but straightforward local image priors [43]. In contrast, low-rank approximation (LRA) models offer a global image prior by revealing the underlying global structure present in seismic data. Matrix-based low-rank approximation (LRA) relies on the assumption that M-dimensional seismic data can be rearranged into a matrix with a low rank. When a seismic image is affected by noise, the rank of the corresponding matrix tends to increase. Thus, denoising can be accomplished by reducing the matrix rank using rank-reduction techniques. These methods include both general unstructured LRA approaches [44–47] and structured Hankel matrix-based LRA techniques [48–51].

*The deep-learning methods*: Instead of relying on manually designed image priors, deep prior learning focuses on directly learning a compact mapping function from noisy observations to the underlying clean seismic images, without making explicit assumptions about the learning process. Heavily influenced by traditional deep image denoising techniques, deep learning (DL)-based methods treat M-dimensional seismic data as matrices or images and utilize various DL models for denoising. These models include convolutional neural networks (CNNs) [52–55], generative adversarial networks (GANs) [56], autoencoders (AEs) [57, 58], and U-Nets [59].

The main limitation of matrix-based methods lies in their ability to capture only two-dimensional information, which is inadequate for comprehensively exploring the intricate, high-dimensional relationships inherent in seismic data.

### 1.1.3 Tensor-Based Seismic Data Processing

As an alternative to constrained matrix-based methods, tensor-based approaches model M-D seismic data processing tasks as problems of recovering a global low-rank tensor structure from incomplete or noisy data. Currently, there are three main types of tensor decomposition techniques used to identify this low-rank structure: canonical polyadic decomposition (CPD), Tucker decomposition (TKD), and tensor singular value decomposition (t-SVD). The CPD method poses significant challenges when completing tensors with missing data, as it requires specialized mathematical solutions (for more details, refer to [60]). This limitation makes t-SVD and TKD models more viable for seismic data interpolation. t-SVD is commonly employed to extract low-tubal-rank structures in several rank-reduction methods, such as multichannel singular spectrum analysis (MSSA, also known as Cadzow filtering) [61–63], damped MSSA (DMSSA) [64–67], structured tensor completion [10], and tensor nuclear norm (TNN) [68]. TKD is used to decompose seismic data tensors into a core tensor and factor matrices in various decomposition models, such as parallel matrix factorization (PMF) [7, 69], higher-order SVD (HOSVD) [9, 70], and hierarchical Tucker (HT) [71].

## 1.2 Practical Application

### *1.2.1 Seismic Data Interpolation*

Seismic data interpolation is crucial for enabling a wide range of subsequent applications involving extensive processing, such as multichannel deconvolution [72], the attenuation of surface-related multiples [20, 73], simultaneous-source separation [74–76], and amplitude variation with offset analysis [77, 78]. Therefore, a variety of interpolation methods, for instance, compressed sensing [28, 79–86] and low-rank recovery [87–90], have been used to achieve promising results. Nevertheless, a summary of these methods indicates that the essential challenge continuing to face seismic data interpolation is how to relate unknown traces to known traces in 3-D physical space. In this book, we proposed a structured low-tubal-rank tensor completion and a DTAE model for capturing the high-dimensional mapping to discover the complicated relationship among seismic traces; details are found in Chaps. 3 and 5.

*Structured low-tubal-rank tensor completion*: We propose a structured low-tubal-rank tensor completion approach using random tubal sampling to effectively integrate the Hankel structure constraint, the low-tubal-rank condition, and the tubal sampling assumption. The benefits of our completion model can be summarized as follows:

(1) *More accurate*: By combining both low-tubal-rank and Hankel structure constraints, the Hankel structure is preserved throughout each iteration of the low-tubal-rank approximation process. This approach inherently leads to more accurate results compared to earlier methods.
(2) *Easier to solve*: By applying Hankel constraints to each frontal slice as a linear matrix equation related to the vectorized slice matrix, the tubal-sampling assumption enables us to effectively determine low-tubal-rank solutions for structured tensor completion (STC) through matrix completion (MC) for each slice. In contrast, element-wise sampling poses challenges in solving these constraints, as Hankel constraints cannot be expressed as linear tensor equations.

(3) *More adaptable*: To accommodate different noise environments, we can create a noise-specific STC model using almost any matrix completion (MC) technique, without being limited by the constraints of tensor algebra [2].

*Deep tensor autoencoder (DTAE)*: Building on the tensor neural network (t-NN) framework [91, 92], we propose a deep tensor autoencoder (DTAE) model, which is designed to establish corrected relationships among seismic traces, even in complex geological structures. Unlike conventional autoencoders (AEs) [93, 94] that utilize matrix products, the DTAE employs the tensor-tensor product (t-product) [95] in its network architecture. This distinction allows the DTAE to offer several advantages in capturing relationships among correlated seismic traces:

(1) *Data-driven*: Our interpolation method is fully data-driven; as a supervised learning model, DTAE operates purely based on data, being trained on a large volume of seismic data. Since it does not depend on any prior assumptions, such as the low-rank assumption, our approach is expected to produce more accurate results compared to tensor decomposition techniques [62, 71].
(2) *High-dimensional*: By utilizing the t-product operator, DTAE can reveal high-dimensional relationships among seismic traces in 3-D space that may be overlooked by traditional matrix-based deep learning (DL) methods [96, 97].
(3) *Nonlinear*: As a representative deep learning method, the DTAE is more capable of capturing nonlinear relationships compared to dictionary learning, as shown by recent research [98]. Although dictionary learning has shown good practical performance, it is restricted by its shallow linear structure [29, 99].

### *1.2.2 Seismic Footprint Removal*

Acquisition footprints are well-known as periodic amplitude stripes that are artifacts caused by the acquisition geometry rather than representing actual geological features. These footprints frequently appear on time or horizon slices of 3-D seismic data [100]. Several factors contribute to footprint noise during both acquisition and processing, with two significant influences being the design of the acquisition and the emphasis on specific features during processing [101, 102]. The presence of footprint noise can obscure seismic images, making it difficult to accurately identify and interpret the true subsurface response, thereby increasing uncertainty in seismic image analysis [103, 104]. As a result, removing footprints is a vital step in seismic data processing [105, 106] and is one of the key focuses of this book.

In this book, we present an innovative tensor model for reducing 3-D seismic random and footprint noise, based on unidirectional total variation regularized low-rank tensor approximation (UTV-LRTA). The advantages of our UTV-LRTA model are as follows:

(1) *Excellent mixed noise removal capability*: By integrating low-rank tensor approximation with unidirectional total variation (UTV) constraints, our UTV-LRTA model effectively employs LRTA to minimize random noise, while utilizing UTV regularization to effectively eliminate footprint artifacts.
(2) *Three-dimensional low-rank structure*: By leveraging the t-product operator, the UTV-LRTA model is capable of identifying the three-dimensional low-rank structure concealed within data contaminated by footprint artifacts. This underlying structure often goes unnoticed in traditional matrix-based methods for footprint removal [35, 107].
(3) *Low computational complexity*: The split Bregman iteration algorithms are known for their rapid convergence, requiring fewer iterations to reach results [108–110]. Consequently, our UTV-LRTA model exhibits significantly lower computational complexity compared to current state-of-the-art (SOTA) removal techniques, including those based on deep learning and low-rank approximation.

### *1.2.3 Seismic Noise Attenuation*

Suppressing seismic random noise is essential for enabling various subsequent processes, such as attribute analysis [111, 112], seismic inversion [113, 114], and oil and gas exploration [115, 116]. In the field of applied geophysics, a variety of denoising methods have been developed, yielding promising results. Key approaches include low-rank approximation (LRA) [44, 46–51] and deep learning (DL) techniques [52–55, 57, 117–123]. One major challenge in these methods is effectively distinguishing the well-defined structures in true signals from random noise. To address these challenges, this book proposes a novel denoising framework based on Stein's unbiased risk estimate (SURE), combined with a tensor convolutional neural network (TCNN) and an order-$p$ tensor neural network (tNN) framework utilizing framelet transform, known as FPTNN. These frameworks are designed to capture the complicated image structures hidden within clean 3-D and multi-dimensional seismic images, respectively.

*SURE-based tensor CNN (SURE-TCNN) denoising method*: By combining the tensor neural network (t-NN) framework [91, 92] with Stein's unbiased risk estimate (SURE) [124, 125], we propose a SURE-based tensor convolutional neural network (SURE-TCNN) for denoising 3-D seismic data. This method is trained entirely on noisy data samples and does not require paired noisy and ground truth data. The advantages of our denoising approach are as follows.

(1) *Ground truth-free*: The ground truth-free denoising solution is achieved by using the Stein's unbiased risk estimate (SURE) cost function instead of the mean squared error (MSE) cost function. Essentially, the SURE function can be calculated directly from noisy 3-D seismic images, whereas the MSE requires access to ground truth images, which is typically used in supervised learning.

(2) *Data-driven*: SURE-TCNN enables a comprehensive mapping from a noisy seismic image to a cleaner version without requiring prior information. Since it does not rely on specific assumptions, such as the low-rank assumption, SURE-TCNN is expected to achieve better noise reduction performance compared to non-deep learning (DL) methods [21, 51].
(3) *High-dimensional capability*: By leveraging the tensor-tensor product (t-product) operator, the SURE-based tensor CNN (SURE-TCNN) can uncover high-dimensional image structures hidden within noisy 3-D seismic images, which would typically remain inactive in traditional matrix-based deep learning (DL) methods [119, 123].
(4) *Exceptional noise reduction capability*: As Stein's unbiased risk estimate (SURE) provides an accurate unbiased estimate of the mean squared error (MSE), training the SURE-based tensor CNN (SURE-TCNN) using the SURE loss function can achieve results comparable to those obtained by using the MSE with ground truth images in a supervised learning setting [124, 125].

*The framelet-based order-$p$ tNN (FPTNN) model*: To effectively capture the prior knowledge of the underlying M-D image structure, it is important to combine image priors from various models, such as sparse representation models, tensor low-rank models, and deep learning (DL) models. Specifically, we propose using both the framelet transform [126, 127] and the order-$p$ t-product [128] within the tensor neural network (tNN) framework [91, 92], resulting in the development of a framelet-based order-$p$ tNN (FPTNN) model. This model demonstrates promising capabilities for denoising M-D seismic data through training. The benefits of our FPTNN model include:

(1) *Suited for high dimensions*: By utilizing the order-$p$ t-product, the FPTNN is designed to establish a true M-D inference function that links noisy and clean seismic images without the need for flattening or vectorization operations. This approach is particularly beneficial as it avoids the limitations of conventional 2-D or 3-D deep learning (DL) denoising methods.
(2) *Superior denoising quality*: Since the framelet basis is well-established in the time-frequency domain and incorporated into a specifically designed deep learning (DL) denoising model, the FPTNN is expected to provide enhanced noise reduction performance compared to non-DL methods [20, 129].
(3) *Simple to implement*: As the order-$p$ t-product is converted into matrix multiplication in the framelet domain (as outlined in Definition 7.2), the FPTNN can efficiently derive a solution by applying deep learning (DL) to the set of transformed matrix frontal slices.
(4) *An open framework*: Our FPTNN provides a versatile framework that integrates matrix deep learning (DL) and linear time-frequency transformations with tensor deep learning (TDL). By utilizing the framelet-based order-$p$ t-product, the framelet transform can be replaced with any linear time-frequency transformation. Moreover, as mentioned in point 3), FPTNN can be easily combined with any matrix DL model, whether supervised or unsupervised, although this paper primarily focuses on an unsupervised approach.

### 1.2.4 Seismic Erratic Noise Suppression

Noise is an inevitable aspect of seismic field acquisition and can significantly degrade the quality of the obtained seismic images. Removing noise from these images is a critical step in seismic data processing and is essential for various interpretation tasks, such as seismic inversion [113, 130], facies recognition [131, 132], and attribute analysis [133]. Typically, noise is categorized into two primary types: coherent noise and incoherent noise, depending on its predictability [134]. Coherent noise has a consistent phase across multiple traces, such as ground roll and multiples. In contrast, incoherent noise varies across traces, including Gaussian noise and erratic noise [135].

By integrating a sparse tensor norm [136] with a tensor neural network (tNN) framework [91, 92, 137, 138], we introduce a robust tensor deep learning (RTDL) approach designed for unsupervised attenuation of 3-D seismic erratic noise. This method leverages the strengths of both strategies, and the key features of our RTDL technique are outlined below:

(1) *Data-Driven Modeling*: In the RTDL framework, the data-driven tensor deep learning (TDL) model offers an effective means to represent clear 3-D seismic data, especially in situations where the low-rank assumption may not hold true. Conversely, the assumption regarding erratic noise in three-dimensional space, which can be characterized by tensor sparsity norms, is consistently valid. Consequently, our RTDL methodology fully leverages the strengths of both model-driven and data-driven approaches to enhance robustness and generalizability throughout the denoising process.
(2) *High Dimensional*: In our RTDL method, the tensor-tensor product (t-product) operator is added to a deep tNN that can capture complete 3-D image structures without flattening operations. These structures typically remain undiscovered when employing conventional image or matrix-based denoising techniques [135, 139, 140].
(3) *Ground-Truth-Free Denoising*: The approach to ground-truth-free denoising is based on the integration of unsupervised tensor deep learning (TDL) with the tensor sparse norm. Consequently, the noise reduction process is performed in an unsupervised manner, removing the requirement for ground-truth field data, similar to the method described in [58].

### 1.2.5 Seismic Data Super-Resolution

Seismic data are generally sparsely sampled along spatial dimensions, resulting in insufficient lateral resolution in seismic recordings [141]. Seismic trace interpolation is an essential task to achieve high-resolution imaging of seismic data. A variety of traditional methods have been proposed for this purpose. For example, Spitz [141] introduced an interpolation technique using predictive filtering in the $f-x$ domain.

Claerbout [142] investigated prediction error filtering in the $t - x$ domain, enabling data reconstruction, albeit with possible false frequencies. Sacchi et al. [143] implemented a five-dimensional interpolation method within an inversion-based framework. More recently, Jia and Ma [144] proposed a machine learning approach for seismic data interpolation. Despite these advances, many of these methods often fail to capture the structural features inherent in three-dimensional (3D) seismic data.

Super-resolution (SR) is a well-recognized problem in image signal processing, and numerous effective dictionary learning-based methods have been developed to generate high-resolution data [145–147]. More recently, Ledig et al. [148] introduced the application of generative adversarial networks (GAN) for image super-resolution. Given that formation parameters exhibit lateral continuity and seismic data across different traces exhibit similarities, it is possible to derive a sparse dictionary from seismic data [149], suggesting that seismic data can be sparsely represented [150]. Moreover, compressed sensing theory has been applied to seismic data [80]. However, many existing SR methods transform data into vectors, which neglects the spatial correlations in multi-dimensional datasets [150]. These spatial correlations contain critical geophysical information that is instrumental in achieving higher resolution.

To effectively capture the spatial correlations present in multidimensional data, we propose a seismic trace interpolation method based on tensor sparse coding (TSC) [151], specifically developed for 3D seismic data. This approach involves training two dictionaries: one for high resolution and the other for low resolution. Initially, tensor sparse coding is applied to the seismic data. Subsequently, we recognize that structural features are similarly represented in both the low-resolution and high-resolution seismic data. We assert that these data pairs share the same sparse representation, enabling us to use the sparse coefficients obtained from the low-resolution data along with the high-resolution dictionary to reconstruct the high-resolution data. Finally, experimental results using real field seismic data show that our proposed method significantly improves the resolution of seismic imaging.

## 1.3 Coverage of This Book

The book consists of 10 chapters and 1 appendix, with the first part comprising 2 chapters that cover the fundamentals of seismic data processing and tensor computation.

- This chapter provides an overview of the motivation behind the study and discusses various practical applications related to seismic data processing.
- Chapter 2 presents a comprehensive overview of various tensor computation methods, including tensor decomposition, tensor completion, tensor deep learning, and tensor dictionary learning.

Part II focuses on the technical elements of tensor-based seismic data analysis methods, covering both theoretical foundations and practical applications, and comprises eight chapters.

- Chapter 3 presents a structured low-tubal-rank tensor completion framework that efficiently exploits the combination of the Hankel structure and the low-tubal-rank constraint to enhance performance [10].
- Chapter 4 presents an innovative tensor model designed for suppressing 3-D seismic noise, targeting both random noise and footprint noise. This model is based on unidirectional total variation (UTV) regularized low-rank tensor approximation (UTV-LRTA) [152].
- Chapter 5 presents a basic deep tensor autoencoder (DTAE) along with two variations, intended to implicitly learn a data-driven, nonlinear, high-dimensional mapping. This approach seeks to reveal the complex relationships among seismic traces without imposing any underlying assumptions [8].
- Chapter 6 presents a data denoising framework called SURE-TCNN, which is based on a tensor convolutional neural network (TCNN) utilizing Stein's unbiased risk estimate (SURE). This method aims to learn the intrinsic high-dimensional structures of seismic data without requiring ground truth information [153].
- Chapter 7 presents the framelet-based order-tensor neural network model called FPTNN. This model aims to implicitly learn priors that capture the typical features of clear multidimensional seismic images in a data-driven manner [154].
- Chapter 8 presents a robust tensor deep learning (RTDL) denoising approach aimed at unsupervised suppression of 3-D seismic erratic noise. This method utilizes a strategic combination of tensor sparse representation (SR) and a tensor neural network (tNN) to achieve its objectives [155].
- Chapter 9 presents a novel technique for generating high-resolution three-dimensional (3D) seismic data from low-resolution data using tensor sparse coding (TSC). This approach involves simultaneously training both a high-resolution dictionary and a low-resolution dictionary [11].
- Chapter 10 encapsulates the key concepts and findings discussed throughout this monograph while also emphasizing potential avenues for future research.

At the conclusion of this book, the appendix provides a summary of the available software for tensor computation, which can be accessed online along with the corresponding user manual.

### Suggestions for Further Reading

Our book complements other modern monographs, such as those of [156–162], each of these publications focuses on distinct facets of tensor computation or seismic data processing. Thus, a curated selection of references is provided for further reading.

(1) Nanda NC (2021) Seismic data interpretation and evaluation for hydrocarbon exploration and production. Springer Nature Switzerland AG: Springer International Publishing.
(2) Liu Y (ed) (2021) Tensors for data processing: theory, methods, and applications. Academic Press.
(3) Liu Y, Liu J, Long Z et al (2022) Tensor computation for data analysis. Springer, Berlin.

(4) Zhou HW (2014) Practical seismic data analysis. Cambridge University Press.
(5) Al-Shuhail A, Al-Dossary S (2020) Attenuation of incoherent seismic noise. Springer.
(6) Tiwari RK, Rekapalli R (2020) Modern singular spectral-based denoising and filtering techniques for 2D and 3D reflection seismic data. Springer Nature.
(7) Yilmaz Ö (2001) Seismic data analysis: processing, inversion, and interpretation of seismic data. Society of exploration geophysicists.

## References


1. De Lathauwer L, Vandewalle J (2004) Dimensionality reduction in higher-order signal processing and rank-(r1, r2,..., rn) reduction in multilinear algebra. Linear Algebra Its Appl 391:31–55
2. De Lathauwer L, De Moor B (1998) From matrix to tensor: multilinear algebra and signal processing. In: Institute of mathematics and its applications conference series, vol 67, pp 1–16. Citeseer
3. Panagakis Y, Kossaifi J, Chrysos GG, Oldfield J, Nicolaou MA, Anandkumar A, Zafeiriou S (2021) Tensor methods in computer vision and deep learning. Proce IEEE 109(5):863–890
4. Shashua A, Hazan T (2005) Non-negative tensor factorization with applications to statistics and computer vision. In: Proceedings of the 22nd international conference on machine learning, pp 792–799
5. Papalexakis EE, Faloutsos C, Sidiropoulos ND (2016) Tensors for data mining and data fusion: models, applications, and scalable algorithms. ACM Trans Intell Syst Technol (TIST) 8(2):1–44
6. Morten M (2011) Applications of tensor (multiway array) factorizations and decompositions in data mining. Wiley Interdisc Rev: Data Min Knowl Disc 1(1):24–40
7. Gao J, Cheng J, Sacchi MD (2016) Five-dimensional seismic reconstruction using parallel square matrix factorization. IEEE Trans Geosci Remote Sens 55(4):2124–2135
8. Feng Q, Zhangbo L, Yan W, Songjie L, Shengli P, Hu G (2021) DTAE: deep tensor autoencoder for 3-d seismic data interpolation. IEEE Trans Geosci Remote Sens 60:1–19
9. Kreimer N, Stanton A, Sacchi MD (2013) Tensor completion based on nuclear norm minimization for 5d seismic data reconstruction. Geophysics 78(6):V273–V284
10. Feng Q, Cangcang Z, Lingtian F, Lu C, Gulan Z, Hu G (2020) Tubal-sampling: bridging tensor and matrix completion in 3-d seismic data reconstruction. IEEE Trans Geosci Remote Sens 59(1):854–870
11. Liao S, Liu X-Y, Qian F, Yin M, Hu G-M (2019) Tensor super-resolution for seismic data. In: ICASSP 2019—2019 IEEE international conference on acoustics, speech and signal processing (ICASSP). IEEE, pp 8598–8602
12. Sheriff RE, Geldart LP (1995) Exploration seismology. Cambridge University Press
13. Raynolds MK, Jorgenson JC, Torre Jorgenson M, Kanevskiy M, Liljedahl AK, Nolan M, Sturm M, Walker DA (2020) Landscape impacts of 3d-seismic surveys in the arctic national wildlife refuge. Alaska. Ecol Appl 30(7):e02143
14. Liner CL (2016) Elements of 3D seismology. Society of exploration geophysicists
15. Fehmers GC, Höcker CFW (2003) Fast structural interpretation with structure-oriented filtering. Geophysics 68(4):1286–1293
16. Qingbao Z, Jinghuai G, Zhiguo W, Kexue L (2015) Adaptive variable time fractional anisotropic diffusion filtering for seismic data noise attenuation. IEEE Trans Geosci Remote Sens 54(4):1905–1917

17. Shuwei G, Shoudong W, Yangkang C, Xiaohong C, Kui X (2016) Separation of simultaneous sources using a structural-oriented median filter in the flattened dimension. Comput Geosci 86:46–54
18. David B, Mauricio S (2012) Denoising seismic data using the nonlocal means algorithm. Geophysics 77(1):A5–A8
19. Huang W, Wang R, Zu S, Chen Y (2017) Low-frequency noise attenuation in seismic and microseismic data using mathematical morphological filtering. Geophys J Int 222(3):1728–1749
20. Wei C, Jianyong X, Zu S, Shuwei G, Yangkang C (2016) Multiple-reflection noise attenuation using adaptive randomized-order empirical mode decomposition. IEEE Geosci Remote Sens Lett 14(1):18–22
21. Jian M, Jinghuai G, Wenchao C (2006) On the denoising method of prestack seismic data in wavelet domain. In: 2006 SEG annual meeting. OnePetro
22. Hao S, Jianwei M, Huizhu Y (2009) Comparisons of wavelets, contourlets and curvelets in seismic denoising. J Appl Geophys 69(2):103–115
23. Li J-H, Zhang Y-J, Qi R, Liu QH (2017) Wavelet-based higher order correlative stacking for seismic data denoising in the curvelet domain. IEEE J Sel Top Appl Earth Obs Remote Sens 10(8):3810–3820
24. Andrzej G, Anna A, Michał M (2014) Application of curvelet denoising to 2d and 3d seismic data-practical considerations. J Appl Geophys 105:78–94
25. Dehui K, Zhenming P (2015) Seismic random noise attenuation using shearlet and total generalized variation. J Geophys Eng 12(6):1024–1035
26. Liu C, Wang D, Wang T, Feng F, Cheng H, Meng G (2014) Random seismic noise attenuation based on the shearlet transform. Acta Petrolei Sinica 35(4):692–699
27. Yang P, Gao J, Chen W (2012) Noise attenuation of the geophysical data using the framelet transform. In: 2012 IEEE international geoscience and remote sensing symposium. IEEE, pp 2300–2303
28. Weilin H, Wu R-S, Runqiu W (2018) Damped dreamlet representation for exploration seismic data interpolation and denoising. IEEE Trans Geosci Remote Sens 56(6):3159–3172
29. Jingwei L, Jianwei M, Xiaoqun Z (2014) Seismic data restoration via data-driven tight frame. Geophysics 79(3):V65–V74
30. Liu Y, Fomel S, Liu C (2015) Signal and noise separation in prestack seismic data using velocity-dependent seislet transform. Geophysics 80(6):WD117–WD128
31. Li J, Sacchi MD (2021) An LP-space matching pursuit algorithm and its application to robust seismic data denoising via time-domain radon transforms. Geophysics 86(3):V171–V183
32. Wang H, Chen W, Zhang Q, Liu X, Zu S, Chen Y (2020) Fast dictionary learning for high-dimensional seismic reconstruction. IEEE Trans Geosci Remote Sens 59(8)
33. Yangkang C (2017) Fast dictionary learning for noise attenuation of multidimensional seismic data. Geophys J Int 209(1):21–31
34. Zhou Y, Yang J, Wang H, Huang G, Chen Y (2020) Statistics-guided dictionary learning for automatic coherent noise suppression. IEEE Trans Geosci Remote Sens 60
35. Chen W, Saad OM, Wang H, Chen Y (2021) Statistics-guided residual dictionary learning for footprint noise removal. IEEE Trans Geosci Remote Sens 60:1–11
36. Chen W, Saad OM, Innocent Oboué YAS, Yang L, Chen Y (2021) Retrieving the leaked signals from noise using a fast dictionary learning method. Geophysics 87(1):1–86
37. Lina L, Jianwei M (2018) Structured graph dictionary learning and application on the seismic denoising. IEEE Trans Geosci Remote Sens 57(4):1883–1893
38. Turquais P, Asgedom EG, Söllner W (2017) A method of combining coherence-constrained sparse coding and dictionary learning for denoising. Geophysics 82(3):V137–V148
39. Xiaojing W, Jianwei M (2019) Adaptive dictionary learning for blind seismic data denoising. IEEE Geosci Remote Sens Lett 17(7):1273–1277
40. Siahsar MAN, Gholtashi S, Roshandel Kahoo A, Chen W, Chen Y (2017) Data-driven multitask sparse dictionary learning for noise attenuation of 3d seismic data. Geophysics 82(6):V385–V396

41. Zhu L, Liu E, McClellan JH (2015) Seismic data denoising through multiscale and sparsity-promoting dictionary learning. Geophysics 80(6):WD45–WD57
42. Yangkang C, Jianwei M, Sergey F (2016) Double-sparsity dictionary for seismic noise attenuation. Geophysics 81(2):V103–V116
43. Lei Z, Wangmeng Z (2017) Image restoration: from sparse and low-rank priors to deep priors [lecture notes]. IEEE Signal Process Mag 34(5):172–179
44. Anvari R, Siahsar MAN, Gholtashi S, Kahoo AR, Mohammadi M (2017) Seismic random noise attenuation using synchrosqueezed wavelet transform and low-rank signal matrix approximation. IEEE Trans Geosci Remote Sens 55(11):6574–6581
45. Yuxing Z, Yue L, Xintong D, Baojun Y (2018) Low-frequency noise suppression method based on improved DNCNN in desert seismic data. IEEE Geosci Remote Sens Lett 16(5):811–815
46. Xingye L, Xiaohong C, Jingye L, Yangkang C (2020) Nonlocal weighted robust principal component analysis for seismic noise attenuation. IEEE Trans Geosci Remote Sens 59(2):1745–1756
47. Ouyang S, Lingling W, Hu X, Zhidan L (2021) Seismic denoising via truncated nuclear norm minimization. Geophysics 86(2):V153–V169
48. Chong W, Zhihui Z, Gu H, Wu X, Shuaiqi L (2018) Hankel low-rank approximation for seismic noise attenuation. IEEE Trans Geosci Remote Sens 57(1):561–573
49. Chong W, Zhihui Z, Gu H (2020) Low-rank seismic denoising with optimal rank selection for Hankel matrices. Geophys Prospect 68(3):892–909
50. Yushu Z, Hongbo L, Yue L, Haitao M (2019) Seismic signal enhancement and noise suppression using structure-adaptive nonlinear complex diffusion. IEEE Trans Geosci Remote Sens 58(3):2198–2211
51. Yangkang C, Yatong Z, Wei C, Zu S, Weilin H, Dong Z (2017) Empirical low-rank approximation for seismic noise attenuation. IEEE Trans Geosci Remote Sens 55(8):4696–4711
52. Siwei Yu, Jianwei M, Wenlong W (2019) Deep learning for denoising. Geophysics 84(6):V333–V350
53. Wang E, Nealon J (2019) Applying machine learning to 3d seismic image denoising and enhancement. Interpretation 7(3):SE131–SE139
54. Wenda L, Hong L, Jian W (2021) A deep learning method for denoising based on a fast and flexible convolutional neural network. IEEE Trans Geosci Remote Sens 60:1–13
55. Liuqing Y, Wei C, Hang W, Yangkang C (2021) Deep learning seismic random noise attenuation via improved residual convolutional neural network. IEEE Trans Geosci Remote Sens 59(9):7968–7981
56. Yuan Y, Si X, Zheng Y (2020) Ground-roll attenuation using generative adversarial networksground-roll attenuation using gans. Geophysics 85(4):WA255–WA267
57. Saad OM, Chen Y (2020) Deep denoising autoencoder for seismic random noise attenuation. Geophysics 85(4):V367–V376
58. Feng Q, Wei G, Zhangbo L, Yu H, Gulan Z, Hu G (2022) Unsupervised erratic seismic noise attenuation with robust deep convolutional autoencoders. IEEE Trans Geosci Remote Sens 60:1–16
59. Sun J, Slang S, Elboth T, Greiner TL, McDonald S, Gelius L-J (2020) Attenuation of marine seismic interference noise employing a customized u-net. Geophys Prospect 68(3):845–871
60. Mikael S, De Lathauwer L (2019) Fiber sampling approach to canonical polyadic decomposition and application to tensor completion. SIAM J Matrix Anal Appl 40(3):888–917
61. Trickett S (2008) F-xy cadzow noise suppression. In: SEG technical program expanded abstracts 2008. Society of Exploration Geophysicists, pp 2586–2590
62. Vicente O, Mauricio S (2011) Simultaneous seismic data denoising and reconstruction via multichannel singular spectrum analysis. Geophysics 76(3):V25–V32
63. Dong Z, Yatong Z, Hanming C, Wei C, Zu S, Yangkang C (2017) Hybrid rank-sparsity constraint model for simultaneous reconstruction and denoising of 3d seismic data. Geophysics 82(5):V351–V367
64. Weilin H, Runqiu W, Yangkang C, Huijian L, Shuwei G (2016) Damped multichannel singular spectrum analysis for 3d random noise attenuation. Geophysics 81(4):V261–V270

65. Siahsar MAN, Gholtashi S, Torshizi EO, Chen W, Chen Y (2017) Simultaneous denoising and interpolation of 3-d seismic data via damped data-driven optimal singular value shrinkage. IEEE Geosci Remote Sens Lett 14(7):1086–1090
66. Yangkang C, Weilin H, Dong Z, Wei C (2016) An open-source matlab code package for improved rank-reduction 3d seismic data denoising and reconstruction. Comput Geosci 95:59–66
67. Yangkang C, Dong Z, Zhaoyu J, Xiaohong C, Shaohuan Z, Weilin H, Shuwei G (2016) Simultaneous denoising and reconstruction of 5-d seismic data via damped rank-reduction method. Geophys J Int 206(3):1695–1717
68. Ely G, Aeron S, Hao N, Kilmer ME (2015) 5d seismic data completion and denoising using a novel class of tensor decompositions. Geophysics 80(4):V83–V95
69. Gao J, Stanton A, Sacchi MD (2015) Parallel matrix factorization algorithm and its application to 5d seismic reconstruction and denoising. Geophysics
70. Kreimer N, Sacchi MD (2012) A tensor higher-order singular value decomposition for prestack seismic data noise reduction and interpolation. Geophysics 77(3):V113–V122
71. Da Silva C, Herrmann F (2013) Hierarchical tucker tensor optimization-applications to 4d seismic data interpolation. In: 75th EAGE conference & exhibition incorporating SPE EUROPEC 2013
72. Kazemi N, Bongajum E, Sacchi MD (2016) Surface-consistent sparse multichannel blind deconvolution of seismic signals. IEEE Trans Geosci Remote Sens 54(6):3200–3207
73. Yanghua W (2004) Multiple prediction through inversion: a fully data-driven concept for surface-related multiple attenuation. Geophysics 69(2):547–553
74. Weilin H, Runqiu W, Xiangbo G, Yangkang C (2017) Iterative deblending of simultaneous-source seismic data with structuring median constraint. IEEE Geosci Remote Sens Lett 15(1):58–62
75. Shuwei G, Shoudong W, Yangkang C, Xiaohong C (2015) Simultaneous-source separation using iterative seislet-frame thresholding. IEEE Geosci Remote Sens Lett 13(2):197–201
76. Yanhui Z, Jinghuai G, Wenchao C, Pascal F (2016) Seismic simultaneous source separation via patchwise sparse representation. IEEE Trans Geosci and Remote Sens 54(9):5271–5284
77. Mahmoudian F, Margrave GF, Wong J, Henley DC (2015) Azimuthal amplitude variation with offset analysis of physical modeling data acquired over an azimuthally anisotropic mediumazimuthal amplitude analysis. Geophysics 80(1):C21–C35
78. Fang X, Zheng Y, Fehler MC (2017) Fracture clustering effect on amplitude variation with offset and azimuth analysesavoaz fracture characterization. Geophysics 82(1):N13–N25
79. Sacchi MD, Kaplan S, Naghizadeh M (2009) Fx gabor seismic data reconstruction. In: 71st EAGE conference and exhibition incorporating SPE EUROPEC 2009
80. Felix J Herrmann and Gilles Hennenfent. Non-parametric seismic data recovery with curvelet frames. Geophys J Int 173(1):233–248
81. Qiao T, Ren J, Wang Z, Zabalza J, Sun M, Zhao H, Li S, Benediktsson JA, Dai Q, Marshall S (2016) Effective denoising and classification of hyperspectral images using curvelet transform and singular spectrum analysis. IEEE Trans Geosci Remote Sens 55(1):119–133
82. Yaru X, Meixi M, Shaohuan Z, Fanglan C, Yangkang C (2017) Amplitude-preserving iterative deblending of simultaneous source seismic data using high-order radon transform. J Appl Geophys 139:79–90
83. Shuwei G, Shoudong W, Yangkang C, Yizhuo Z, Zhaoyu J (2015) Dealiased seismic data interpolation using seislet transform with low-frequency constraint. IEEE Geosci Remote Sens Lett 12(10):2150–2154
84. Wei L, Siyuan C, Shuwei G, Yangkang C, Shaohuan Z, Zhaoyu J (2016) One-step slope estimation for dealiased seismic data reconstruction via iterative seislet thresholding. IEEE Geosci Remote Sens Lett 13(10):1462–1466
85. Trad DO, Ulrych TJ, Sacchi MD (2002) Accurate interpolation with high-resolution time-variant radon transforms. Geophysics 67(2):644–656
86. Latif A, Mousa WA (2016) An efficient undersampled high-resolution radon transform for exploration seismic data processing. IEEE Trans Geosci Remote Sens 55(2):1010–1024

87. Jianwei M (2013) Three-dimensional irregular seismic data reconstruction via low-rank matrix completion. Geophysics 78(5):V181–V192
88. Kumar R, Da Silva C, Akalin O, Aravkin AY, Mansour H, Recht B, Herrmann FJ (2015) Efficient matrix completion for seismic data reconstruction. Geophysics 80(5):V97–V114
89. Aravkin A, Kumar R, Mansour H, Recht B, Herrmann FJ (2014) Fast methods for denoising matrix completion formulations, with applications to robust seismic data interpolation. SIAM J Sci Comput 36(5):S237–S266
90. Yi Y, Jianwei M, Stanley O (2013) Seismic data reconstruction via matrix completion. Inverse Prob Imaging 7(4):1379–1392
91. Novikov A, Podoprikhin D, Osokin A, Vetrov D (2015) Tensorizing neural networks. In: Proceedings of the 28th international conference on neural information processing systems, vol 1, pp 442–450
92. Horesh L, Newman E, Kilmer ME, Avron H (2020) Generating and managing deep tensor neural networks. US Patent App. 16/189,744, May 14, 2020
93. Vincent P, Larochelle H, Lajoie I, Bengio Y, Manzagol P-A, Bottou L (2010) Stacked denoising autoencoders: learning useful representations in a deep network with a local denoising criterion. J Mach Learn Res 11(12)
94. Vincent P, Larochelle H, Bengio Y, Manzagol PA (2008) Extracting and composing robust features with denoising autoencoders. In: Proceedings of the 25th international conference on Machine learning, pp 1096–1103
95. Kilmer ME, Martin CD (2011) Factorization strategies for third-order tensors. Linear Algebra Its Appl 435(3):641–658
96. Benfeng W, Ning Z, Wenkai L, Jialin W (2019) Deep-learning-based seismic data interpolation: a preliminary result. Geophysics 84(1):V11–V20
97. Xintao C, Genyang T, Shangxu W, Ronghua P, Wei C, Jingnan L (2020) Deep learning for regularly missing data reconstruction. IEEE Trans Geosci Remote Sens 58(6):4406–4423
98. Xie J, Xu L, Chen E (2012) Image denoising and inpainting with deep neural networks. In: Advances in neural information processing systems, pp 341–349
99. Yu S, Ma J, Zhang X, Sacchi MD (2015) Interspolation and denoising of high-dimensional seismic data by learning a tight frame. Geophysics 80(5):V119–V132
100. Satinder C, Glen L (2000) Acquisition footprint-its detection and removal. CSEG Rec 25(8):16–20
101. Stephen H, Michael S, Joel B (1999) Acquisition footprint and fold-of-stack plots. Lead Edge 18(6):686–695
102. Li J, Gao J, Mei J, Chen X (2008) The footprint simulation and analysis for offshore seismic tow-cable acquisition. In: 2008 SEG annual meeting. OnePetro
103. Marfurt KJ, Scheet RM, Sharp JA, Harper MG (1998) Suppression of the acquisition footprint for seismic sequence attribute mapping. Geophysics 63(3):1024–1035
104. Cooper JK (2010) Seismic acquisition footprint: modelling and mitigation. Master's thesis, Graduate Studies
105. Alali A, Machado G, Marfurt KJ (2018) Attribute-assisted footprint suppression using a 2d continuous wavelet transform. Interpretation 6(2):T457–T470
106. Sahai SK, Soofi KA (2006) Use of simple 2-d filters to reduce footprint noise in seismic data. Geohorizons 7:14–17
107. Dawei L, Lei G, Xiaokai W, Wenchao C (2021) A dictionary learning method with atom splitting for seismic footprint suppression. Geophysics 86(6):V509–V523
108. Tom G, Stanley O (2009) The split bregman method for l1-regularized problems. SIAM J Imaging Sci 2(2):323–343
109. Yi C, Luxin Y, Houzhang F, Chunan L (2015) Anisotropic spectral-spatial total variation model for multispectral remote sensing image destriping. IEEE Trans Image Process 24(6):1852–1866
110. Yi C, Houzhang F, Luxin Y, Hai L (2013) Robust destriping method with unidirectional total variation and framelet regularization. Opt Express 21(20):23307–23323

111. Wu X, Shi Y, Fomel S, Liang L, Zhang Q, Yusifov AZ (2019) Faultnet3d: predicting fault probabilities, strikes, and dips with a single convolutional neural network. IEEE Trans Geosci Remote Sens 57(11):9138–9155
112. Xiaokai W, Wenchao C, Zhenyu Z (2018) Robust seismic volumetric dip estimation combining structure tensor and multiwindow technology. IEEE Trans Geosci Remote Sens 57(1):395–405
113. Yuqing W, Qiang G, Wenkai L, Xinfei Y (2020) Well-logging constrained seismic inversion based on closed-loop convolutional neural network. IEEE Trans Geosci Remote Sens 58(8):5564–5574
114. Zhaoqi G, Zhibin P, Jinghuai G (2016) Multimutation differential evolution algorithm and its application to seismic inversion. IEEE Trans Geosci Remote Sens 54(6):3626–3636
115. Rapoport MB, Rapoport LI, Ryjkov VI (2004) Direct detection of oil and gas fields based on seismic inelasticity effect. Lead Edge 23(3):276–278
116. Hu X, Chen Y, Liang X, Lang K (2005) New technology for direct hydrocarbon reservoir detection using seismic information. In: SEG technical program expanded abstracts 2005. Society of Exploration Geophysicists, pp 1735–1738
117. Liu D, Wang W, Chen W, Wang X, Zhou Y, Shi Z (2018) Random noise suppression in seismic data: What can deep learning do? In: SEG technical program expanded abstracts 2018. Society of Exploration Geophysicists, pp 2016–2020
118. Dawei L, Wei W, Xiaokai W, Cheng W, Jiangyun P, Wenchao C (2019) Poststack seismic data denoising based on 3-d convolutional neural network. IEEE Trans Geosci Remote Sens 58(3):1598–1629
119. Mi Z, Yang L, Yangkang C (2019) Unsupervised seismic random noise attenuation based on deep convolutional neural network. IEEE Access 7:179810–179822
120. Yangkang C, Mi Z, Min B, Wei C (2019) Improving the signal-to-noise ratio of seismological datasets by unsupervised machine learning. Seismol Res Lett 90(4):1552–1564
121. Saad OM, Chen Y (2021) A fully unsupervised and highly generalized deep learning approach for random noise suppression. Geophys Prospect 69(4):709–726
122. Yang L, Wang S, Chen X, Saad OM, Chen W, Innocent Oboué YAS, Chen Y (2021) Unsupervised 3-d random noise attenuation using deep skip autoencoder. IEEE Trans Geosci Remote Sens 60:1–16
123. Qiu C, Wu B, Liu N, Zhu X, Ren H (2021) Deep learning prior model for unsupervised seismic data random noise attenuation. IEEE Geosci Remote Sens Lett 19
124. Soltanayev S, Chun SY (2018) Training deep learning based denoisers without ground truth data. In: Proceedings of the 32nd international conference on neural information processing systems, pp 3261–3271
125. Nguyen HV, Ulfarsson MO, Sveinsson JR (2020) Hyperspectral image denoising using sure-based unsupervised convolutional neural networks. IEEE Trans Geosci Remote Sens 59(4):3369–3382
126. Yan S, Xiaoyuan Y, Yuhua G (2013) Translation invariant directional framelet transform combined with Gabor filters for image denoising. IEEE Trans Image Process 23(1):44–55
127. Tai-Xiang J (2020) Michael K Ng, Xi-Le Zhao, and Ting-Zhu Huang. Framelet representation of tensor nuclear norm for third-order tensor completion, IEEE Trans Image Process, p 29
128. Martin CD, Shafer R, LaRue B (2013) An order-p tensor factorization with applications in imaging. SIAM J Sci Comput 35(1):A474–A490
129. Feng J, Li X, Liu X, Chen C, Chen H (2021) Seismic data denoising based on tensor decomposition with total variation. IEEE Geosci Remote Sens Lett 18(7)
130. Zhaoqi G, Zhibin P, Chen Z, Jinghuai G, Zongben X (2019) An optimized deep network representation of multimutation differential evolution and its application in seismic inversion. IEEE Trans Geosci Remote Sens 57(7):4720–4734
131. Liu X, Chen X, Li J, Zhou X, Chen Y (2020) Facies identification based on multikernel relevance vector machine. IEEE Trans Geosci Remote Sens 58(10):7269–7282
132. Feng Q, Miao Y, Xiao-Yang L, Yao-Jun W, Cai L, Guang-Min H (2018) Unsupervised seismic facies analysis via deep convolutional autoencoders. Geophysics 83(3):A39–A43

133. Kuijie C, Jianwei M (2018) Robust estimation of multiple local dips via multidirectional component analysis. IEEE Trans Geosci Remote Sens 57(5):2798–2810
134. Qiang Z, Qizhen D, Xufei G, Yangkang C (2018) Signal-preserving erratic noise attenuation via iterative robust sparsity-promoting filter. IEEE Trans Geosci Remote Sens 56(6):3547–3560
135. Chen K, Sacchi MD (2015) Robust reduced-rank filtering for erratic seismic noise attenuation. Geophysics 80(1):V1–V11
136. Zhang Z, Ely G, Aeron S, Hao N, Kilmer M (2014) Novel methods for multilinear data completion and de-noising based on tensor-SVD. In: Proceedings of the IEEE conference on computer vision and pattern recognition, pp 3842–3849
137. Jen-Tzung C, Yi-Ting B (2017) Tensor-factorized neural networks. IEEE Trans Neural Netw Learn Syst 29(5):1998–2011
138. Newman E, Horesh L, Avron H, Kilmer M (2018) Stable tensor neural networks for rapid deep learning. arXiv preprint arXiv:1811.06569
139. Weilin H (2023) A two-step singular spectrum analysis method for robust low-rank approximation of seismic data. Geophysics 88(1):V9–V19
140. Trickett S, Burroughs L, Milton A (2012) Robust rank-reduction filtering for erratic noise. In: SEG international exposition and annual meeting, pp SEG–2012
141. Simon S (1991) Seismic trace interpolation in the FX domain. Geophysics 56(6):785–794
142. Claerbout JF (1991) and DAVE Nichols. Interpolation beyond aliasing by (t, x)-domain PEFs. In: 53rd EAEG meeting. European Association of Geoscientists & Engineers, pp cp–42
143. Sacchi MD, Gao J, Stanton A, Cheng J (2015) Tensor factorization and its application to multidimensional seismic data recovery. In: SEG technical program expanded abstracts 2015. Society of Exploration Geophysicists, pp 4827–4831
144. Yongna J, Jianwei M (2017) What can machine learning do for seismic data processing? An interpolation application. Geophysics 82(3):V163–V177
145. Spyrou L, Sanei S (2016) Coupled dictionary learning for multimodal data: an application to concurrent intracranial and scalp EEG. In: 2016 IEEE international conference on acoustics, speech and signal processing (ICASSP). IEEE, pp 2349–2353
146. Huang J-J, Dragotti PL (2018) A deep dictionary model for image super-resolution. In: 2018 IEEE international conference on acoustics, speech and signal processing (ICASSP). IEEE, pp 6777–6781
147. Yang J, Wright J, Huang TS, Ma Y (2010) Image super-resolution via sparse representation. IEEE Trans Image Process 19(11):2861–2873
148. Ledig C, Theis L, Huszár F, Caballero J, Cunningham A, Acosta A, Aitken A, Tejani A, Totz J, Wang Z et al (2017) Photo-realistic single image super-resolution using a generative adversarial network. In: Proceedings of the IEEE conference on computer vision and pattern recognition, pp 4681–4690
149. Bin S, Yaojun W, Jiandong L, Zhining L, Chengyun S, Guangmin H (2018) A data-driven amplitude variation with offset inversion method via learned dictionaries and sparse representation. Geophysics 83(6):R725–R748
150. Wang Z, Long Z, AlRegib G (2016) Tensor-based subspace learning for tracking salt-dome boundaries constrained by seismic attributes. In: 2016 ieee international conference on acoustics, speech and signal processing (ICASSP). IEEE, pp 1125–1129
151. Jiang F, Liu X-Y, Lu H, Shen R (2017) Efficient two-dimensional sparse coding using tensor-linear combination. arXiv preprint arXiv:1703.09690
152. Feng Qian Yu, Yue HY, Yingjie Z, Bo W, Guangmin H (2023) Improved low-rank tensor approximation for seismic random plus footprint noise suppression. IEEE Trans Geosci Remote Sens 61:1–19
153. Feng Q, Zhangbo L, Yan W, Yingjie Z, Guangmin H (2022) Ground truth-free 3-d seismic random noise attenuation via deep tensor convolutional neural networks in the time-frequency domain. IEEE Trans Geosci Remote Sens 60:1–17
154. Feng Q, Yan W, Bingwei Z, Zhangbo L, Yingjie Z, Guangmin H (2022) Multidimensional seismic data denoising using framelet-based order-p tensor deep learning. IEEE Trans Geosci Remote Sens 60:1–18

155. Qian F, Hua H, Wen Y, Pan S, Zhang G, Hu G (2024) Unsupervised 3-d seismic erratic noise attenuation with robust tensor deep learning. IEEE Trans Geosci Remote Sens 62
156. Nanda NC (2021) Seismic data interpretation and evaluation for hydrocarbon exploration and production. Springer
157. Liu Y (2021) Tensors for data processing: theory, methods, and applications. Academic Press
158. Liu Y, Liu J, Long Z, Zhu C (2022) Tensor computation for data analysis. Springer
159. Zhou H-W (2014) Practical seismic data analysis. Cambridge University Press
160. Öz Yilmaz (2001) Seismic data analysis: processing, inversion, and interpretation of seismic data. Society of exploration geophysicists
161. Al-Shuhail A, Al-Dossary S (2020) Attenuation of incoherent seismic noise. Springer
162. Tiwari RK, Rekapalli R (2020) Modern singular spectral-based denoising and filtering techniques for 2D and 3D reflection seismic data. Springer Nature

# Chapter 2
# The Foundations of Tensor Computation

## 2.1 Tensor Notations

In this book, we adopt the following notational conventions: tensors are represented by bold calligraphic letters, such as $\mathcal{T}$; matrices are represented by bold uppercase letters, such as $T$; vectors are represented by bold lowercase letters, such as $\mathbf{t}$; and scalars are represented by lowercase letters, such as $t$, as illustrated in Fig. 2.1. Our focus is on real-valued third-order seismic data tensors in the space $\mathbb{R}^{N_x \times N_y \times N_t}$, as depicted in Fig. 2.2. The indices $i$, $j$, and $k$ are used to denote the x-line, y-line, and time dimensions of a tensor, respectively, where $i \in [N_x]$, $j \in [N_y]$, and $k \in [N_t]$, with $[N]$ representing the set $1, 2, \ldots, N$. For a given tensor $\mathcal{T} \in \mathbb{R}^{N_x \times N_y \times N_t}$, the $(i, j, k)$th element is written as $\mathcal{T}(i, j, k)$, or more compactly as $\mathcal{T}_{ijk}$. The $i$th horizontal and $j$th lateral slices are denoted by $\mathcal{T}(i, :, :)$ and $\mathcal{T}(:, j, :)$, respectively. The $k$th frontal slice is represented by $\mathcal{T}(:, :, k)$, or concisely as $\mathcal{T}^{(k)}$.

We utilize the following norms and operators for matrices and tensors in this work: the Frobenius norm of a matrix is defined as $\|T\|_F = \sqrt{\sum_{i=1}^{m} \sum_{j=1}^{n} T_{ij}^2}$, while for a tensor it is given by $\|\mathcal{T}\|_F = \sqrt{\sum_{ijk} |\mathcal{T}_{ijk}|^2}$. The nuclear norm of a matrix is expressed as $|T|_* = \sum_{i=1}^{L} \sigma_i$, where $\sigma_i$ represents the $i$th singular value of $T$ and $L$ is the total number of singular values of $X$. The inner product operator is defined as $\langle T_1, T_2 \rangle = \sum T_1(i, j) T_2(i, j)$. The transpose of a matrix is indicated by $\top$, and the conjugate transpose by $H$. For conciseness, we introduce three primary tensor products, which are used in Sects. 3.3 and 8.3.

**Definition 2.1** (*Element-wise tensor product* [1]) The element-wise tensor product $\mathcal{T} = \mathcal{T}_1 \odot \mathcal{T}_2$ of two tensors $\mathcal{T}_1, \mathcal{T}_2 \in \mathbb{R}^{N_x \times N_y \times N_t}$, having identical dimensions, results in a tensor $\mathcal{T}$ of the same size. The $(i, j, k)$th entry of $\mathcal{T}$ is defined by $\mathcal{T}(i, j, k) = \mathcal{T}_1(i, j, k) \mathcal{T}_2(i, j, k)$.

 
F. Qian et al., *Tensor Computation for Seismic Data Processing*,
Earth Systems Data and Models 6, https://doi.org/10.1007/978-3-031-78900-7_2

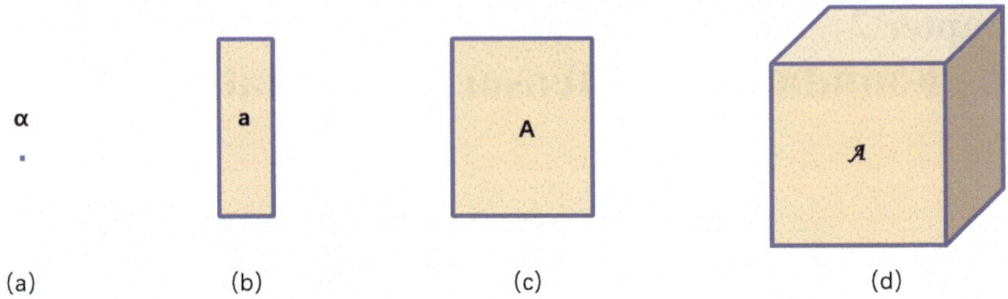


**Fig. 2.1** A visual depiction of **a** a scalar, **b** a vector, **c** a matrix, and **d** a tensor

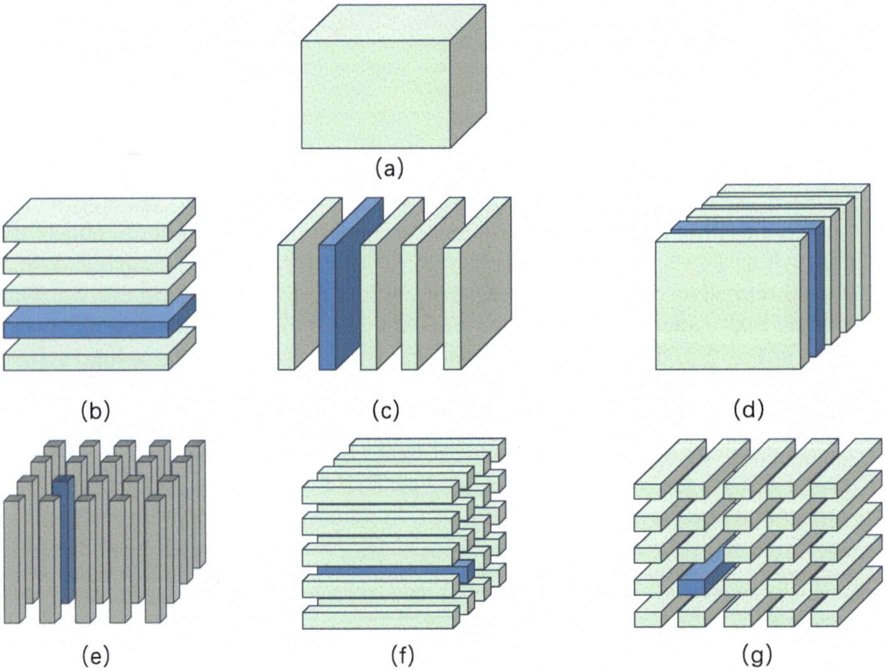


**Fig. 2.2** A graphical illustration of the fibers and slices of a third-order tensor $\mathcal{A}$. **a** Visualization of the third-order tensor $\mathcal{A}$. **b** A horizontal slice denoted as $\mathcal{A}(2, :, :)$ . **c** A lateral slice represented by $\mathcal{A}(:, 2, :)$ . **d** A frontal slice depicted as $\mathcal{A}(:, :, 2)$. **e** A Mode-1 fiber, given by $\mathcal{A}(:, 2, 2)$. **f** A Mode-2 fiber expressed as $\mathcal{A}(2, :, 2)$. (g) A Mode-3 fiber identified as $\mathcal{A}(2, 2, :)$

**Definition 2.2** (*Tube-wise circular convolution* [1]) The tube-wise circular convolution $\mathcal{T} = \mathcal{T}_1 \cdot \circledast \mathcal{T}_2$ of two tensors $\mathcal{T}_1, \mathcal{T}_2 \in \mathbb{R}^{N_x \times N_y \times N_t}$ of identical size results in a tensor $\mathcal{T}$ with the same dimensions. For each $(i, j)$th tube, $\mathcal{T}(i, j, :)$ is defined by $\mathcal{T}(i, j, :) = \mathcal{T}_1(i, j, :) * \mathcal{T}_2(i, j, :)$, where $*$ represents the circular convolution operator.

**Definition 2.3** (*Frontal-slice-wise tensor product* [1]) The frontal slice-wise tensor product $\mathcal{T} = \mathcal{T}_1 \cdot \S\mathcal{T}_2$ is defined by performing matrix multiplication on the corresponding frontal slices of two tensors. Specifically, for tensors $\mathcal{T} \in \mathbb{R}^{N_x \times N_y \times N_t}$, $\mathcal{T}_1 \in \mathbb{R}^{N_x \times N_y \times N_t}$, and $\mathcal{T}_2 \in \mathbb{R}^{N_y \times N_y \times N_t}$, the $(k)$th frontal slice of $\mathcal{T}$ is calculated as $\mathcal{T}^{(k)} = \mathcal{T}_1^{(k)} * \mathcal{T}_2^{(k)}$ for $k \in [N_t]$.

## 2.2 Tensor-Tensor Product (t-Prouduct)

### 2.2.1 *Order-3 and Oder-p t-Product*

In general, the processes of folding and unfolding are defined as follows.

$$\text{unfold}(\mathcal{X}) := \begin{bmatrix} \mathcal{X}^{(1)} \\ \mathcal{X}^{(2)} \\ \vdots \\ \mathcal{X}^{(n_p)} \end{bmatrix}, \quad \text{fold}(\text{unfold}(\mathcal{X})) = \mathcal{X}, \tag{2.1}$$

where $\text{unfold}(\mathcal{X}) \in \mathbb{R}^{n_1 n_p \times n_2 \times \cdots \times n_{p-1}}$. Using block-based operators, we derive several key definitions based on [2, 3] below.

**Definition 2.4** (*Order-3 t-product* [3]) Let $\mathcal{A} \in \mathbb{R}^{n_1 \times r \times n_3}$ and $\mathcal{B} \in \mathbb{R}^{r \times n_2 \times n_3}$ be two tensors. Their t-product, denoted by $\mathcal{A} * \mathcal{B}$, results in a third-order tensor of dimensions $n_1 \times n_2 \times n_3$.

$$\mathcal{A} * \mathcal{B} = \text{fold}(\text{circ}(\mathcal{A}) \cdot \text{unfold}(\mathcal{B})). \tag{2.2}$$

Equation (2.2) clearly relies on standard matrix multiplication. By substituting equation (2.1) into the definition given in (2.2), the t-product can be generalized to tensors of orders higher than three, as outlined in the following. This generalization expands the applicability of the t-product, enabling efficient computations across higher-order tensor structures.

**Definition 2.5** (*Order-p t-product in the time domain* [2, 4]) Consider the tensors $\mathcal{A}$ with dimensions $n_1 \times r \times n_3 \times \cdots \times n_p$ and $\mathcal{B}$ with dimensions $r \times n_2 \times n_3 \times \cdots \times n_p$. Their t-product, denoted by $\mathcal{A} * \mathcal{B}$, produces an order-$p$ tensor of size $n_1 \times n_2 \times n_3 \times \cdots \times n_p$.

$$\mathcal{A} * \mathcal{B} = \text{fold}(\text{circ}(\mathcal{A}) * \text{unfold}(\mathcal{B})). \tag{2.3}$$

Definition 2.2 is computed recursively by applying the $t$-product between two order-$(p-1)$ tensors.

**Algorithm 2.1:** t-SVD for 3-way data

**Input**: $\mathcal{A} \in \mathbb{R}^{I_1 \times I_2 \times I_3}$
**Output**: $\mathcal{U}, \mathcal{S}, \mathcal{V}$
$\mathcal{D} \leftarrow \text{fft}(\mathcal{A}, [], 3)$.
**for** $i_3 = 1, \cdots, I_3$ **do**
  $[\mathbf{U}, \mathbf{S}, \mathbf{V}] = \text{SVD}(\mathcal{D}(:, :, i_3))$,
  $\hat{\mathcal{U}}(:, :, i_3) = \mathbf{U}, \hat{\mathcal{S}}(:, :, i_3) = \mathbf{S}, \quad \hat{\mathcal{V}}(:, :, i_3) = \mathbf{V}$
**end**
$\mathcal{U} \leftarrow \text{ifft}(\hat{\mathcal{U}}, [], 3), \mathcal{S} \leftarrow \text{ifft}(\hat{\mathcal{S}}, [], 3), \mathcal{V} \leftarrow \text{ifft}(\hat{\mathcal{V}}, [], 3)$

**Definition 2.6** (*Order-p t-product in the frequency domain* [4, 5]) According to the convolution theorem, the $t$-product can be transformed (2.3) into matrix multiplication in the frequency domain as

$$\overline{\mathcal{C}}(:, :, \ell) = \overline{\mathcal{A}}(:, :, \ell) \cdot \overline{\mathcal{B}}(:, :, \ell), \ \ell \in \left[n_3 n_4 \cdots n_p\right]. \tag{2.4}$$

**Definition 2.7** (*Tensor SVD* [3]) The tSVD of $\mathcal{T} \in \mathbb{R}^{n_1 \times n_2 \ldots \times n_p}$ is expressed as $\mathcal{T} = \mathcal{U} * \mathcal{S} * \mathcal{V}^\dagger$ (see Algorithm 7.2 and Fig. 2.3), where $\mathcal{U} \in \mathbb{R}^{n_1 \times r \ldots \times n_p}$ and $\mathcal{V} \in \mathbb{R}^{r \times n_2 \ldots \times n_p}$ are orthogonal tensors and $\mathcal{S} \in \mathbb{R}^{r \times r \ldots \times n_p}$ is a rectangular $f$-diagonal tensor whose frontal slice is a diagonal matrix.

The t-SVD exhibits similarities to matrix SVD, particularly in maintaining the orthogonality of $\mathcal{U}$ and $\mathcal{V}$, as well as the diagonal form of $\mathcal{S}$. Each frontal slice of the t-SVD corresponds to a matrix SVD in the frequency domain. This formulation of t-SVD introduces the notion of tensor tubal rank, which is defined as follows.

**Definition 2.8** (*Tensor tubal-rank* [3, 6]) The tensor multi-rank of $\mathcal{T}$ is denoted by a vector $\mathbf{r} \in \mathbb{R}^{N_t \times 1}$, where the $i$th element represents the rank of the $i$th frontal slice of $\widetilde{\mathcal{T}}$ in the Fourier domain. The tensor tubal-rank $r$ of $\mathcal{T}$ is defined as the maximum rank among all the frontal slices, given by $r = \max\{\mathbf{r}\}$.

### 2.2.2 L Transform-Based t-Product

In tensor operations, the standard t-product functions similarly to a convolution-like operation, enabling efficient computation using the fast Fourier transform (FFT). However, research in [7, 8] expands the definition of the t-product to include any invertible linear transform $L$. In our study, we focus on developing a ground truth-free data denoising model that utilizes a t-product involving an invertible linear transform. Given the linear transformation $L : \mathbb{R}^{n_1 \times n_2 \times n_3} \rightarrow \mathbb{R}^{n_1 \times n_2 \times n_3}$, we obtain $\widetilde{\mathcal{T}}$

**Algorithm 2.2:** $L$ transform-based t-product [8]

**Input:** $\mathcal{T}_1 \in \mathbb{R}^{n_1 \times n_2 \times n_3}$, $\mathcal{T}_2 \in \mathbb{R}^{n_2 \times l \times n_3}$, and $L : \mathbb{R}^{n_1 \times n_2 \times n_3} \to \mathbb{R}^{n_1 \times n_2 \times n_3}$.
**Output:** $\mathcal{T} = \mathcal{T}_1 *_L \mathcal{T}_2 \in \mathbb{R}^{n_1 \times l \times n_3}$.
1: $\widetilde{\mathcal{T}}_1 = L(\mathcal{T}_1)$, $\widetilde{\mathcal{T}}_1 = L(\mathcal{T}_2)$.
2: **for** $k = 1, \cdots, n_3$
3: $\widetilde{\mathcal{T}}^{(k)} = \widetilde{\mathcal{T}}_1^{(k)} \widetilde{\mathcal{T}}_2^{(k)}$. **do**
**end**
4: $\mathcal{T} = L^{-1}(\widetilde{\mathcal{T}})$.

by applying the linear transformation $L$ to each tube fiber $\mathcal{T}(i, j, :)$. Alternatively, we can express this as follows:

$$\widetilde{\mathcal{T}} = L(\mathcal{T}) = \mathcal{T} \times_3 L, \tag{2.5}$$

where $\times_3$ represents the mode-3 (matrix) product between a tensor $\mathcal{T}$ and an arbitrary invertible matrix $L \in \mathbb{R}^{n_3 \times n_3}$. Similarly, we can articulate the following inverse transform:

$$L^{-1}(\mathcal{T}) = \mathcal{T} \times_3 L^{-1}. \tag{2.6}$$

According to the standard definition of the $t$-product, the $L$ transform-based $t$-product is defined in the following manner.

**Definition 2.9** (*$L$ transform-based t-product* [8]) Given an invertible linear transform $L$ and two third-order tensors $\mathcal{T}_1 \in \mathbb{R}^{n_1 \times n_2 \times n_3}$ and $\mathcal{T}_2 \in \mathbb{R}^{n_2 \times l \times n_3}$, the $L$ transform-based $t$-product is defined as $C = \mathcal{T}_1 *_L \mathcal{T}_2$, such that $L(\mathcal{T}) = L(\mathcal{T}_1) \odot L(\mathcal{T}_2)$. Here, $\odot$ represents the facewise product of the two third-order tensors, which is defined as follows.

**Definition 2.10** (*Facewise product* [7]) The facewise product of two third-order tensors $\mathcal{T}_1$ and $\mathcal{T}_2$, denoted as $\mathcal{T}_1 \odot \mathcal{T}_2$, is defined as follows:

$$(\mathcal{T}_1 \odot \mathcal{T}_2)^{(k)} = \mathcal{T}_1^{(k)} \mathcal{T}_2^{(k)}, \quad k \in [n]. \tag{2.7}$$

Following this step, $L(\mathcal{T}) = L(\mathcal{T}_1) \odot L(\mathcal{T}_2)$ is equivalent to $\widetilde{\mathcal{T}}^{(k)} = \widetilde{\mathcal{T}}_1^{(k)} \widetilde{\mathcal{T}}_2^{(k)}$, which shows that the $L$ transform-based $t$-product is analogous to the matrix–matrix product of two frontal slices in the transform domain. Using these properties, an algorithm for computing the $t$-product is provided in Algorithm 2.2.

**Definition 2.11** (*Tensor transpose* [8]) Given a invertible linear transform $L$ and a third-order tensor $\mathcal{T} \in \mathbb{R}^{n_1 \times n_2 \times n_3}$, the tensor transpose is defined as $\mathcal{T}^{\dagger}$, which satisfies $L\left(\mathcal{T}^{\dagger}\right)^{(k)} = \left(L(\mathcal{T})^{(k)}\right)^{\dagger}, k \in [n_3]$.

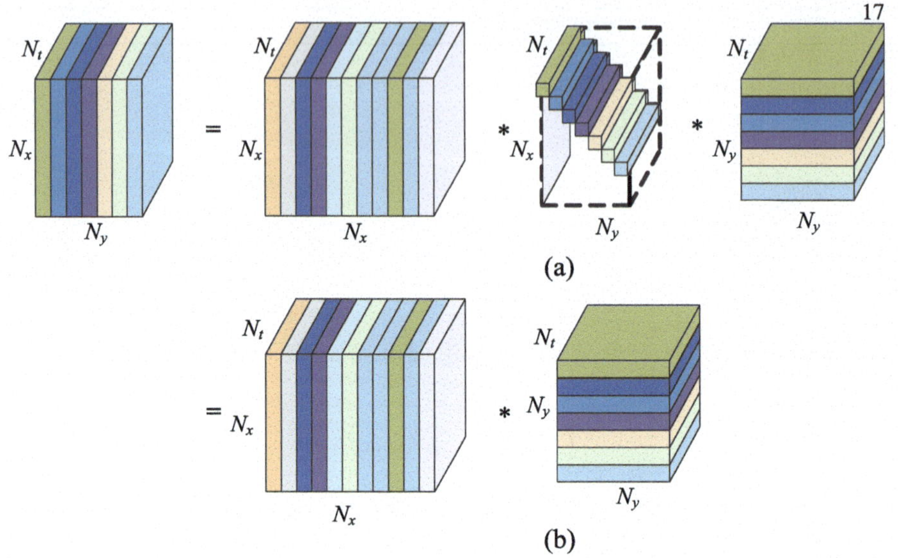


**Fig. 2.3** t-SVD of an $N_x \times N_y \times N_t$ tensor. **a** Trilinear decomposition. **b** Bilinear decomposition

## 2.3 Tensor Completion

### *2.3.1 Rank Minimization Model*

This section illustrates the recovery process of seismic data using the nuclear norm minimization method when the tensor data is either randomly sampled or downsampled [9]. Let us consider an unknown seismic data tensor $\mathcal{M}$ of size $n_1 \times n_2 \times n_3$, which is assumed to have a low tubal-rank. Given a subset of entries $\{\boldsymbol{\mathcal{M}}_{ijk} : (i, j, k) \in \boldsymbol{\Omega}\}$, where $\boldsymbol{\Omega}$ is an indicator tensor of size $n_1 \times n_2 \times n_3$, our objective is to recover the complete tensor $\mathcal{M}$. This is accomplished by solving the following complexity penalized algorithm, defined as follows.

$$\begin{aligned} &\min \|\mathcal{X}\|_* \\ &\text{s.t. } P_{\boldsymbol{\Omega}}(\mathcal{X}) = P_{\boldsymbol{\Omega}}(\mathcal{M}) \end{aligned} \tag{2.8}$$

where $P_\Omega$ is an orthogonal projection operator that projects the seismic data tensor onto the seismic observation set $\boldsymbol{\Omega}$, such that

$$\left[\mathcal{P}_\Omega(\mathcal{M})\right]_{ij\kappa} = \begin{cases} \mathcal{M}_{ij\kappa}, & \text{if } (i, j, \kappa) \in \Omega, \\ 0, & \text{otherwise.} \end{cases} \tag{2.9}$$

Assuming $\boldsymbol{y}$ represents the observed (or sampled) 3D seismic data: $\boldsymbol{y} = P_{\boldsymbol{\Omega}}(\mathcal{M})$, we define $\mathcal{G} = \mathcal{F}_3 P_\Omega \mathcal{F}_3^{-1}$, where $\mathcal{F}_3$ and $\mathcal{F}_3^{-1}$ denote the operators for the FFT and inverse FFT along the third dimension of the seismic data tensor. Consequently, we have $\widehat{\mathbf{y}} = \mathcal{G}(\widehat{\mathcal{M}})$, with $\widehat{\mathbf{y}}$ and $\widehat{\mathcal{M}}$ being the FFT transforms of $\mathbf{y}$ and $\mathcal{M}$ along the third dimension. Thus, Eq. (2.8) can be expressed in the following form:

$$
\begin{aligned}
&\min \|\operatorname{blkdiag}(\widehat{\mathcal{X}})\|_* \\
&\text{s.t. } \widehat{\mathbf{y}} = \mathcal{G}(\widehat{\mathbf{x}})
\end{aligned}
\tag{2.10}
$$

where $\widehat{\mathcal{X}}$ denotes the FFT of $\mathcal{X}$ along the third dimension of the 3D seismic data, and $\operatorname{blkdiag}(\widehat{\mathcal{X}})$ represents a block diagonal matrix defined as follows:

$$
\operatorname{blkdiag}(\hat{\mathcal{X}}) = \begin{bmatrix} \hat{\mathcal{X}}^{(1)} & & & \\ & \hat{\mathcal{X}}^{(2)} & & \\ & & \ddots & \\ & & & \hat{\mathcal{X}}^{(n_3)} \end{bmatrix}
\tag{2.11}
$$

It is important to mention that the equivalence above relies on the relationship: $\|\mathcal{X}\|_* = \|\operatorname{blkdiag}(\hat{\mathcal{X}})\|_*$. The optimization problem (2.10) can thus be reformulated equivalently as follows:

$$
\begin{aligned}
&\min \|\operatorname{blkdiag}(\hat{z})\|_* + \mathbf{1}_{\hat{y}=\mathcal{G}(\hat{x})} \\
&\text{s.t. } \hat{\mathbf{x}} - \hat{z} = 0
\end{aligned}
\tag{2.12}
$$

where $\mathbf{1}_{(\cdot)}$ denotes the indicator function. By applying the general framework of the alternating direction method of multipliers (ADMM) [10] to solve problem (2.12), the following recursive relation can be derived:

$$
\begin{aligned}
\mathcal{X}^{k+1} &= \arg\min_{\mathcal{X}} \left\{ \mathbf{1}_{y=P_\Omega(x)} + \mathcal{X}(:)^{\mathrm{T}}\mathbf{Q}^k(:) + \frac{1}{2}\left\|\mathcal{X} - z^k\right\|_F^2 \right\} \\
&= \arg \min_{x:y=P_\Omega(x)} \left\{ \left\|\mathcal{X} - (z^k - \mathbf{Q}^k)\right\|_F^2 \right\} \\
\hat{z}^{k+1} &= \arg\min_{\hat{\mathbf{z}}} \left\{ \frac{1}{\rho}\|\operatorname{blkdiag}(\hat{\mathbf{Z}})\|_* + \frac{1}{2}\left\|\hat{\mathbf{z}} - \left(\hat{\mathbf{x}}^{k+1} + \hat{\mathbf{Q}}^k\right)\right\|_F^2 \right\} \\
\mathbf{Q}^{k+1} &= \mathbf{Q}^k + \left(\mathcal{X}^{k+1} - z^{k+1}\right)
\end{aligned}
\tag{2.13}
$$

In Eq. (2.13), the first recursive formula represents the least-squares projection onto the constraint, while the solution to the second recursive formula is provided by the singular value thresholding method. The notation $\mathcal{X}(:)$ and $\mathbf{Q}^k(:)$ indicates the vectorization of the 3D seismic data tensors, as commonly used in MATLAB.

### 2.3.2 Low-Rank Tensor Factorization Model

This section addresses the seismic data interpolation problem using the tensor factorization method [11]. Compared to the nuclear norm approach, this method provides improved recovery accuracy, although it is less resistant to noise. It is assumed here that the 3D seismic data tensor has a low tubal rank, meaning that the seismic data tensor $\mathcal{T} \in \mathbb{R}^{m\times n\times k}$ has a tubal rank of $r \ll \min(m, n)$. Specifically, by observing elements of $\mathcal{T}$ in the subset $\Omega \subseteq [m] \times [n] \times [k]$, the objective is to recover $\mathcal{T}$. In other words, given the elements $\mathcal{T}_{ij\kappa}$ for $(i, j, \kappa) \in \Omega$, the goal is to estimate the remaining elements outside of $\Omega$ as accurately as possible.

Assuming that $\mathcal{T}$ is a low-tubal-rank tensor, the estimated tensor $\widehat{\mathcal{T}}$ should closely approximate $\mathcal{T}$ on the 3D seismic data observation set $\Omega$. Therefore, estimating the low-tubal-rank tensor can be framed as the following optimization problem:

$$\begin{aligned} \widehat{\mathcal{T}} = &\mathop{\arg\min}_{\mathcal{Z}\in\mathbb{R}^{m\times n\times k}} \|\mathcal{P}_\Omega(\mathcal{Z}) - \mathcal{P}_\Omega(\mathcal{T})\|_F \\ &\text{s.t.} \quad \operatorname{rank}(\mathcal{Z}) \le r, \end{aligned} \tag{2.14}$$

where $\mathcal{Z} \in \mathbb{R}^{m\times n\times k}$ is the decision variable, and the function $\operatorname{rank}(\cdot)$ denotes the tensor tubal-rank. The target tensor $\widehat{\mathcal{T}} \in \mathbb{R}^{m\times n\times k}$ can be decomposed as $\widehat{\mathcal{T}} = \mathcal{X} * \mathcal{Y}^\dagger$, where $\mathcal{X} \in \mathbb{R}^{m\times r\times k}$ and $\mathcal{Y} \in \mathbb{R}^{n\times r\times k}$, with $r$ as the target tubal-rank. Based on this decomposition, problem (2.14) reduces to:

$$\widehat{\mathcal{T}} = \mathop{\arg\min}_{\mathcal{X}\in\mathbb{R}^{m\times r\times k}, \mathcal{Y}\in\mathbb{R}^{n\times r\times k}} \left\|\mathcal{P}_\Omega(\mathcal{T}) - \mathcal{P}_\Omega\left(\mathcal{X} * \mathcal{Y}^\dagger\right)\right\|_F^2, \tag{2.15}$$

The objective of Eq. (2.15) is to determine the tensor $\widehat{\mathcal{T}} = \mathcal{X} * \mathcal{Y}^\dagger$. Simplifying $\mathcal{T}_\Omega = \mathcal{P}_\Omega(\mathcal{T})$ gives $\mathcal{T}_\Omega = \mathcal{P}_\Omega \odot \mathcal{T}$. By applying the Convolution Theorem, which states that elementwise multiplication in the time domain equates to circular convolution in the frequency domain, the tensor least squares minimization problem can be reformulated in the frequency domain as follows:

$$\widehat{\mathcal{Y}} = \mathop{\arg\min}_{,\tilde{\mathcal{Y}}\in\mathbb{R}^{r\times n\times k}} \left\|\widetilde{\mathcal{T}}_\Omega - \widetilde{\mathcal{P}}_\Omega \cdot \otimes(\tilde{\mathcal{X}} \cdot \S\widetilde{\mathcal{Y}})\right\|_F^2, \tag{2.16}$$

For simplicity, the tensor transpose is omitted. To solve (4.6), an alternating minimization algorithm is proposed. Equation (4.6) is thereby transformed into $n$ separate standard least squares minimization subproblems:

$$\widehat{\mathcal{Y}}(:, j, :) = \mathop{\arg\min}_{\mathcal{Y}::,j:)\in\mathbb{R}^{r\times 1\times k}} \left\|\widetilde{\mathcal{T}}_\Omega(:, j, :) - \widetilde{\mathcal{P}}_\Omega(:, j, :) \cdot \otimes(\widetilde{\mathcal{X}} \cdot \S\widetilde{\mathcal{Y}}(:, j, :))\right\|_F^2, \tag{2.17}$$

where each subproblem involves estimating a lateral slice $\widetilde{\mathcal{Y}}(:, j, :),\ j \in [n]$.

## 2.4 Tensor Deep Learning

### *2.4.1 Tensor Neural Networks*

Building on the t-product description from Definition 2.4, Newman et al. [12] proposed the t-NN framework, presenting an innovative approach for constructing neural networks capable of handling tensor data. Utilizing the t-NN framework, we can effectively extend an autoencoder (AE) to create a tensor autoencoder. However, prior to formulating the tensor AE, it is essential to first review several key concepts inherent to the t-NN framework.

**Definition 2.12** (*Matrix representation of a neural network* [12, 13]). Consider a fully connected neural network (FNN) shown in Fig. 2.4a. Let $W_{\ell-1}$ be the weight matrix with dimensions $n_\ell \times n_{\ell-1}$, and let $X_{\ell-1}$ represent the output from layer $\ell - 1$, sized $n\ell - 1 \times n_2$. The input data matrix $X_0$ has dimensions $n_1 \times n_2$, and $\mathbf{b}_\ell$ is the bias vector of length $n_\ell$. The forward propagation is then defined as follows.

$$X_\ell = \sigma\left(W_{\ell-1}X_{\ell-1} + \mathbf{b}_\ell\right), \quad \ell \in [K], \tag{2.18}$$

where $\sigma(\cdot)$ represents an elementwise, nonlinear activation function[1] and $K$ denotes the total number of layers in the network, as depicted in Fig. 2.4b. Subsequently, it is logical to explore a tensor extension of this type of neural network. Unlike the matrix case, in the framework of the t-product, the previously mentioned fully connected layers are substituted with layers of the following form.

**Definition 2.13** (*Tensor representation of a neural network* [12, 14]) Assume we have tensors $\mathcal{X}_{\ell-1} \in \mathbb{R}^{n_{\ell-1}\times n_2 \times n_3}$, $\mathcal{W}_\ell \in \mathbb{R}^{n_\ell \times n_{\ell-1} \times n_3}$, and $\overrightarrow{\mathcal{B}}_\ell \in \mathbb{R}^{n_\ell \times 1 \times n_3}$. The tensor forward propagation can then be defined as follows.

$$\mathcal{X}_\ell = \sigma\left(\mathcal{W}_{\ell-1} * \mathcal{X}_{\ell-1} + \overrightarrow{\mathcal{B}}_\ell\right), \qquad \ell \in [K]. \tag{2.19}$$

It is important to note that $\mathcal{W}_{\ell-1}$ represents the mapping from layer $\ell - 1$ to layer $\ell$. The summation operator "+" combines $\overrightarrow{\mathcal{B}}_\ell$ with each lateral slice of $\mathcal{W}_{\ell-1} * \mathcal{X}_{\ell-1}$, as illustrated in Fig. 2.4c.

Once the tensor deep learning network is established as per Definition 2.13, the next goal is to minimize the following objective function $\mathcal{F}$:

$$\mathcal{F} \equiv \mathcal{L}\left(f\left(\mathcal{W}_N * \mathcal{X}_N\right), C\right) \tag{2.20}$$

This objective function $\mathcal{F}$ characterizes a standard classification problem, where $C$ denotes the true classification and is comprised of two components: $\mathcal{L} : \mathbb{R}^{p\times m} \to \mathbb{R}$,

---

[1] In this study, we utilize the ReLU function $\sigma(x) = \max(0, x)$.

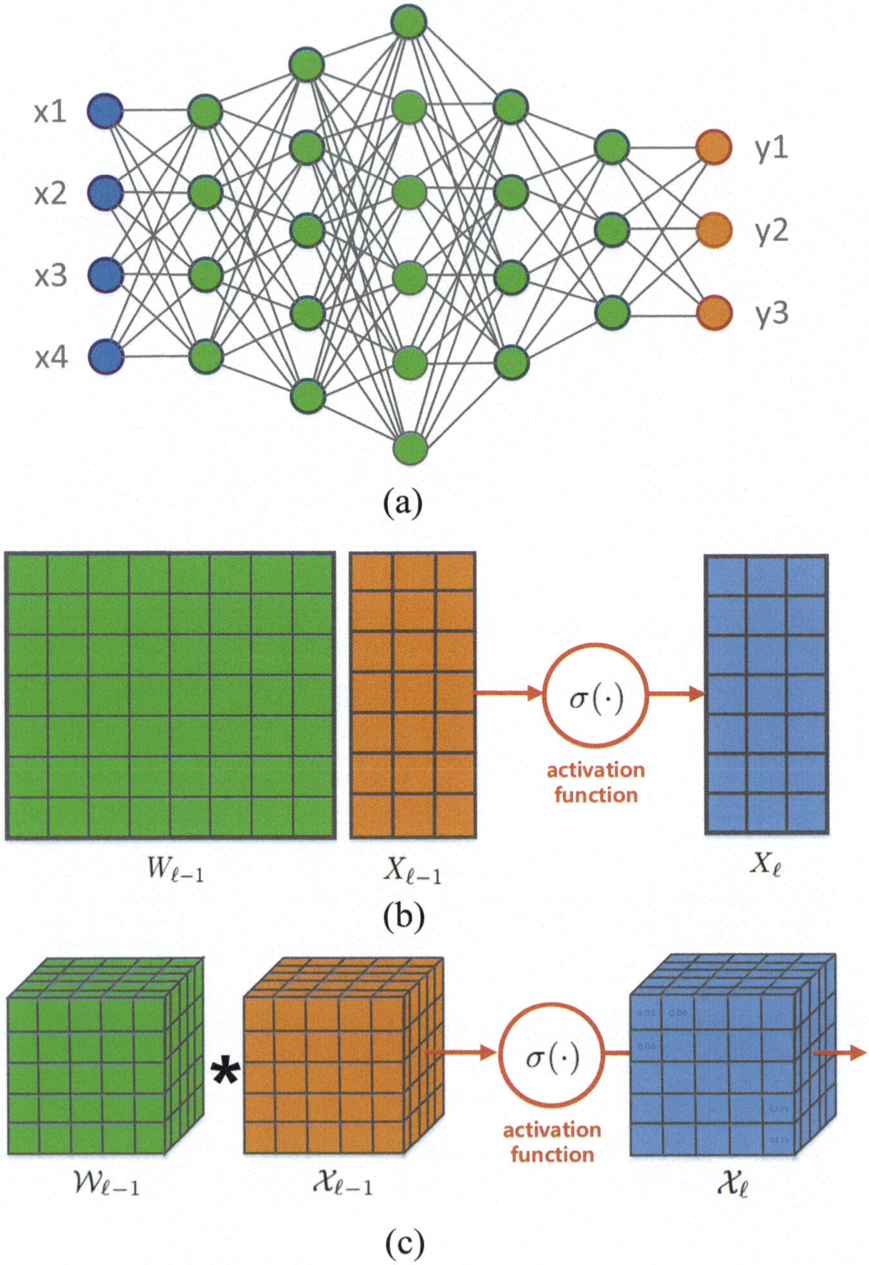


**Fig. 2.4** A fully connected neural network (FNN) and its representations in matrix and tensor forms are illustrated as follows: **a** the structure of the FNN; **b** the matrix-based representation of the FNN; and **c** the tensor-based representation of the FNN

representing the tensor loss function, and $f : \mathbb{R}^{p \times m \times n} \to \mathbb{R}^{p \times m}$, which is the scalar tubal function. The representation of the true classification $C$ varies according to the selected loss function $\mathcal{L}$. For example, if $\mathcal{L}$ is defined as the cross-entropy function, then $C$ is expressed as a $1 \times m$ vector where $C_j = i$ indicates that the $j$th sample belongs to the $i$th class.

### 2.4.2 *Tensor Back-Propagation*

Once the network architecture and objective function are defined, tensor deep learning modifies its network parameters using a layer evolution method known as back-propagation. The detailed formulas for tensor backpropagation are presented as follows:

$$\delta \mathcal{X}_N = W_N^{\top} * \partial \mathcal{F} / \partial \mathcal{X}_N \tag{2.21}$$

$$\delta \mathcal{X}_j = \mathcal{W}_j^{\top} * \left(\delta \mathcal{X}_{j+1} \odot \sigma' \left(\mathcal{Z}_{j+1}\right)\right) \tag{2.22}$$

$$\delta \mathcal{W}_j = \left(\delta \mathcal{X}_{j+1} \odot \sigma' \left(\mathcal{Z}_{j+1}\right)\right) * \mathcal{X}_j^{\top} \tag{2.23}$$

$$\delta \overrightarrow{\mathcal{B}}_j = \operatorname{sum}\left(\delta \mathcal{X}_{j+1} \odot \sigma' \left(\mathcal{Z}_{j+1}\right), 2\right) \tag{2.24}$$

where $\odot$ signifies the Hadamard element-wise product, $\delta \mathcal{X}_j = \partial \mathcal{F} / \partial \mathcal{X}_j$ indicates the error at the $j$th layer, $\mathcal{Z}_{j+1} = \mathcal{W}_j * \mathcal{X}_j + \overrightarrow{\mathcal{B}}_j$, $\sigma'(\cdot)$ represents the derivative of the activation function, and $\operatorname{sum}(\cdot, 2)$ refers to summing along the second dimension, which corresponds to the summation of lateral slices.

The derivation of Eq. (2.21) is intricate and is based on the matrix backpropagation formulas outlined in [15]. The detailed derivation of Eq. (2.21) proceeds as follows:

$$\frac{\partial \mathcal{F}}{\partial \overrightarrow{\mathcal{X}}_N} = \frac{\partial \mathcal{L}}{\partial \mathbf{y}} \cdot \frac{\partial \mathbf{y}}{\partial \overrightarrow{\mathcal{X}}_N} = \frac{\partial \mathcal{L}}{\partial \mathbf{y}} \cdot \frac{\partial f\left(\mathcal{W}_N * \overrightarrow{\mathcal{X}}_N\right)}{\partial \overrightarrow{\mathcal{X}}_N} \tag{2.25}$$

where $\mathbf{y} = f\left(\mathcal{W}_N * \overrightarrow{\mathcal{X}}_N\right)$. To derive equation (2.22), we first rewrite Equation (2.19) in matrix form, and then apply the following matrix backpropagation formula:

$$\operatorname{unfold}\left(\delta \mathcal{X}_j\right) = \operatorname{bcirc}\left(\mathcal{W}_j\right)^{\top} \cdot \left(\operatorname{unfold}\left(\delta \mathcal{X}_{j+1}\right) \odot \sigma'\left(\operatorname{unfold}\left(\mathcal{Z}_{j+1}\right)\right)\right) \tag{2.26}$$

Given that $\operatorname{bcirc}\left(\mathcal{W}_j\right)^{\top} = \operatorname{bcirc}\left(\mathcal{W}_j^{\top}\right)$, we can express equation (2.26) in tensor form, resulting in Eq. (2.22). The proof for Eq. (2.24) follows a similar approach, with the summation along the second dimension arising from the application of the chain rule. To derive equation (2.23), we will rewrite the forward propagation formula for a specific frontal slice, as shown in Eq. (2.19):

$$X_{j+1}^{(k)} = \sigma\left(W_j^{(k)} \cdot X_j^{(1)} + \sum_{i=1}^{k-1} W_j^{(i)} \cdot X_j^{(k-i+1)} + \sum_{i=k+1}^{n} W_j^{(i)} \cdot X_j^{(n-i+k+1)} + \mathbf{B}_j^{(i)}\right) \tag{2.27}$$

Since frontal slices are matrices, differentiating equation (2.27) with respect to $W_j^{(i)}$ results in $\delta W_j^{(i)}$. Assuming that $\delta\tilde{\mathcal{X}}_{j+1} = \delta\mathcal{X}_{j+1} \odot \sigma'\left(\mathcal{Z}_{j+1}\right)$, $\delta W_j^{(i)}$ can be calculated using the following formula:

$$\begin{aligned}\delta W_j^{(i)} &= \sum_{k=1}^{n} \frac{\partial \mathcal{F}}{\partial X_{j+1}^{(k)}} \cdot \frac{\partial X_{j+1}^{(k)}}{\partial W_j^{(i)}} \\ &= \delta\tilde{X}_{j+1}^{(i)} \cdot \left(X_j^{(1)}\right)^{\top} + \sum_{k=1}^{i-1} \delta\tilde{X}_{j+1}^{(k)} \cdot \left(X_j^{(n-i+k+1)}\right)^{\top} \\ &+ \sum_{k=i+1}^{n} \delta\tilde{X}_{j+1}^{(k)} \cdot \left(X_j^{(k-i+1)}\right)^{\top}\end{aligned} \tag{2.28}$$

In Eq. (2.28), the first summation includes $X^{(n-i+k+1)}$ instead of $X^{(i-k+1)}$, while the second summation includes $X^{(k-i+1)}$ instead of $X^{(n-k+i+1)}$. This precisely corresponds to the frontal slice mapping of the tensor transpose as defined in Definition 2.11. The specific result is as follows:

$$\delta\mathcal{W}_j = \left(\delta\mathcal{X}_{j+1} \odot \sigma\left(\mathcal{Z}_{j+1}\right)\right) * \mathcal{X}_j^{\top}. \tag{2.29}$$

Equations (2.22), (2.23), and (2.24) closely resemble their matrix counterparts. This similarity arises from the fact that in the $t$-product framework, tensors function as $t$-linear operators, much like matrices operate as linear operators. Consequently, tensor backpropagation (TBP) serves as a natural high-dimensional extension of backpropagation (BP) algorithms derived from matrix theory.

## 2.5 Tensor Dictionary Learning

Given a three-dimensional seismic dataset of dimensions $m \times k \times n$, it can be expressed as a third-order tensor $\mathcal{X} \in \mathbb{K}^{m\times n}$. This tensor can be approximated through the tensor product of $\mathcal{D} \in \mathbb{K}^{m\times r}$ and $C \in \mathbb{K}^{r}$.

$$\mathcal{X} = \mathcal{D} * C, \tag{2.30}$$

where $*$ denotes the tensor-tensor product. Let $\mathcal{D} \in \mathbb{K}^{m\times r}$ be a tensor dictionary, with each lateral slice $\mathcal{D}(:, j, :)$ representing a tensor basis. Additionally, let $\mathcal{S} \in \mathbb{K}^{r\times n}$

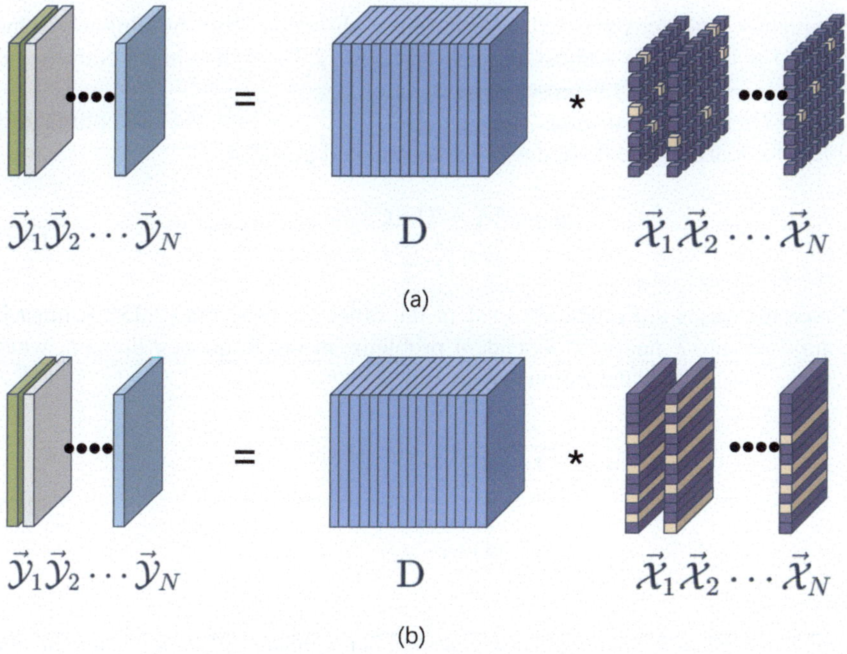


**Fig. 2.5** Sparse tensor representation utilizing the t-product, represented either **a** with an $\ell_1$ constraint or **b** with tubal sparsity imposed through the $\ell_{1,1,2}$ norm

denote the corresponding tensor representation. This indicates that each seismic image $\mathcal{X}(:, j, :)$ can be approximately expressed as a sparse linear combination of these tensor bases. Therefore, the objective function for the tubular sparse coding (TubSC) model (see Fig. 2.5) [16] can be formulated as follows:

$$\begin{array}{ll} \min_{\mathcal{D},\mathcal{C}} & \frac{1}{2}\|\mathcal{X} - \mathcal{D} * \mathcal{C}\|_F^2 + \beta\|\mathcal{C}\|_1 \\ \text{s.t.} & \left\|\vec{\mathcal{D}}_j\right\|_F^2 \leq 1,\ j \in [r] \end{array} \tag{2.31}$$

where $\beta$ is the sparsity regularizer. The objective function's non-convex nature and the involvement of tensor product operators make solving problem (2.31) quite challenging. In [16], an efficient algorithm was introduced that alternates between optimizing $\mathcal{C}$ and $\mathcal{D}$ while keeping the other variable fixed. This approach is specifically divided into the following two steps.

*Tensor Coefficient Learning*: Given the dictionary $\mathcal{D}$, we aim to determine the sparse coefficients $\mathcal{C}$ by defining the sub-optimization problem as follows:

$$\min_{\mathcal{C}} \frac{1}{2}\|\mathcal{X} - \mathcal{D} * \mathcal{C}\|_F^2 + \beta\|\mathcal{C}\|_1. \tag{2.32}$$

Subsequently, the tensor-based fast iterative shrinkage thresholding algorithm (TFISTA) is utilized to address the sub-problem (2.32), achieving a convergence rate of $O\left(1/t^2\right)$, where $t$ represents the iteration count.

*Tensor Dictionary Learning*: Then, when $\mathcal{C}$ is held constant, the sub-optimization problem for learning the dictionary $\mathcal{D}$ is formulated as follows:

$$\begin{aligned} &\min_{\mathcal{D}} \ \|\mathcal{X} - \mathcal{D} * \mathcal{C}\|_F^2 \\ &\text{s.t.} \quad \left\|\mathcal{D}_j\right\|_F^2 \le 1,\ j \in [r] \end{aligned} \tag{2.33}$$

Since the bases are interconnected in the objective function (2.33), it must be decomposed into $k$ nearly independent problems in the frequency domain, where $k = n_3 n_4 \cdots n_p$. Their definitions are as follows:

$$\begin{aligned} &\min_{\widehat{\mathcal{D}}^{(\ell)}, \ell \in [k]} \sum_{\ell=1}^{k} \left\|\widehat{\mathcal{X}}^{(\ell)} - \widehat{\mathcal{D}}^{(\ell)} \widehat{\mathcal{C}}^{(\ell)}\right\|_F^2 \\ &\text{s.t.} \sum_{\ell=1}^{k} \left\|\widehat{\mathcal{D}}^{(\ell)}(:, j)\right\|_F^2 \le k,\ j \in [r]. \end{aligned} \tag{2.34}$$

Subsequently, the Lagrange dual method [17] is employed to solve for the dual variables using Newton's algorithm. An additional benefit of the Lagrange dual is that the number of optimization variables is $r$, which is significantly smaller than $n_1 n_3 \cdots n_p r$ in the primal problem (2.34).

## References


1. Liu X-Y, Aggarwal S, Wang X (2016). Low-tubal-rank tensor completion using alternating minimization. CoRR arXiv:1610.01690
2. Martin CD, Shafer R, LaRue B (2013) An order-p tensor factorization with applications in imaging. SIAM J Sci Comput 35(1):A474–A490
3. Kilmer ME, Braman K, Hao N, Hoover RC (2013) Third-order tensors as operators on matrices: a theoretical and computational framework with applications in imaging. SIAM J Matrix Anal Appl 34(1):148–172
4. Jiang F, Liu X-Y, Lu H, Shen R (2018) Efficient multi dimensional tensor sparse coding using t-linear combination. In: Proceedings of the AAAI conference on artificial intelligence, vol 32
5. Wang J-L, Huang T-Z, Zhao X-L, Jiang T-X, Ng MK (2021) Multi-dimensional visual data completion via low-rank tensor representation under coupled transform. IEEE Trans Image Process 30:3581–3596
6. Zhang Z, Aeron S (2017) Exact tensor completion using t-SVD. IEEE Trans Signal Process 65(6):1511–1526
7. Kernfeld E, Kilmer M, Aeron S (2015) Tensor-tensor products with invertible linear transforms. Linear Algebra Its Appl 485:545–570
8. Lu C, Peng X, Wei Y (2019) Low-rank tensor completion with a new tensor nuclear norm induced by invertible linear transforms. In: Proceedings of the IEEE/CVF conference on computer vision and pattern recognition, pp 5996–6004

9. Zhang Z, Ely G, Aeron S, Hao N, Kilmer M (2014) Novel methods for multilinear data completion and de-noising based on tensor-SVD. In: Proceedings of the IEEE conference on computer vision and pattern recognition, pp 3842–3849
10. Boyd S, Parikh N, Chu E, Peleato B, Eckstein J et al (2011) Distributed optimization and statistical learning via the alternating direction method of multipliers. Found Trends® Mach Learn 3(1):1–122
11. Liu X-Y, Aeron S, Aggarwal V, Wang X (2019) Low-tubal-rank tensor completion using alternating minimization. IEEE Trans Inform Theory 66(3):1714–1737
12. Newman E, Horesh L, Avron H, Kilmer M (2018) Stable tensor neural networks for rapid deep learning (2018). arXiv preprint arXiv:1811.06569
13. Denil M, Shakibi B, Dinh L, Ranzato M, De Freitas N (2013) Predicting parameters in deep learning. In: Advances in neural information processing systems, pp 2148–2156
14. Horesh L, Newman E, Kilmer ME, Avron H (2020) Generating and managing deep tensor neural networks. US Patent App. 16/189744, May 14 2020
15. Rumelhart DE, Hinton GE, Williams RJ (1986) Learning representations by back-propagating errors. Nature 323(6088):533–536
16. Jiang F, Liu X-Y, Lu H, Shen R (2017) Graph regularized tensor sparse coding for image representation. In: 2017 IEEE International Conference on Multimedia and Expo (ICME). IEEE, pp 67–72
17. Lee H, Battle A, Raina R, Ng A (2006) Efficient sparse coding algorithms. In: Advances in neural information processing systems, vol 19

# Chapter 3
# Tensor Completion for Seismic Data Reconstruction

## 3.1 Introduction

Techniques for low-rank matrix and tensor completion have demonstrated notable success across a wide range of domains, including recommendation systems [1], machine learning [2], data mining [3, 4], and computer vision [5, 6]. Similarly, in seismic data reconstruction, the fully sampled data is represented as a low-rank matrix or tensor, where the presence of missing values and noise causes a rank increase. Thus, the main challenge in seismic data reconstruction is the design of efficient rank-reduction techniques that not only recover absent traces but also improve the signal quality in the presence of varying noise levels, tackling the complexities of different noise environments and SNR conditions.

In seismic data processing, current approaches to rank reduction are typically divided into two categories: unstructured low-rank approximations [7, 8] and structured low-rank approximations [9, 10]. The first category views seismic data as unorganized multidimensional arrays and utilizes various rank-reduction methods for completing and recovering missing data. These methods typically include nuclear norm-based approaches [8, 11–14], SVD-free factorization techniques [15–17], and traditional matrix or tensor factorization methods [7, 18]. In contrast, the second category advocates for the use of structured Hankel matrices or tensors to more effectively capture the characteristics of seismic data, as seismic signals generally have lower ranks, while the observed data tends to have higher ranks [19]. This category focuses on matrix-based low-rank approximation, with notable techniques including singular spectrum analysis (SSA) [20–22], multichannel SSA (MSSA) [9, 23, 24], and damped multichannel SSA (DMSSA) [10, 25, 26]. Nonetheless, in practical scenarios, the predefined Hankel structure is often disrupted during the iterative process by operations such as truncated singular value decomposition (SVD) (see Fig. 3.3c), resulting in the Hankel matrix failing to achieve the desired low-rank form in the final outcome (see Fig. 3.3d and e). This issue is particularly evident in noise attenuation tasks, as further confirmed by the study in [19].



F. Qian et al., *Tensor Computation for Seismic Data Processing*,
Earth Systems Data and Models 6, https://doi.org/10.1007/978-3-031-78900-7_3

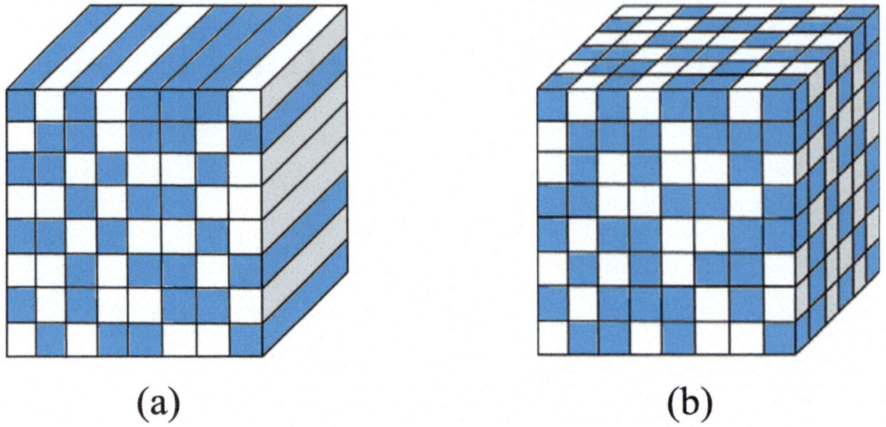


**Fig. 3.1** Illustrating the differences between tubal sampling and element-wise sampling for 3D tensor: **a** Tubal sampling. **b** Element-wise sampling

To address this challenge, we introduce a new method grounded in the random tubal-sampling assumption [27, 28], assuming that the seismic data exhibit low-rank characteristics, in order to more accurately capture the behavior of missing traces [29], as illustrated in Fig. 3.1a. In contrast to traditional element-wise random sampling methods [9, 10] (shown in Fig. 3.1b), the tubal-sampling assumption offers significant advantages. Specifically, it provides two key benefits: first, this assumption allows the seismic tensor completion problem to be decomposed into a series of individual matrix completion (MC) tasks, each related to an individual frontal or temporal slice, with sampling taking place in the frequency space [28]; second, matrix completion techniques are not only applicable to interpolation and denoising in seismic data but can also be easily extended to various tensor completion tasks, thus better addressing the diverse and complex needs of seismic data processing.

This chapter introduces an innovative STC framework designed to address interpolation and noise suppression in seismic data under various noise conditions. The framework utilizes different strategies depending on the noise level: for low-noise scenarios, structured tensor decomposition is used; for moderate noise levels, a singular value decomposition (SVD)-free tensor factorization technique is adopted; and for high-noise conditions, a weighted tensor nuclear norm method is employed. Additionally, we propose an efficient alternating minimization (AltMin) algorithm to solve the least squares problem for both structured matrix factorization and SVD-free matrix factorization. To enhance the stability and convergence of the solution process, we also employ the alternating direction method of multipliers (ADMM) to manage the structured weighted matrix nuclear norm. Ultimately, through experiments on both synthetic and real seismic data, we show that the proposed approach achieves superior performance compared to three leading methods, excelling in both data recovery quality and computational speed.

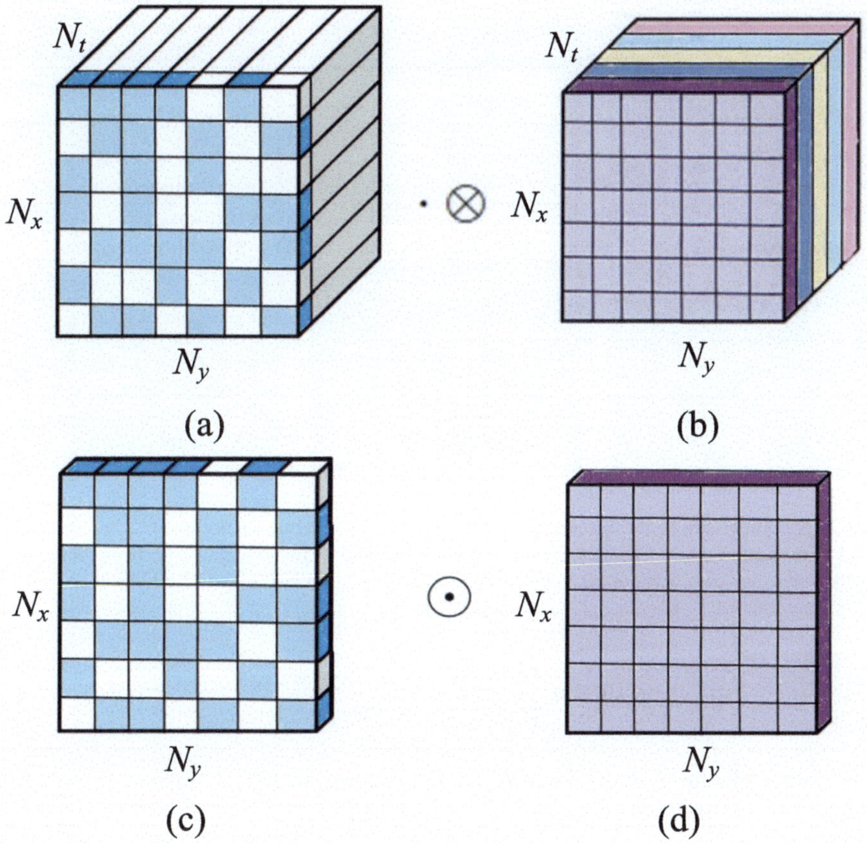


**Fig. 3.2** Illustrating the relationship between tubal sampling in tensor reconstruction and matrix completion: **a** The linear operator $\widetilde{\mathcal{P}}_{\Omega}$ depicted in the Fourier space. **b** The representation of seismic tensor data within the Fourier space. **c** The first frontal slice $\widetilde{\mathcal{P}}_{\Omega}^{(1)}$ of the linear operator $\widetilde{\mathcal{P}}_{\Omega}$. **d** The representation of seismic data in the Fourier domain across different frequency slices

## 3.2 Preliminaries in Seismic Data Reconstruction

This section introduces a novel 3D seismic data reconstruction method based on the STC model. Unlike traditional seismic completion approaches, our approach considers missing seismic traces as instances of random tubal sampling within a 3D domain, illustrated in Fig. 3.1a. Specifically, the complete seismic data set, denoted as $\mathcal{T}$, is acquired along the temporal or depth axis, with the sampling locations determined by the index set $\Omega$. To simulate the observed missing data, the sampling operator $\mathcal{P}_{\Omega}(\cdot)$ maps the complete data tensor to the observed subset. $\Omega$, defined as follows:

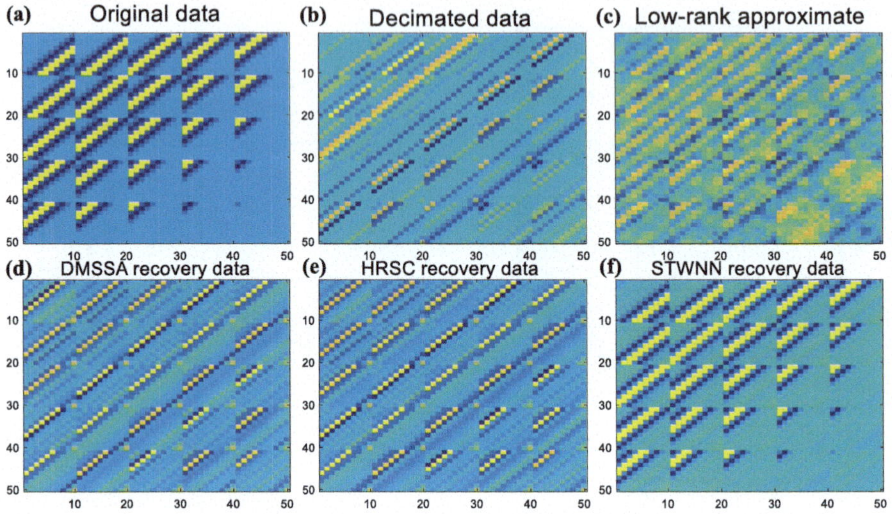


**Fig. 3.3** A comparison of the Hankel structure for the 5th crossline section from a 3D synthetic dataset is presented under heavy noise conditions, including: **a** the original data, **b** incompletely sampled and noisy data, **c** low-rank approximation, **d** DMSSA, **e** HRSC, and **f** STWNN

$$\mathcal{P}_\Omega(\mathcal{T}) = \begin{cases} \mathcal{T}(i, j, :), & \text{if } (i, j) \in \Omega, \\ 0, & \text{otherwise}, \end{cases} \tag{3.1}$$

In this definition, the $(i, j)$th tube of $\mathcal{P}_\Omega(\mathcal{T})$ is assigned to $\mathcal{T}(i, j, :)$ when $(i, j) \in \Omega$, and is set to zero in other cases..

The seismic data tensor, $\mathcal{T}_\Omega$, is derived from the complete tensor, $\mathcal{T}$, using the sampling operator $\mathcal{P}_\Omega(\cdot)$, producing a sparse set of observations, $\mathcal{P}_\Omega(\mathcal{T})$, with the addition of noise $\mathcal{N}$. This relationship is expressed as follows:

$$\mathcal{T}_\Omega = \mathcal{P}_\Omega(\mathcal{T}) + \mathcal{N}. \tag{3.2}$$

Since the observed data points are much fewer than the total entries in the tensor $\mathcal{T}$, the problem described in Eq. (3.2) becomes highly ill-conditioned and cannot be solved directly unless additional regularization is applied to $\mathcal{T}$. To address this issue, we impose a combination of low-rank assumptions and structure Hankel constraints on the solution, and solve the problem using SSA-based method [9, 10, 21–26, 30]. However, these existing methods do not fully leverage the Hankel structure constraints, leading to suboptimal performance in noise reduction and data recovery. To overcome this limitation, we propose an innovative STC model, which enforces both low-rank and Hankel structure constraints in the process of low-tubal-rank tensor completion. Specifically, we postulate that the target Hankel tensor $\mathcal{M} \in \mathbb{R}^{I_x \times I_y \times N_t}$ has a tubal rank $r \ll \min(I_x, I_y)$. Given a set $\Omega \in [N_x] \times \left[N_y\right]$ of observed elements from the tensor $\mathcal{M}$, the objective is to reconstruct the full tensor $\mathcal{M}$. That is,

**Algorithm 3.1:** Structured Low-tubal-rank Tensor Completion with Tubal Sampling

**Input:** $\mathcal{T}_\Omega, \mathcal{P}_\Omega, r, N_{ite}, P_{MC}$.
**Output:** $\widehat{\mathcal{T}}$.
$\mathcal{M}_\Omega \leftarrow \mathcal{H}(\mathcal{T}_\Omega)$, $\mathcal{P}_\Omega \leftarrow \mathcal{H}(\mathcal{P}_\Omega)$.
$\widetilde{\mathcal{M}}_\Omega \leftarrow \text{fft}(\mathcal{M}_\Omega, [\,], 3)$, $\widetilde{\mathcal{P}}_\Omega \leftarrow \text{fft}(\mathcal{P}_\Omega, [\,], 3)$.
**for** $k = 1, \cdots, N_t$ **do**
 $\widetilde{\mathcal{M}}^{(k)} \leftarrow \text{LRMC}(\widetilde{\mathcal{M}}_\Omega^{(k)}, \widetilde{\mathcal{P}}_\Omega^{(1)}, r, N_{ite}, P_{MC})$.
**end**
$\widehat{\mathcal{M}} \leftarrow \text{ifft}(\widetilde{\mathcal{M}}, [\,], 3)$. $\widehat{\mathcal{T}} \leftarrow \mathcal{A}(\widehat{\mathcal{M}})$.

from the observed elements $\mathcal{M}_\Omega(i, j, :)$ for $(i, j) \in \Omega$, our goal is to reconstruct the missing elements outside of $\Omega$ with the highest possible accuracy. The mathematical formulation of the proposed STC approach can be presented below.

$$\begin{aligned} &\min_{\mathcal{M}} \ \text{rank}(\mathcal{M}), \\ &\text{s.t. } \mathcal{M}_\Omega = \mathcal{P}_\Omega(\mathcal{M}), \ \mathcal{B}(\mathcal{M}) = 0, \end{aligned} \tag{3.3}$$

where rank$(\cdot)$ represents the tubal rank constraint of the tensor $\mathcal{M}$, as defined in Definition 2.8. The operator $\mathcal{B}(\cdot)$ performs the Hankel constraint transformation, which converts the input linear structure into a Hankel structure constraint for each frontal slice of $\mathcal{M}$. This transformation ensures that the condition $\mathcal{T}_{1,2,3} = \mathcal{T}_{2,1,3}$ is satisfied, similar to the micro Hankel matrix structure defined in Eq. (3.42). Mathematically, this transformation can be expressed as the linear equation $\mathcal{B}(\mathcal{M}) = B\overline{M} = 0$, where $B \in \mathbb{R}^{N_1(N_2-1)(N_y-N_2)(N_x-N_1+1)+N_y(N_1-1)(N_x-N_1)\times I_x I_y}$ is a matrix which captures the structure of each frontal slice in $\mathcal{M}$. Detailed derivations of this process can be found in the appendix of [31]. Furthermore, $\overline{M} \in \mathbb{R}^{I_x I_y \times N_t}$ is defined as $\overline{M} = [\mathbf{m}_1; \mathbf{m}_2; \ldots; \mathbf{m}_{N_t}]$, where each $\mathbf{m}_k \in \mathbb{R}^{I_x I_y}$ is a vector that reshapes each frontal slice into a column vector through a linear transformation, as described below:

$$\mathbf{m}_k = [\text{squeeze}(\mathcal{M}(:, 1, k)); \text{squeeze}(\mathcal{M}(:, 2, k)); \ldots; \text{squeeze}(\mathcal{M}(:, I_y, k))], \tag{3.4}$$

The operation squeeze$(\mathcal{M}(i, j, k))$ compresses the $j$th tube of the $k$th frontal slice of $\mathcal{M}$ into a column vector of dimension $I_y \times 1$. Unlike models based on singular spectrum analysis (SSA), the proposed model incorporates a Hankel constraint into the goal function to ensure that the solution adheres to the structured Hankel format at each iteration. This constraint creates dependencies between the elements of the reconstructed frontal slice, effectively limiting the solution's flexibility.

## 3.3 The Bridge Between Tensor and Matrix Completion

In this section, we explore the connection between tensor completion and matrix completion (MC) by leveraging tubal sampling techniques. Specifically, this connection consists of two main components: first, we propose a tensor completion approach by mapping the tensor problem to the matrix space, as detailed in Sect. 3.3.1; second, we advance the study of tensor completion by providing a pathway from matrix completion, further driving the development of this field, as discussed in Sect. 3.3.2.

### *3.3.1 From Tensor to Matrix Completion*

Due to the complexity of problem (3.3), which may not be solvable directly, we employ tubal sampling to transform the intractable STC problem into a manageable matrix completion (MC) problem. Initially, we simplify the expression $\mathcal{M}_\Omega = \mathcal{P}_\Omega(\mathcal{M})$ by employing $\mathcal{T}_\Omega = \mathcal{P}_\Omega \odot \mathcal{T}$. According to the convolution theorem (which states that element-wise multiplication. According to the convolution theorem (which asserts that element-wise multiplication in the time domain is equivalent to circular convolution in the Fourier space), one can transform the objective function (3.3) into its frequency domain representation as follows:

$$\begin{aligned} &\min_{\widetilde{\mathcal{M}}} \ \operatorname{rank}(\widetilde{\mathcal{M}}), \\ &\text{s.t.} \ \widetilde{\mathcal{M}}_\Omega = \widetilde{\mathcal{P}}_\Omega \cdot \otimes \widetilde{\mathcal{M}}, \ \mathcal{B}(\widetilde{\mathcal{M}}) = 0. \end{aligned} \tag{3.5}$$

Based on the core properties of Fourier transforms, the transform of a constant produces a Dirac delta function, while the transform of a Dirac delta function results in a constant. The operator $\mathcal{P}_\Omega(\cdot)$ contains two distinct types of fibers, where each element is assigned a value of either 0 or 1, as illustrated in Fig. 3.2. By applying the Fourier transform to $\mathcal{P}_\Omega(\cdot)$, we obtain $\widetilde{\mathcal{P}}_\Omega(\cdot)$, where the characteristics of the fibers are represented by Dirac delta functions. The specific form of $\widetilde{\mathcal{P}}_\Omega(\cdot)$ is described as follows:

$$\widetilde{\mathcal{P}}_\Omega(\cdot) = \begin{cases} \mathcal{P}_\Omega^{(k)}(\cdot), & \text{if } k = 1, \\ 0, & \text{otherwise.} \end{cases} \tag{3.6}$$

The characteristics of the Dirac delta function greatly streamline the resolution of the initial problem (3.3). When there are $k$ slices, Eq. (3.5) can be effectively decomposed into $k$ independent subproblems, as outlined below:

$$\begin{aligned} &\min_{\widetilde{\mathcal{M}}^{(k)}} \ \operatorname{rank}(\widetilde{\mathcal{M}}^{(k)}), \\ &\text{s.t.} \ \widetilde{\mathcal{M}}_\Omega^{(k)} = \widetilde{\mathcal{P}}_\Omega^{(1)} \odot \widetilde{\mathcal{M}}^{(k)}, \ \mathcal{B}(\widetilde{\mathcal{M}}^{(k)}) = 0. \end{aligned} \tag{3.7}$$

**Algorithm 3.2:** AltMin-based Matrix Completion

**Input:** $M_\Omega, P_\Omega, r, N_{ite}, \mu, \lambda, \mu_0, \epsilon_0$.
**Output:** $\widehat{M}$.
**Function** LRMC ($M_\Omega, P_\Omega, r, N_{ite}, \mu, \lambda, \mu_0, \epsilon_0$):
  Initialize: $Y_0 \leftarrow$ the top $r$ singular vectors of $M_\Omega$.
  **for** $\ell = 1, \cdots, N_{ite}$ **do**
    $X_\ell \leftarrow \text{LS}(M_\Omega, P_\Omega, Y_{\ell-1}, r, \mu, \lambda)$;
    $X_\ell \leftarrow \text{QR}(X_\ell, \epsilon_0, \mu_0)$;
    $Y_\ell \leftarrow \text{LS}(M_\Omega, P_\Omega, X_\ell, r, \mu, \lambda)$;
    $Y_\ell \leftarrow \text{QR}(Y_\ell, \epsilon_0, \mu_0)$.
  **end**
  $\widehat{M} = X_\ell Y_\ell$.
**return**

Thus, the STC framework with tubal sampling can essentially be viewed as matrix completion (MC) with element-wise sampling of every frontal slice in the frequency domain. This process is described in detail in Algorithm 3.1. In this procedure, the user applies LRMC$(\cdot)$ to perform rank-reduction on each frontal slice in the frequency space, where $r$ denotes the rank of $\widetilde{\mathcal{M}}_\Omega^{(k)}$. Furthermore, the user must apply $\mathcal{H}(\cdot)$ to transform the original tensor into its Hankel form to generate the input data, and use $\mathcal{A}(\cdot)$ to reverse the Hankel transformation for each frontal slice of the Hankel tensor. In addition, $P_{MC}$ refers to the specialized parameter set for each matrix completion (MC) problem.

### 3.3.2 *From Matrix to Tensor Completion*

Tubal sampling provides an effective approach to solve the low-tubal-rank tensor completion problem by decomposing the complex tensor completion task into multiple simpler matrix completion (MC) subproblems. In other words, matrix completion techniques can be applied in the Fourier domain to achieve tensor recovery. This occurs because the t-product of two third-order tensors in the Fourier space is analogous to matrix multiplication between their frontal slices. For instance, performing matrix singular value decomposition (SVD) on each frequency slice in the Fourier domain results in the t-SVD for tensor data, thereby offering a pathway for efficient tensor completion via the tubal sampling technique.

Our main proposition is that the effectiveness of matrix completion (MC) methods is significantly influenced by the noise conditions [32, 33], and that it is possible to develop a corresponding tensor formulation that achieves similar performance to the noise-specific MC under such conditions. The noise environment here refers to varying levels of noise, including light, moderate, and heavy noise. For a more precise quantitative characterization, we define a moderate noise environment as one with a signal-to-noise ratio (SNR) ranging from 10 to 19 dB. Based on this, we classify the

**Algorithm 3.3:** ADMM-based Matrix Completion

**Input:** $M_\Omega, \mathcal{P}_\Omega, r, N_{ite}, \rho, \mu, \delta, \epsilon$.
**Output:** $\widehat{M}$.
**Function** LRMC ($M_\Omega, \mathcal{P}_\Omega, r, N_{ite}, \rho, \mu, \delta, \epsilon$) **:**
  Initialize: $W_0 = M_\Omega$, $M_0 = W_0$, and $Z_0 = W_0$.
  **for** $\ell = 1, \cdots, N_{ite}$ **do**
    Update the weights $\omega_\ell$ by Eq. (3.36);
    Update $W_\ell$ by Eq. (3.34);
    Update $\mathbf{m}_\ell$ by Eq. (3.38);
    Update $Z_\ell$ by Eq. (3.39).
  **end**
  $\widehat{M} = \text{reshape}(\mathbf{m}_\ell, I_x, I_y)$.
**return**

noise-specific matrix completion problems into light noisy MC, medium noisy MC, and heavy noisy MC, with detailed descriptions of each category's characteristics.

#### 3.3.2.1 Light Noisy Matrix Completion

When noise is not present, the primary challenge in matrix completion (MC) lies in accurately filling in the missing entries, rather than addressing noise issues. Matrix factorization [34, 35] provides an effective solution by decomposing the matrix into the product of two low-dimensional factor matrices, which enables the recovery of the full matrix. This approach constrains the rank of the unknown matrix within the dimensional limits of the factor matrices, thereby ensuring accurate recovery, as expressed in the following:

$$\min_{X,Y} \|M_\Omega - P_\Omega \odot (XY)\|_F^2, \tag{3.8}$$

Here, $P_\Omega = \mathcal{P}_\Omega^{(1)}$, $M_\Omega = \widetilde{\mathcal{M}}_\Omega^{(k)}$, and $X = \widetilde{\mathcal{X}}^{(k)}$ and $Y = \widetilde{\mathcal{Y}}^{(k)}$ represent the factor matrices. By reducing the approximation error and avoiding the computationally expensive singular value decomposition (SVD), matrix factorization achieves superior reconstruction accuracy and faster convergence compared to other methods. However, since noise reduction in matrix completion (MC) relies primarily on exploiting the low-rank structure of the data matrix, matrix factorization has limited noise suppression capabilities. This limitation arises because no explicit rank constraint is imposed on the factor matrices $X$ and $Y$, hindering its effectiveness in high-noise environments.

#### 3.3.2.2 Medium Noisy Matrix Completion

In moderate noise environments, SVD-free matrix factorization [33] improves denoising performance by introducing a rank regularization constraint on the two factor matrices, effectively preventing overfitting of the reconstruction error. This approach can be expressed as follows:

$$\begin{aligned} &\min_{X,Y} \|X\|_F^2 + \|Y\|_F^2, \\ &\text{s.t. } M_\Omega = P_\Omega \odot (XY). \end{aligned} \tag{3.9}$$

#### 3.3.2.3 Heavy Noisy Matrix Completion

Under high noise conditions, it becomes difficult for matrix factorization methods to consistently maintain the low-rank properties of the factor matrices during each iteration [32]. To address this issue, the nuclear norm is commonly introduced as a regularization term to guide the solution towards a low-rank matrix. The expression is given as follows:

$$\begin{aligned} &\min_{M} \|M\|_*, \\ &\text{s.t. } M_\Omega = P_\Omega \odot M, \end{aligned} \tag{3.10}$$

However, this approach inevitably affects the data reconstruction performance. A summary of the performance comparison for the three matrix completion (MC) methods is provided in Table 3.1. It is clear that the choice of the MC method plays a critical role in determining performance across different noise environments. This is the reason why tensor decomposition, SVD-free tensor decomposition, and the tensor nuclear norm (TNN) have been introduced to simultaneously perform data recovery and noise reduction on 3D seismic datasets, tailored to varying noise intensities.

**Table 3.1** ComparISON OF PERFORMANCE ACROSS VArious MC models

| Methods | Recovery | Denoising | Convergence |
|---|---|---|---|
| Matrix Factorization (MF) | High | Weak | Fast |
| SVD-free MF | Medium | Medium | Medium |
| Matrix Nuclear Norm | Low | Strong | Slow |

## 3.4 Structured Tensor Completion (STC)

Based on the previous discussion, we have proposed three distinct tensor models within the structured tensor completion (STC) framework, each tailored to different noise environments. These models include: structured tensor factorization for low noise conditions (detailed in Sect. 3.4.3), SVD-free tensor factorization for moderate noise conditions (explained in Sect. 3.4.2), and the weighted tensor nuclear norm for severe noise conditions (outlined in Sect. 3.4.3).

### *3.4.1 Model 1: Structured Tensor Factorization*

Inspired by the works of [31, 36], we broaden the use of low-tubal-rank tensor decomposition by incorporating Hankel constraints on the multiplication of two considerably smaller tensors that exhibit low-tubal-rank properties [37, 38]. The resulting tensor formulation is termed structured tensor factorization (STF), which is defined as follows:

$$\min_{\mathcal{X},\mathcal{Y}} \|\mathcal{M}_\Omega - \mathcal{P}_\Omega(\mathcal{X} * \mathcal{Y})\|_F^2 + \frac{\mu}{2} \|\mathcal{B}(\mathcal{X} * \mathcal{Y})\|_F^2 , \tag{3.11}$$

where $\mathcal{X} \in \mathbb{R}^{I_x \times r \times N_t}$ and $\mathcal{Y} \in \mathbb{R}^{r \times I_y \times N_t}$ are factor tensors that satisfy $\text{rank}(\mathcal{X}) = \text{rank}(\mathcal{Y})$, with $\mu > 0$ serving as a regularization parameter to control the error in the observed constraints. Using the convolution theorem , one can reformulate problem (3.11) into the following form in the frequency domain:

$$\min_{\widetilde{\mathcal{X}},\widetilde{\mathcal{Y}}} \|\widetilde{\mathcal{M}}_\Omega - \widetilde{\mathcal{P}}_\Omega \cdot \otimes(\widetilde{\mathcal{X}} \cdot \S\widetilde{\mathcal{Y}})\|_F^2 + \frac{\mu}{2} \left\|B(\widetilde{\mathcal{X}} \cdot \S\widetilde{\mathcal{Y}})\right\|_F^2 . \tag{3.12}$$

Utilizing the bridge introduced in Sect. 3.3, we can decompose problem (3.12) into $N_t$ distinct standard least squares (LS) minimization subproblems, as shown below:

$$\min_{\widetilde{\mathcal{X}}^{(k)},\widetilde{\mathcal{Y}}^{(k)}} |\widetilde{\mathcal{M}}_\Omega^{(k)} - \mathcal{P}_\Omega^{(1)} \odot (\widetilde{\mathcal{X}}^{(k)} \cdot \S\widetilde{\mathcal{Y}}^{(k)})\|_F^2 + \frac{\mu}{2} \left\|B(\widetilde{\mathcal{X}}^{(k)} \cdot \S\widetilde{\mathcal{Y}}^{(k)})\right\|_F^2 , \tag{3.13}$$

In this framework, the primary task of each subproblem is to approximate the factor matrices $\widetilde{\mathcal{X}}^{(k)}$ and $\widetilde{\mathcal{Y}}^{(k)}$ for the frontal slice $\mathcal{M}^{(k)}$, where $k \in [N_t]$. Therefore, problem (3.13) can be viewed as a structured matrix rank optimization problem. Given that AltMin is both computationally effective and memory-efficient [35], we opt to use AltMin to solve this structured matrix completion problem (3.13), with the implementation details provided in Algorithm 3.2. The algorithm begins with truncated singular value decomposition (SVD) (see Step 2). The function $\text{LS}(\cdot)$ (refer to Steps 4 and 6) essentially solves a least squares (LS) problem in an alternating manner to progressively recover $\widetilde{\mathcal{X}}^{(k)}$ and $\widetilde{\mathcal{Y}}^{(k)}$. Finally, orthogonalization is ensured at each iteration through QR decomposition (see Steps 5 and 7) [34]. The following steps outline the procedure for solving the LS problem LS$(\cdot)$.

1. A frontal slice $\widetilde{\mathcal{M}}_{\Omega}^{(k)}$ with dimensions $I_x \times I_y \times 1$ is transformed into a vector $\mathbf{m}_{\Omega}$ of size $I_x I_y \times 1$ through the following process:

$$\mathbf{m}_{\Omega} = [\text{squeeze}(\widetilde{\mathcal{M}}_{\Omega}(:, 1, k)); \text{squeeze}(\widetilde{\mathcal{M}}_{\Omega}(:, 2, k)); \ldots; \text{squeeze}(\widetilde{\mathcal{M}}_{\Omega}(:, I_y, k))]. \tag{3.14}$$

Likewise, when estimating $\widetilde{\mathcal{Y}}$, the slice $\widetilde{\mathcal{Y}}(:, :, k)$ is converted into a vector $y$ of size $rI_y \times 1$ using the following method:

$$\mathbf{y} = [\text{squeeze}(\widetilde{\mathcal{Y}}(:, 1, k)); \text{squeeze}(\widetilde{\mathcal{Y}}(:, 2, k)); \ldots; \text{squeeze}(\widetilde{\mathcal{Y}}(:, I_y, k))]. \tag{3.15}$$

2. $\widetilde{\mathcal{X}}^{(k)}$ is converted into a block diagonal matrix of dimensions $I_x I_y \times rI_y$ using the following process:

$$A_1 = \begin{bmatrix} \widetilde{\mathcal{X}}^{(k)} & & & \\ & \widetilde{\mathcal{X}}^{(k)} & & \\ & & \ddots & \\ & & & \widetilde{\mathcal{X}}^{(k)} \end{bmatrix}_{I_x I_y \times r I_y}. \tag{3.16}$$

3. The slice $\mathcal{P}_{\Omega}^{(1)}$ is initially converted into a vector of size $I_x I_y \times 1$, and subsequently transformed into a matrix $A_2$ with dimensions $I_x I_y \times I_x I_y$ using the following method:

$$P_i = \text{diag}(\mathcal{P}_{\Omega}^{(1)}(i, :)), \; i \in [I_x], \tag{3.17}$$

$$A_2 = \begin{bmatrix} P_1 & & & \\ & P_2 & & \\ & & \ddots & \\ & & & I_x \end{bmatrix}_{I_x I_y \times I_x I_y}, \tag{3.18}$$

Here, the operator diag$(\cdot)$ is used to transform a tube into a diagonal matrix by arranging its elements along the diagonal.

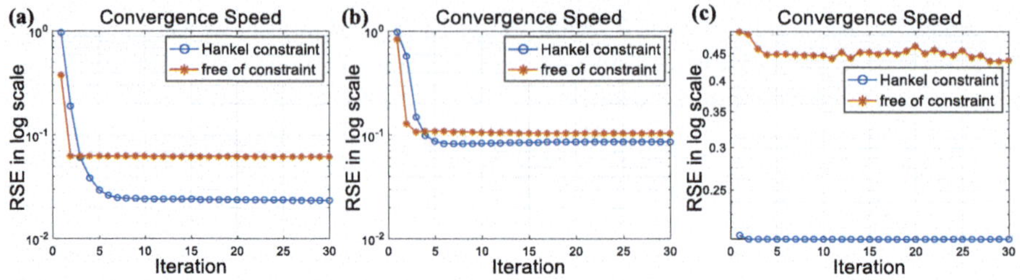


**Fig. 3.4** A comparison of the recovery error, measured by RSE, on a logarithmic scale against the iteration number is presented for methods both with and without the Hankel constraint under **a** light noise, **b** medium noise, and **c** heavy noise conditions, with 50% of the traces missing. The symbols "+", "∗", "×", and "o" correspond to our STC framework, MSSA, HRSC, DMSSA, and proposed STC, respectively

Consequently, the problem in (3.13) transforms into the following least squares (LS) problem:

$$\min_{\mathbf{y}} \frac{1}{2} \|\mathbf{m}_\Omega - A_2 A_1 \mathbf{y}\|_F^2 + \frac{\mu}{2} \|B A_1 \mathbf{y}\|_F^2 . \tag{3.19}$$

A straightforward closed-form solution for (3.19) is available, given by:

$$y = (\mu (BA_1)^H (BA_1) + (A_2 A_1)^H (A_2 A_1))^{-1} (A_2 A_1)^H \mathbf{m}_\Omega . \tag{3.20}$$

In MATLAB notation, $Y$ can be calculated using $Y = \text{reshape}(\mathbf{y}, r, I_y)$, the reshape$(\cdot)$ function reorganizes the vector into a matrix by arranging its elements into columns. It is possible to compute an estimate of $x$ by following similar steps, as outlined below:

$$\min_{\widetilde{\mathcal{X}}^{(k)}} \|\widetilde{\mathcal{M}}_\Omega^{(k)H} - \mathcal{P}_\Omega^{(1)} \odot (\widetilde{\mathcal{X}}^{(k)H} \cdot \S \widetilde{\mathcal{Y}}^{(k)H})\|_F^2 + \frac{\mu}{2} \left\| B(\widetilde{\mathcal{X}}^{(k)H} \cdot \S \widetilde{\mathcal{Y}}^{(k)H}) \right\|_F^2 . \tag{3.21}$$

## 3.4.2 *Model 2: SVD-Free Structured Tensor Factorization*

SVD-free structured tensor factorization (SSTF) is a specialized tensor factorization technique that primarily focuses on minimizing the rank of the objective function (3.22) for tensor decomposition. Unlike traditional tensor factorization methods, SSTF incorporates a variational representation of the trace norm [39] during the optimization process, effectively constraining the rank while preserving the tensor structure, as expressed in the following:

$$\begin{aligned} &\min_{\mathcal{X},\mathcal{Y}} \frac{\lambda}{2} (\|\mathcal{X}\|_F^2 + \|\mathcal{Y}\|_F^2) + \frac{\mu}{2} \|\mathcal{B}(\mathcal{X} * \mathcal{Y})\|_F^2 , \\ &\text{s.t. } \mathcal{M}_\Omega = \mathcal{P}_\Omega (\mathcal{X} * \mathcal{Y}), \end{aligned} \tag{3.22}$$

where $\lambda > 0$ regulates the level of regularization. Following a Fourier transform, we implement tubal sampling in a comparable way to break down the $N_t$ independent subproblems as shown below:

$$\begin{aligned}\min_{\widetilde{\mathcal{X}}^{(k)},\widetilde{\mathcal{Y}}^{(k)}} \frac{1}{2} \left\| \widetilde{\mathcal{M}}_{\Omega}^{(k)} - \mathcal{P}_{\Omega}^{(1)} \odot (\widetilde{\mathcal{X}}^{(k)} \cdot \S\widetilde{\mathcal{Y}}^{(k)}) \right\|_F^2 \\ + \frac{\mu}{2} \left\| \mathcal{B}(\widetilde{\mathcal{X}}^{(k)} \cdot \S\widetilde{\mathcal{Y}}^{(k)}) \right\|_F^2 + \frac{\lambda}{2} (\|\widetilde{\mathcal{X}}^{(k)}\|_F^2 + \left\|\widetilde{\mathcal{Y}}^{(k)}\right\|_F^2).\end{aligned} \tag{3.23}$$

The problem (3.23) essentially represents a structured SVD-free matrix factorization problem. Within the framework of matrix factorization, we can apply the previously mentioned alternating minimization method to address the optimization challenge in (3.23). However, when dealing with the least squares problem outlined in Algorithm 3.2, it is necessary to incorporate a rank minimization constraint, reformulating it into the form presented in problem (3.19), as follows:

$$\min_{\mathbf{y}} \frac{1}{2} \|\mathbf{m}_{\Omega} - A_2 A_1 \mathbf{y}\|_F^2 + \frac{\mu}{2} \|B A_1 \mathbf{y}\|_F^2 + \frac{\lambda}{2} \|\mathbf{y}\|_F^2 . \tag{3.24}$$

Similar to (3.19), the Eq. (3.24) also possesses a straightforward closed-form solution.

$$\begin{aligned}\mathbf{y} =& (\mu(BA_1)^H(BA_1) + (A_2A_1)^H(A_2A_1) + \lambda I)^{-1} \\ &(A_2A_1)^H \mathbf{m}_{\Omega}.\end{aligned} \tag{3.25}$$

In a manner akin to structured tensor factorization (STF), we can also deduce $x$ as follows:

$$\begin{aligned}\min_{\widetilde{\mathcal{X}}^{(k)}} \frac{1}{2} \left\| \widetilde{\mathcal{M}}_{\Omega}^{(k)H} - \mathcal{P}_{\Omega}^{(1)H} \odot (\widetilde{\mathcal{X}}^{(k)H} \cdot \S\widetilde{\mathcal{Y}}^{(k)H}) \right\|_F^2 \\ + \frac{\mu}{2} \left\| \mathcal{B}(\widetilde{\mathcal{X}}^{(k)H} \cdot \S\widetilde{\mathcal{Y}}^{(k)H}) \right\|_F^2 + \frac{\lambda}{2} (\|\widetilde{\mathcal{X}}^{(k)H}\|_F^2 + \left\|\widetilde{\mathcal{Y}}^{(k)H}\right\|_F^2).\end{aligned} \tag{3.26}$$

As highlighted in previous studies [35, 38], the least squares (LS) problems (3.19) and (3.24) in Algorithm 3.2 can be effectively addressed by computing the inverses of the matrices in equations (3.20) and (3.25). Thus, compared to the singular value decomposition (SVD) operation typically used in baseline algorithms, our method may require more computational resources (see Sect. 3.5.1). However, as shown in Fig. 3.5, the proposed method converges faster compared to the reference algorithm, demonstrating one of the main benefits of the AltMin framework [31]. The reduced number of iterations helps alleviate the overall computational burden. Moreover, the intrinsic symmetric structure and frequency-limited nature of real seismic data considerably decrease the number of frontal slices requiring processing through low-rank matrix completion (LRMC). Ultimately, the convergence of Algorithm 3.2 for the problems (3.13) and (3.26) is guaranteed by the AltMin framework [31].

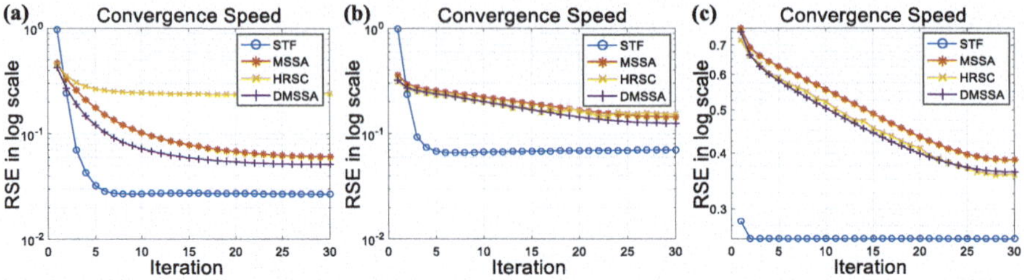


**Fig. 3.5** A comparison of the recovery error, indicated by RSE, on a logarithmic scale concerning the iteration number is shown for various methods under **a** light noise, **b** medium noise, and **c** heavy noise conditions, with 50% of the traces missing. The symbols "+", "∗", "×", and "o" correspond to our STC framework, MSSA, HRSC, DMSSA, and proposed STC, respectively

### *3.4.3 Model 3: Structured Tensor Weighted Nuclear Norm*

Building on the methods proposed by [40, 41], we propose a structured tensor weighted nuclear norm (STWNN) esigned for seismic data reconstruction in the presence of significant noise. The method consists of two main components: first, a low-tubal-rank approximation using the weighted tensor nuclear norm, which imposes a low-tubal-rank structure on the reconstructed data. Compared to the traditional tensor nuclear norm, the weighted tensor nuclear norm provides a more accurate rank representation, thereby enhancing the completion performance. Second, structured regularization with Hankel constraints is employed, which effectively preserves the Hankel structure during each iteration, analogous to the previously discussed models. These two components can be expressed as follows:

$$\begin{aligned} &\min_{\mathcal{M}} \|\mathcal{M}_\Omega - \mathcal{P}_\Omega(\mathcal{M})\|_F^2 + \frac{\lambda}{2}\|\mathcal{W}\|_{*,\omega} + \frac{\mu}{2}\|\mathcal{B}(\mathcal{M})\|_F^2 . \\ &\text{s.t. } \mathcal{W} = \mathcal{M},\ \mathcal{W}_\Omega = \mathcal{M}_\Omega, \end{aligned} \tag{3.27}$$

Here, $\|\mathcal{W}\|_{*,\omega}$ represents the tensor weighted nuclear norm, as defined in [41], and is formulated as follows:

$$\|\mathcal{W}\|_{*,\omega} = \sum_{i=1}^{N} \|\mathcal{W}_{(i)}\|_{*,\omega}, \tag{3.28}$$

In this context, $\|W\|_{*,\omega}$ refers to the weighted nuclear norm of the matrix, as defined in [40], which is expressed as follows:

$$\|W\|_{*,\omega} = \sum_{j=1}^{J} \sigma_j w(j), \tag{3.29}$$

In this context, $\sigma_j$ denotes the $j$th largest singular value of the matrix $X$, and $J$ represents the complete set of singular values. The vector $\omega$ is defined as $\omega = \delta/(\sigma(W) + \epsilon)$, where $\delta \in \mathbb{R}^+$ is a positive constant, $\sigma(W) \in \mathbb{R}^J$ is the vector of singular values of $W$, and $\epsilon$ is a constant introduced to prevent division by zero, as discussed in [40].

The weighted nuclear norm can be regarded as a particular case of the nuclear norm. Similarly, the tensor-weighted nuclear norm (as shown in Eq. 3.27) is a special instance of the tensor nuclear norm defined in [28], and can be expressed as:

$$\begin{aligned} &\min_{\widetilde{\mathcal{T}}^{(k)}} \frac{1}{2} \left\| \widetilde{\mathcal{M}}_{\Omega}^{(k)} - \mathcal{P}_{\Omega}^{(1)} \odot \widetilde{\mathcal{M}}^{(k)} \right\|_F^2 + \frac{\lambda}{2} \|\widetilde{\mathcal{W}}^{(k)}\|_{*,\omega}^2 + \frac{\mu}{2} \left\| \mathcal{B}(\widetilde{\mathcal{M}}^{(k)}) \right\|_F^2, \\ &\text{s.t. } \widetilde{\mathcal{W}}^{(k)} = \widetilde{\mathcal{M}}^{(k)}, \ \widetilde{\mathcal{W}}_{\Omega}^{(k)} = \widetilde{\mathcal{M}}_{\Omega}^{(k)}. \end{aligned} \tag{3.30}$$

To enhance clarity, we introduce two auxiliary variables: $W$ is defined as $\widetilde{\mathcal{W}}^{(k)}$, and $W_\Omega$ as $W_\Omega = \widetilde{\mathcal{W}}_{\Omega}^{(k)}$. With these definitions, Eq. (3.30) can be reformulated as:

$$\begin{aligned} &\min_{\mathbf{m}} \|\mathbf{m}_\Omega - A_2 \mathbf{m}\|_F^2 + \frac{\lambda}{2} \|W\|_{*,\omega} + \frac{\mu}{2} \|B\mathbf{m}\|_F^2, \\ &\text{s.t } W = M, \ W_\Omega = M_\Omega. \end{aligned} \tag{3.31}$$

To solve the minimization problem in (3.30), we apply the alternating direction method of multipliers (ADMM). Following the fundamental framework of ADMM [42], we transform (3.30) as:

$$\begin{aligned} L(M, Z, W, \rho) = \|\mathbf{m}_\Omega - A_2 \mathbf{m}\|_F^2 + \|W\|_{*,\omega} + \frac{\mu}{2} \|B\mathbf{m}\|_F^2 \\ + \frac{\rho}{2} \|W - M\|_F^2 + \langle Z, W - M \rangle, \end{aligned} \tag{3.32}$$

where $\rho$ is a positive scalar. Starting with the initial assignments $W = M_\Omega$, $M = W$, and $Z = W$, the l lagrangian formulation in Eq. (3.32) can be solved in three stages, as outlined below.

#### 3.4.3.1 Update $W_\ell$

Fix $M_{\ell-1}$ and $Z_{\ell-1}$ to minimize $L(M_{\ell-1}, Z_{\ell-1}, W_\ell, \rho)$ for $W_\ell$ as:

$$\begin{aligned} W &= \underset{W}{\operatorname{argmin}} \|W\|_{*,\omega} + \frac{\rho}{2} \|W - M\|_F^2 + \langle Z, W - M \rangle \\ &= \underset{W}{\operatorname{argmin}} \|W\|_{*,\omega} + \frac{\rho}{2} \left\| W - (M - \frac{1}{\rho} Z) \right\|_F^2. \end{aligned} \tag{3.33}$$

According to Theorem 2 from [40], the aforementioned problem (3.33) can be addressed as follows:

$$W_\ell = S_{\frac{1}{\rho}\omega}(M_{\ell-1} - \frac{1}{\rho}Z_{\ell-1}), \tag{3.34}$$

Here, $S_\omega(\Sigma)$ refers to the generalized soft-thresholding operator, which is defined as

$$S_\omega(\Sigma)(i,j) = \begin{cases} \max(\Sigma(i,j) - \omega(i), 0), & i = j, \\ 0, & i \neq j, \end{cases} \tag{3.35}$$

with weights $\omega$ given by

$$\omega = \delta/(\sigma(M_{\ell-1}) + \epsilon). \tag{3.36}$$

#### 3.4.3.2 Update $M_\ell$

Fix $W_\ell$ and $Z_{\ell-1}$ to bbtain $M_\ell$ as:

$$\mathbf{m} = \underset{\mathbf{w}}{\text{argmin}} \; \|\mathbf{m}_\Omega - A_2\mathbf{m}\|_F^2 + \frac{\mu}{2}\|B\mathbf{m}\|_F^2 + \frac{\rho}{2}\|W - M\|_F^2 + \langle Z, W - M \rangle, \tag{3.37}$$

This expression represents a quadratic function of $\mathbf{m}_\ell$. Solving it involves taking the derivative with respect to $\mathbf{m}_\ell$ and setting it equal to zero, leading to the following result

$$\mathbf{m}_\ell = (\rho I + \mu B^\top B + A_2^\top A_2)^{-1}(\rho \mathbf{w}_\ell + A_2^\top \mathbf{m}_\Omega + \mathbf{z}_\ell). \tag{3.38}$$

In MATLAB, $M_\ell$ and $Z_\ell$ can be computed by reshaping the vectors $\mathbf{m}_\ell$ and $\mathbf{z}_\ell$ into matrices of dimensions $I_x \times I_y$, respectively, using the reshape function: $M_\ell = \text{reshape}(\mathbf{m}_\ell, I_x, I_y)$ and $Z_\ell = \text{reshape}(\mathbf{z}_\ell, I_x, I_y)$.

#### 3.4.3.3 Update $Z_{\ell+1}$

Fix $W_\ell$ and $M_\ell$ to evaluate $Z_\ell$ as:

$$Z_\ell = Z_{\ell-1} + \rho(W_\ell - M_\ell). \tag{3.39}$$

The detailed implementation of the SWTNN algorithm is provided in Algorithm 3.3. The algorithm's convergence behavior in solving the optimization problem (3.31) is ensured by employing the alternating direction method, with the theoretical foundation outlined in [42].

## 3.5 Field Examples of STC

This part demonstrates the efficacy and advantages of the proposed approach through its application to the completion of 3D seismic data, specifically in situations where traces are missing. We examine two distinct types of 3D seismic data: (1) synthetic data and (2) real field data. In Sect. 3.5.1, we begin by outlining the experimental framework for the evaluation, followed by a comparison of the proposed method with current leading rank-reduction approaches, using both synthetic data (see Sect. 8.4.2) and real datasets (see Sect. 8.4.3) for evaluation.

### *3.5.1 Experimental Setup*

#### 3.5.1.1 Performance Metrics

To assess the enhanced performance of the proposed models, it is crucial to establish metrics for evaluating the quality of the reconstructed results. Given the original fully observed dataset $\mathcal{T}$ (ground truth) and the reconstructed data $\widehat{\mathcal{T}}$ obtained from the incomplete measurements, we introduce two key performance metrics. One of these is the relative square error (RSE) defined as

$$\mathrm{RSE} = \frac{\|\mathcal{T} - \widehat{\mathcal{T}}\|_F}{\|\mathcal{T}\|_F}, \tag{3.40}$$

and another important metric is the SNR, calculated as [43]

$$\mathrm{SNR}\,[\mathrm{dB}] = 20 \log_{10} \frac{1}{\mathrm{RSE}}. \tag{3.41}$$

#### 3.5.1.2 Compared Algorithms

To highlight the enhanced effectiveness of the proposed STC method, we categorize the algorithms into two groups for comparison. The first group consists of the Hankel t-SVD decomposition method discussed in Sect. 3.7 and our STC model. This group leverages various rank-reduction techniques, such as MSSA [9], DMSSA [26], and HRSC [24], to provide a low-tubal-rank solution for Hankel tensors. The second group includes widely used tensor decomposition methods, including tucker decomposition (TKD) and canonical polyadic decomposition (CPD) [44, 45], which are used to evaluate the interpolation and denoising capabilities, providing a comparison with the STC model. In the TKD framework, the recovery of missing traces from a 3D incomplete seismic tensor can be performed directly, without the need for extra optimization procedures, such as the commonly used parallel matrix factorization

(PMF) algorithm [7, 18]. This explains why t-SVD and TKD are the primary methods used for seismic tensor completion.

In contrast, the CPD model faces considerable challenges when applied to tensor completion tasks with tubal or fiber sampling. This complexity necessitates the use of specific strategies for effective resolution, as outlined in the FCPD framework in [46]. Based on the work of [46], tubal sampling in FCPD can be divided into two distinct types: regular tubal sampling (also referred to as systematic sampling) [46–48] and random tubal sampling [46, 49]. As missing seismic traces generally occur in a random manner, the task of completing the seismic tensor naturally aligns with the framework of random tubal sampling. While [49] provides a comprehensive theoretical analysis of random tubal sampling, the algebraic methods discussed in [46] are based on the assumption of noiseless conditions. However, in practice, real seismic data often contain substantial noise [47].

After a comprehensive evaluation of the aforementioned methods, we selected four reference algorithms that have demonstrated significant performance in interpolation and denoising, as reported in existing literature.

- *MSSA* [9]: This method utilizes truncated SVD to decompose the Hankel matrix of the signal, separating it into signal and noise subspaces, thereby effectively extracting the signal and reducing noise.
- *DMSSA* [26]: By employing the damped DMSSA algorithm, this approach further optimizes the decomposition of signal and noise subspaces, significantly improving noise attenuation and enhancing data reconstruction accuracy.
- *HRSC* [24]: This method combines rank-reduction and sparsity-promoting techniques through a hybrid rank-sparsity constraint (HRSC) model, enabling simultaneous seismic data reconstruction and denoising while leveraging the advantages of both approaches.
- *PMF* [7, 18]: This method recovers missing seismic traces by utilizing low-rank matrix decompositions across various tensor modes, which represent the tensor in various matrix formats, thus achieving precise completion of the seismic data.

#### 3.5.1.3 Parameter Settings

To assess the performance of our proposed STC method relative to other reference approaches, we specify the essential parameters required for the model and the numerical algorithms. To identify the optimal low-tubal-rank $r$ for the Hankel tensor $\mathcal{M}$, we test values ranging from 2 to 20, with an increment of 1, in order to achieve the best reconstruction quality for each method. The numerical algorithm parameters are configured as follows: the maximum iteration count is set to $N_{ite} = 30$. In the STC framework, the performance of the STF model is primarily influenced by the penalty parameter $\mu$, which is set to 20 to enhance the effect of matrix $B$. For the SVD-free structured tensor factorization (SSTF) model, we focus on two key parameters, $\mu$ and $\lambda$, which are set to $\mu = 20$ and $\lambda = 3$. In the structured tensor weighted nuclear norm (STWNN) model, the penalty parameters $\mu$ and $\rho$ are similarly crucial, with values set to $\mu = 20$ and $\rho = 3$. To ensure a fair comparison, we use consistent model parameter settings across the MSSAA, DMSSA, and HRSC algorithms, matching those used in the STC framework. Additionally, the parameters for each algorithm have been fine-tuned based on their respective foundational literature [9, 24, 26].

### 3.5.2 Validation on Synthetic Data

To assess the effectiveness of the STC framework, we use a 3D synthetic dataset created with an open-source MATLAB tool, referenced in [26].[1] The dataset comprises three intersecting linear events in the $t - x - y$ space, with a total of $N_x \times N_y$ traces, where $N_x = 20$ and $N_y = 20$. The dataset spans a time duration of 1200 ms, with a sampling step of 4 ms, leading to $N_t = 300$ temporal samples. The 2-fold Hankel tensor formed in this case has dimensions of $121 \times 100 \times 300$. To assess the robustness of the rank-reduction method under noisy conditions, we introduce artificial random noise to the ground truth data, simulating different signal-to-noise ratios (SNRs). Specifically, we test three SNR levels: SNR = 1 dB (high noise), SNR = 10 dB (moderate noise), and SNR = 20 dB (low noise).

We begin by examining how the inclusion of the Hankel constraint enhances performance in two distinct approaches. The first approach involves comparing the Hankel structure of the final estimates obtained using different methods, as illustrated in Fig. 3.3. To enhance the clarity of the comparison, only a portion of the Hankel matrix is shown, enabling a more direct observation of the differences among the methods. The results demonstrate that the STWNN method successfully recovers the low-rank Hankel structure from the Hankel tensor of the incomplete and noisy data, while other methods fail to achieve the same level of restoration. The second approach investigates the convergence behavior of methods both with and without the Hankel constraint across different noise levels—light, medium, and heavy—shown in Fig. 3.4a–c, respectively. This comparison reveals that the method incorporating the Hankel constraint consistently exhibits a lower relative square error (RSE) across all noise levels. These approaches clearly demonstrate the significant advantage of using the Hankel constraint in improving performance.

We then apply five different algorithms to incomplete 3D data generated by randomly sampling image pixels at various sampling rates ranging from 10 to 90%. The algorithms involved include MSSA, DMSSA, HRSC, PMF, and our STC method. The recovery results are evaluated under light, medium, and heavy noise conditions, as illustrated in Figs. 3.6, 3.7, and 3.8. The findings show that the STC method consistently exceeds the performance of the other four techniques across all noise conditions, with notable improvements seen in the accuracy of subsampled data, especially for decimation levels between 10 and 70%, thus validating the effectiveness of our method. It is noteworthy that the performance of the STC algorithm varies

[1] https://www.sciencedirect.com/science/article/pii/S0098300416301650#s0040.

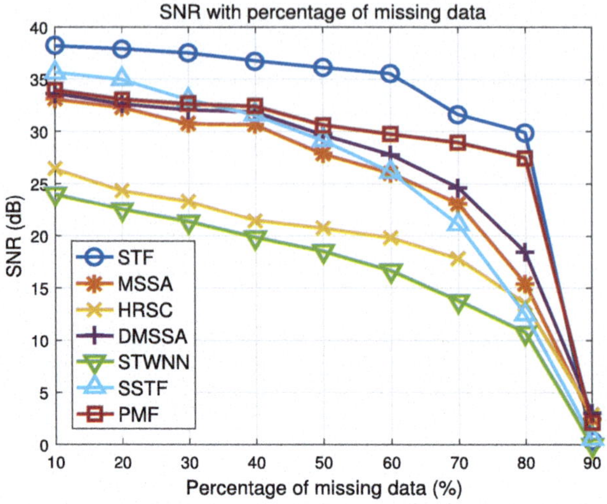


**Fig. 3.6** A comparison of the SNR relative to the proportion of missing data is provided for the proposed methods and other existing techniques under light noise conditions (SNR = 20 dB) is presented

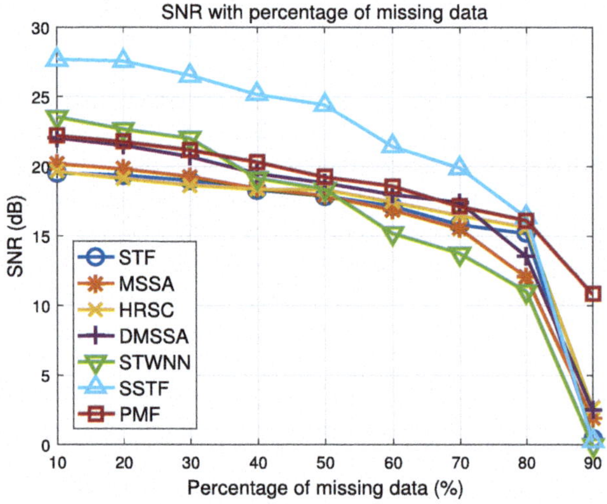


**Fig. 3.7** A comparison of the SNR relative to the proportion of missing data is provided for the proposed methods and other existing techniques under medium noise conditions (SNR = 10 dB) is presented

with the noise environment. For instance, in a light noise scenario, STF performs the best, whereas in a heavy noise scenario, STWNN achieves the highest reconstruction quality. This observation supports the effectiveness of the bridge mechanism discussed in Sect. 3.3.2. However, there are instances where the performance of the STC method is not as satisfactory as the existing methods, as illustrated in Figs. 3.6, 3.7, and 3.8. One possible explanation for this could be that with 90% of the data missing, the AltMin and the ADMM algorithms within the STC framework struggle to perform effectively on highly sparse data, leading to suboptimal recovery results.

To ensure a comprehensive evaluation, we conducted simulation experiments to assess the convergence behavior of the STC framework under different noise scenarios. We used a 3D synthetic dataset with a 50% sampling rate to assess the

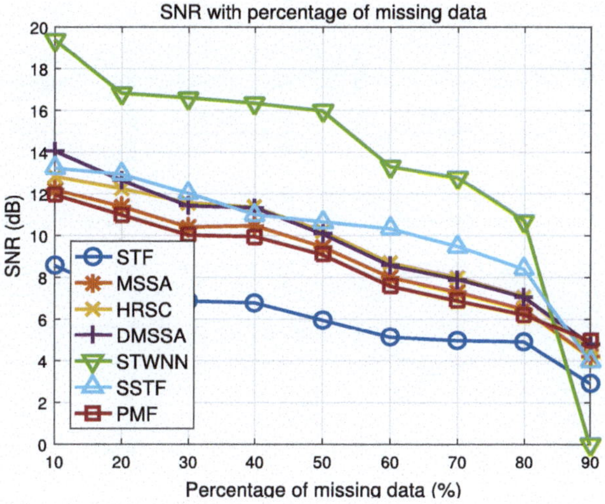


**Fig. 3.8** A comparison of the SNR relative to the proportion of missing data is provided for the proposed methods and other existing techniques under heavy noise conditions (SNR = 1 dB) is presented

stopping criterion and relative square error (RSE) for different methods. The RSE performance is presented against the number of iterations for each method under light, medium, and heavy noise conditions in Fig. 3.5a–c, respectively. To better highlight regions with lower values, Fig. 3.5 employs a semi-logarithmic scale, with a linear scale for the iteration axis and a logarithmic scale for the RSE axis. As depicted in Fig. 3.5, Experiments show that the STC framework delivers the quickest convergence and the smallest final RSE values under all levels of noise. For instance, under moderate noise conditions, methods like MSSA, HRSC, and DMSSA stabilize at an error value of 0.1 after 30 iterations. In contrast, the SSTF model reaches a more rapid convergence, settling at a smaller error value of 0.05 within only 5 iterations. It is important to note that the convergence curve of PMF cannot be monitored here, as it operates iteratively on separate unfolding matrices. This result aligns with our expectations, confirming that the AltMin and the ADMM techniques effectively accelerate convergence within the STC framework. By applying a Hankel constraint to the process noise term, these methods help prevent divergence in the error covariance updates in Eqs. (3.19), (3.24), and (3.38).

We showcase the reconstructed slices of the 3D synthetic dataset, processed using different techniques under varying noise levels in Figs. 3.9, 3.10, and 3.11, corresponding to light, medium, and heavy noise levels, respectively. These plots present a comparative analysis of reconstruction and denoising performance for the third crossline section of the 3D synthetic dataset. Since the order of the sub-images is consistent across Figs. 3.9, 3.10, and 3.11, we focus on Fig. 3.11 for a detailed performance comparison. Figures 3.11a and e display the noisy and decimated data, respectively, where the original data has 50% of the traces missing. The results following the application of DMSSA, HRSC, and STWNN are shown in Fig. 3.11b–d, respectively. The corresponding reconstruction errors for DMSSA, HRSC, and our STWNN are presented in Fig. 3.11f–h. Additionally, we quantitatively compare the

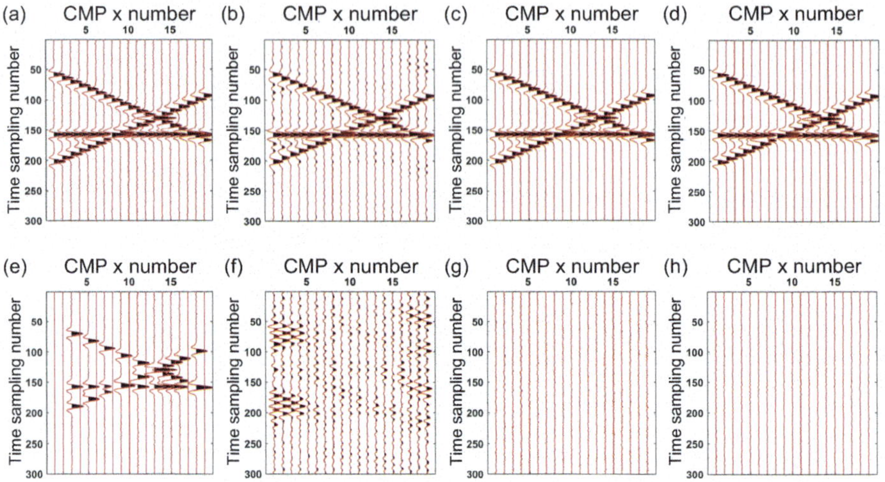


**Fig. 3.9** A comparison of reconstruction and denoising performance is shown for the third crossline section of a 3D synthetic dataset under light noise conditions. Panel (**a**) illustrates the noisy input, while panels (**b**)–(**d**) present the denoised results obtained using HRSC, DMSSA, and STF methods, respectively. Panel (**e**) shows the decimated dataset, and panels **f**–**h** highlight the reconstruction errors for HRSC, DMSSA, and STF

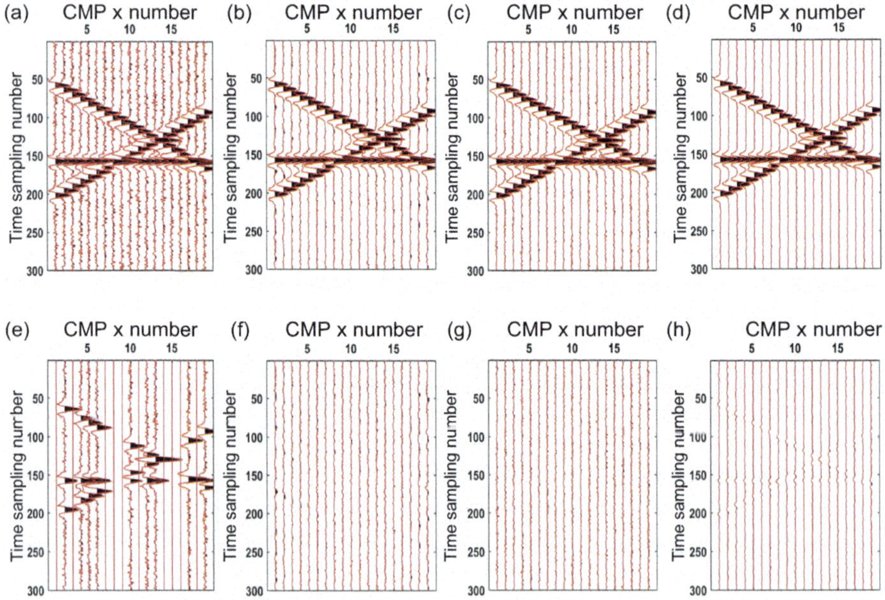


**Fig. 3.10** This figure presents a comparison of the simultaneous reconstruction and noise reduction for the third crossline slice of a 3D synthetic dataset under moderate noise conditions. Panel (**a**) shows the noisy data, while panels (**b**)–(**d**) present the reconstructed and denoised results obtained using HRSC, DMSSA, and SSTF, respectively. The decimated data is displayed in panel (**e**), and panels (**f**)–(**h**) illustrate the reconstruction errors for HRSC, DMSSA, and SSTF, respectively

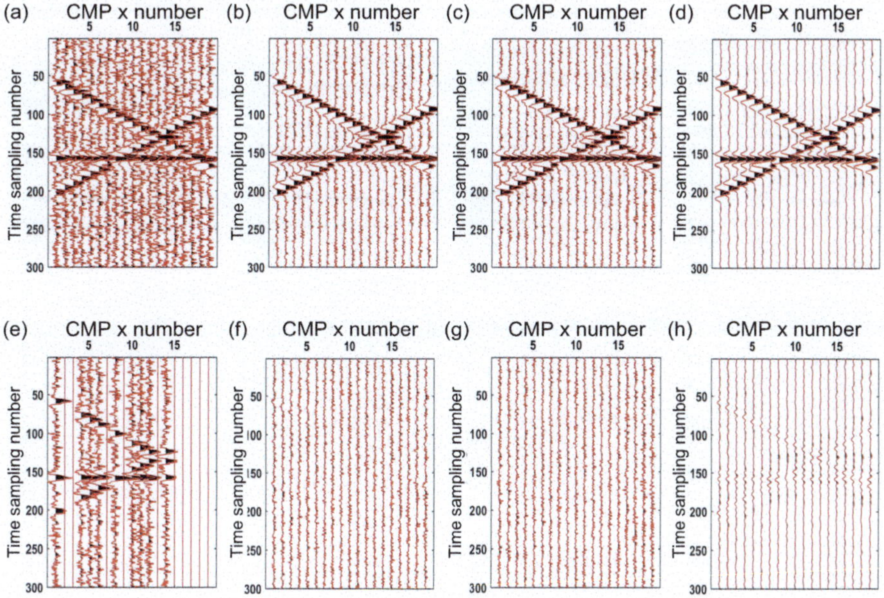


**Fig. 3.11** A comparison of simultaneous reconstruction and noise reduction for the third crossline slice of a 3D synthetic dataset under high noise conditions is presented. Panel (**a**) shows the noisy input, while panels (**b**)–(**d**) display the restored and denoised outputs using HRSC, DMSSA, and STWNN, respectively. Panel (**e**) presents the decimated version of the data, and panels (**f**)–(**h**) show the corresponding reconstruction errors for each method: HRSC, DMSSA, and STWNN

reconstruction performance using SNR and RSE, as shown in Figs. 3.5c and 3.8. As shown in Fig. 3.8, the HRSC and DMSSA methods achieve SNR values of 10.24 and 10.1 dB under the 50% decimation scenario, but the STWNN method produces a higher SNR of 15.99 dB, outperforming the two baseline methods. From Figs. 3.5, 3.6, 3.7, 3.8, 3.9, 3.10 and 3.11, it is evident that while DMSSA and HRSC are effective in reconstructing missing traces and reducing noise, our STC framework delivers superior results, particularly in significantly enhancing the SNR.

### *3.5.3 Validation on Real Field Data*

To validate the effectiveness of our STC framework, we applied it to three real 3D post-stack seismic datasets. For testing, we cropped each dataset into data tensors with dimensions of $N_x \times N_y \times N_t$, where $N_x = 80$, $N_y = 20$, and $N_t = 500$. The resulting 2-fold Hankel tensors for this analysis have dimensions of $451 \times 400 \times 500$. During the processing of these datasets, we employed a noise reduction technique, such as structure-oriented filtering [50], to reduce noise in the original data and generate a set of ground-truth data. We then simulated noisy data by adding synthetic

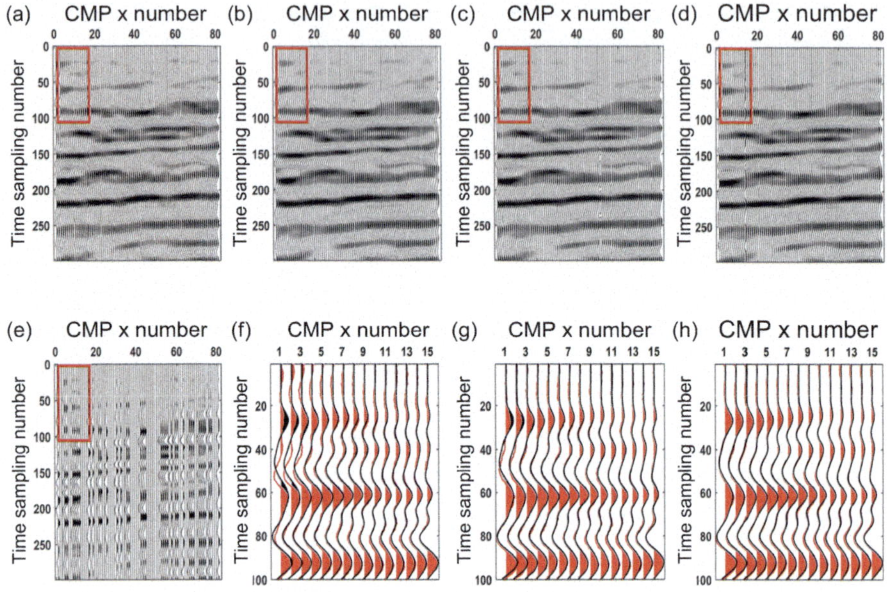


**Fig. 3.12** A comparison of concurrent reconstruction and noise reduction for the fifth crossline slice of a 3D field example under mild noise conditions is provided. (**a**) shows the noisy data, while (**b**)–(**d**) depict the reconstructed and denoised results achieved using HRSC, DMSSA, and STF, respectively. (e) illustrates the decimated data, and plots (**f**)–(**h**) provide close-up views of the results shown in (**b**)–(**d**)

random noise to the ground-truth data, corresponding to light, medium, and heavy noise conditions, with SNR of 20, 10, and 1 dB, respectively. However, it is important to note that the post-stack data are not completely noise-free, so exact SNR values are not provided in this context. The three noisy datasets are illustrated in Figs. 3.12a, 3.13a, and 3.14a. To simulate incomplete data, we randomly omitted half of the traces from the ground-truth dataset, resulting in the decimated datasets shown in Figs. 3.12e, 3.13e, and 3.14e.

Figures 3.12, 3.13, and 3.14 present a comparative analysis of simultaneous reconstruction and denoising for real-world field data under low, medium, and high noise levels, respectively. To further evaluate the robustness of the proposed methods, we focus on data recovery performance in heavily noisy environments, as illustrated in Fig. 3.14. Specifically, Fig. 3.14b–d show the restored and noise-reduced outcomes achieved using DMSSA, HRSC, and STWNN, respectively. Notably, STWNN demonstrates significant improvements in the regions highlighted by red rectangles, showcasing its adaptability and effectiveness under severe noise conditions. To provide a clearer visual comparison, we overlaid the recovered data (in red) and the original ground-truth data (in black) using the seismic analysis toolbox SeisLab.[2]

[2] http://www.mathworks.com/matlabcentral/fileexchange/15674-seislab-3-01.

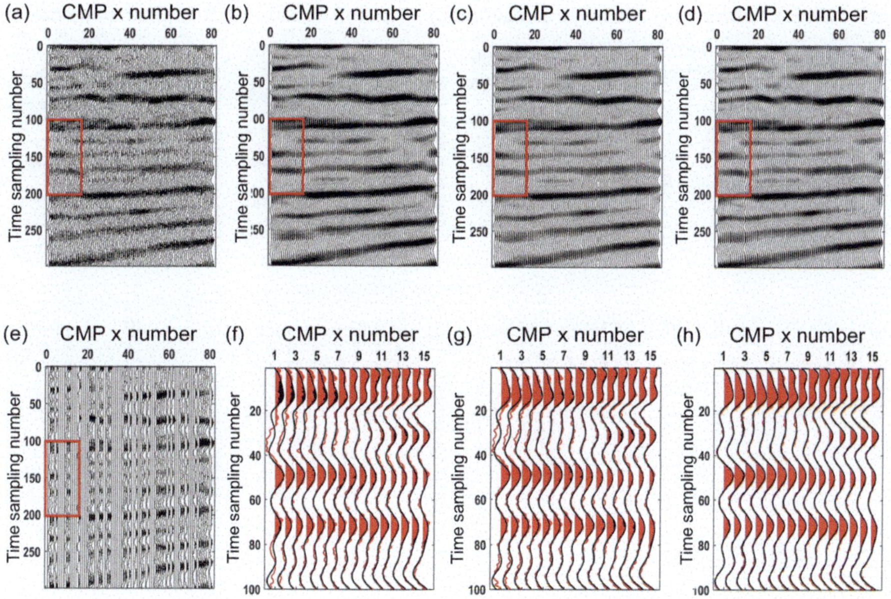


**Fig. 3.13** A comparison of joint reconstruction and noise reduction for the fifth crossline slice of a 3D field dataset under moderate noise levels is presented. Panel (**a**) presents the noisy data, while panels (**b**)–(**d**) showcase the reconstruction and denoising outcomes achieved using HRSC, DMSSA, and SSTF, respectively. Panel (**e**) illustrates the decimated data, and panels (**f**)–(**h**) zoom in on the results shown in (**b**)–(**d**) for a closer examination

These results are displayed in Fig. 3.14f–h. It is evident that HRSC exhibits noticeable deviations from the ground truth across most missing traces, while DMSSA encounters similar limitations. In contrast, the STC framework consistently achieves superior performance across all scenarios, as evidenced by the close alignment between reconstructed (red) and ground-truth (black) curves in Figs. 3.12, 3.13, and 3.14. This consistency underscores the robustness and effectiveness of the proposed framework across varying noise levels.

## 3.6 Summary

This chapter presents an innovative STC framework designed to leverage the inherent structure Hankel and low-tubal-rank constraint of seismic data, addressing the challenges of data recovery under complex noise conditions. By incorporating tubal sampling techniques, the STC framework integrates methods such as structured tensor factorization, SVD-free tensor decomposition, and weighted tensor nuclear norm to achieve effective reconstruction and denoising across light, medium, and heavy

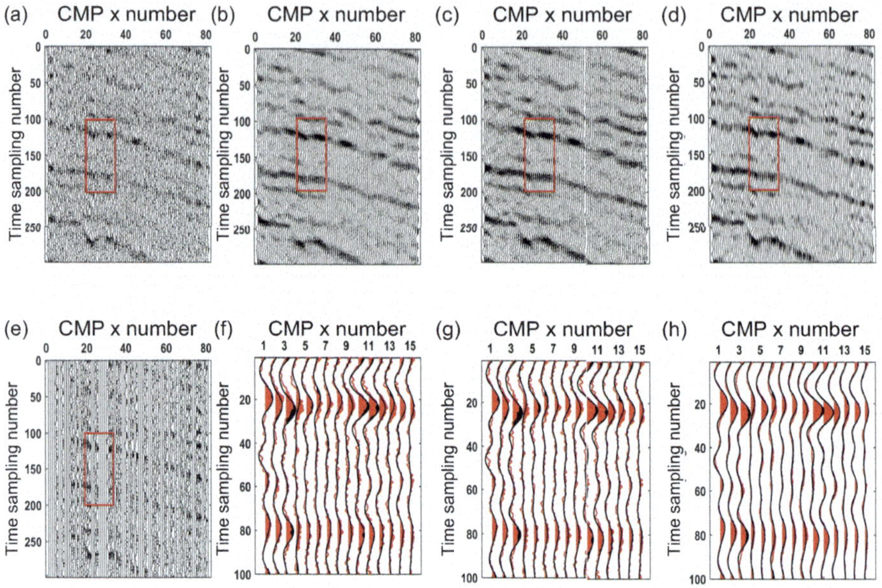


**Fig. 3.14** A comparison of the joint reconstruction and noise reduction performance for the fifth crossline slice of a 3D seismic field dataset under high noise levels is provided. Panel (**a**) presents the noisy data, while panels (**b**)–(**d**) show the results of reconstruction and denoising achieved using the HRSC, DMSSA, and STWNN methods, respectively. Panel (**e**) depicts the subsampled data. Additionally, close-up views of the results in panels (**b**)–(**d**) are provided in plots (**f**)–(**h**)

noise scenarios. Additionally, we designed an optimization strategy based on alternating minimization and employed the ADMM to enhance the efficiency of solving matrix decomposition and structured nuclear norm problems. Compared to existing techniques, including MSSA, DMSSA, and HRSC, the STC framework offers a significant advantage by incorporating the structure Hankel constraint and developing a tensor model optimized for varying noise levels. Experimental results demonstrate its superior performance in both synthetic and real-world datasets, achieving notable improvements in reconstruction and denoising. In particular, the framework excels in enhancing SNR and ensuring data consistency, highlighting its robustness and versatility across diverse scenarios.

In subsequent research, we plan to broaden the scope of the STC framework to encompass a wider range of applications, including prestack seismic data completion [21], tackling the challenges of five-dimensional seismic reconstruction [7, 51], and advancing efficient data recovery in global seismology [52], contributing to further advancements in these fields.

## 3.7 Appendix

### Hankel Tensor Singular Value Decomposition

Based on Definitions 2.4 and 2.7, different variants of SSA can be interpreted as specific implementations of Hankel tensor singular value decomposition (Ht-SVD). Building on this concept, we employ MSSA to construct the Hankel tensor, which serves as the input for the STC framework. In this approach, the three-dimensional seismic data tensor $\mathcal{T}$ is reorganized into a block Hankel tensor $\mathcal{M}$ to better capture the intrinsic structural features of the data. Given the inherent two-level block structure of the Hankel matrix, MSSA processes the tensor $\mathcal{T}$ in two stages. In the first stage, each row of $\mathcal{T}^{(k)}$ is transformed into a corresponding micro Hankel matrix, which acts as a building block for the overall block Hankel matrix. Specifically, for the $i$th row of $\mathcal{T}^{(k)}$, the associated micro Hankel matrix $R_i$ is defined as follows:

$$R_i = \begin{bmatrix} \mathcal{T}_{i1k} & \mathcal{T}_{i2k} & \cdots & \mathcal{T}_{iN_1k} \\ \mathcal{T}_{i2k} & \mathcal{T}_{i3k} & \cdots & \mathcal{T}_{i(N_1+1)k} \\ \vdots & \vdots & \ddots & \vdots \\ \mathcal{T}_{i(N_x-N_1+1)k} & \mathcal{T}_{i(N_x-N_1+2)k} & \cdots & \mathcal{T}_{iN_xk} \end{bmatrix}. \quad (3.42)$$

To ensure that $R_i$ and $\mathcal{M}^{(k)}$ are as nearly to square matrices as possible, the pencil parameters $N_1$ and $N_2$ are typically chosen with balanced dimensions. A common approach is to set $N_1 = N_x - \lfloor N_x/2 \rfloor$ and $N_2 = N_y - \lfloor N_y/2 \rfloor$, ensuring symmetry and facilitating subsequent processing. Here, $\lfloor \cdot \rfloor$ denotes the floor function, which rounds a given value down to the nearest integer. Using $R_i$, a 2-fold Hankel matrix $\mathcal{M}^{(k)}$ is then constructed, as defined below:

$$\mathcal{M}^{(k)} = \begin{bmatrix} R_1 & R_2 & \cdots & R_{N_2} \\ R_2 & R_3 & \cdots & R_{N_2+1} \\ \vdots & \vdots & \ddots & \vdots \\ R_{N_y-N_2+1} & R_{N_y-N_2+2} & \cdots & R_{N_y} \end{bmatrix}, \quad (3.43)$$

Let $\mathcal{M} \in \mathbb{R}^{I_x \times I_y \times N_t}$ represent the Hankel tensor, where its dimensions are defined as $I_x = (N_x - N_1 + 1)(N_y - N_2 + 1)$ and $I_y = N_1 N_2$. The tensor $\mathcal{M}$ is then transformed into the Fourier domain, producing a tensor $\widetilde{\mathcal{M}}$ containing complex values. In the context of Multichannel Singular Spectrum Analysis (MSSA), each frequency component $\widetilde{\mathcal{M}}^{(k)}$ in the Fourier domain is used to construct a low-rank Hankel matrix. This is achieved through rank reduction techniques, including truncated singular value decomposition (Truncated SVD), as outlined in [22, 25]. This process effectively implements the Ht-SVD and extracts the low-rank features of the signal.

## References

1. Boumal N, Absil P-A (2011) RTRMC: a Riemannian trust-region method for low-rank matrix completion. In: Advances in neural information processing systems, pp 406–414
2. Obozinski G, Taskar B, Jordan MI (2010) Joint covariate selection and joint subspace selection for multiple classification problems. Stat Comput 20(2):231–252
3. Kolda TG, Sun J (2008) Scalable tensor decompositions for multi-aspect data mining. In: 2008 Eighth IEEE international conference on data mining. IEEE, pp 363–372
4. Sun J, Papadimitriou S, Lin C-Y, Cao N, Liu S, Qian W (2009) Multivis: content-based social network exploration through multi-way visual analysis. In: Proceedings of the 2009 SIAM international conference on data mining. SIAM, pp 1064–1075
5. Ji T-Y, Yokoya N, Zhu XX, Huang T-Z (2018) Nonlocal tensor completion for multitemporal remotely sensed images' inpainting. IEEE Trans Geosci Remote Sens 56(6):3047–3061
6. Ng MK-P, Yuan Q, Yan L, Sun J (2017) An adaptive weighted tensor completion method for the recovery of remote sensing images with missing data. IEEE Trans Geosci Remote Sens 55(6):3367–3381
7. Gao J, Cheng J, Sacchi MD (2017) Five-dimensional seismic reconstruction using parallel square matrix factorization. IEEE Trans Geosci Remote Sens 55(4):2124–2135
8. Wang Y, Zhou H, Shaohuan Z, Mao W, Chen Y (2017) Three-operator proximal splitting scheme for 3-d seismic data reconstruction. IEEE Geosci Remote Sensing Lett 14(10):1830–1834
9. Oropeza V, Sacchi M (2011) Simultaneous seismic data denoising and reconstruction via multichannel singular spectrum analysis. Geophysics 76(3):V25–V32
10. Chen Y, Zhang D, Jin Z, Chen X, Shaohuan Z, Huang W, Gan S (2016) Simultaneous denoising and reconstruction of 5-d seismic data via damped rank-reduction method. Geophys J Int 206(3):1695–1717
11. Kreimer N, Stanton A, Sacchi MD (2013) Tensor completion based on nuclear norm minimization for 5d seismic data reconstruction. Geophysics 78(6):V273–V284
12. Kreimer N, Sacchi MD (2012) A tensor higher-order singular value decomposition for prestack seismic data noise reduction and interpolation. Geophysics 77(3):V113–V122
13. Yang Y, Ma J, Osher S (2013) Seismic data reconstruction via matrix completion. Inverse Probl Imaging 7(4):1379–1392
14. Ely G, Aeron S, Hao N, Kilmer ME (2015) 5d seismic data completion and denoising using a novel class of tensor decompositions. Geophysics 80(4):V83–V95
15. Kumar R, Aravkin AY, Mansour H, Recht B, Herrmann FJ (2013) Seismic data interpolation and denoising using SVD-free low-rank matrix factorization. In: 75th EAGE conference & exhibition incorporating SPE EUROPEC 2013
16. Kumar R, Da Silva C, Akalin O, Aravkin AY, Mansour H, Recht B, Herrmann FJ (2015) Efficient matrix completion for seismic data reconstruction. Geophysics 80(5):V97–V114
17. Aravkin A, Kumar R, Mansour H, Recht B, Herrmann FJ (2014) Fast methods for denoising matrix completion formulations, with applications to robust seismic data interpolation. SIAM J Sci Comput 36(5):S237–S266
18. Gao J, Stanton A, Sacchi MD (2015) Parallel matrix factorization algorithm and its application to 5d seismic reconstruction and denoising. Geophysics 80(6)
19. Wang C, Zhu Z, Hanming G, Xinming W, Liu S (2018) Hankel low-rank approximation for seismic noise attenuation. IEEE Trans Geosci Remote Sens 57(1):561–573
20. Vautard R, Ghil M (1989) Singular spectrum analysis in nonlinear dynamics, with applications to paleoclimatic time series. Physica D: Nonlinear Phenomena 35(3):395–424
21. Gao J, Sacchi MD, Chen X (2013) A fast reduced-rank interpolation method for prestack seismic volumes that depend on four spatial dimensions. Geophysics 78(1):V21–V30
22. Ma J (2013) Three-dimensional irregular seismic data reconstruction via low-rank matrix completion. Geophysics 78(5):V181–V192
23. Trickett S, Burroughs L, Milton A, Walton L, Dack R (2010) Rank-reduction-based trace interpolation. In: SEG technical program expanded abstracts 2010. Society of Exploration Geophysicists, pp 3829–3833

24. Zhang D, Zhou Y, Chen H, Chen W, Shaohuan Z, Chen Y (2017) Hybrid rank-sparsity constraint model for simultaneous reconstruction and denoising of 3d seismic data. Geophysics 82(5):V351–V367
25. Siahsar MAN, Gholtashi S, Torshizi EO, Chen W, Chen Y (2017) Simultaneous denoising and interpolation of 3-d seismic data via damped data-driven optimal singular value shrinkage. IEEE Geosci Remote Sens Lett 14(7):1086–1090
26. Chen Y, Huang W, Zhang D, Chen W (2016) An open-source matlab code package for improved rank-reduction 3d seismic data denoising and reconstruction. Comput Geosci 95:59–66
27. Liu X-Y, Aeron S, Aggarwal V, Wang X, Min-You W (2016) Adaptive sampling of RF fingerprints for fine-grained indoor localization. IEEE Trans Mob Comput 15(10):2411–2423
28. Zhang Z, Aeron S (2017) Exact tensor completion using t-SVD. IEEE Trans Signal Process 65(6):1511–1526
29. Rouphael TJ, Cruz JR (1998) Multichannel trace interpolation in the presence of spatial aliasing. IEEE Trans Geosci Remote Sens 36(3):1025–1030, May 1998
30. Trickett S, Burroughs L, Milton A (2013) Interpolation using Hankel tensor completion. In: SEG technical program expanded abstracts 2013. Society of Exploration Geophysicists, pp 3634–3638
31. Yu AW, Ma W, Yu Y, Carbonell J, Sra S (2014) Efficient structured matrix rank minimization. In: Advances in neural information processing systems, pp 1350–1358
32. Davenport MA, Romberg J (2016) An overview of low-rank matrix recovery from incomplete observations. IEEE J Sel Top Signal Process 10(4):608–622
33. Li XP, Huang L, So HC, Zhao B (2019) A survey on matrix completion: perspective of signal processing (2019). arXiv preprint arXiv:1901.10885
34. Jain P, Netrapalli P, Sanghavi S (2013) Low-rank matrix completion using alternating minimization. In: Proceedings of the forty-fifth annual ACM symposium on theory of computing. ACM, pp 665–674
35. Hardt M (2014) Understanding alternating minimization for matrix completion. In: 2014 IEEE 55th annual symposium on foundations of computer science (FOCS). IEEE, pp 651–660
36. Fazel M, Pong TK, Sun D, Tseng P (2013) Hankel matrix rank minimization with applications to system identification and realization. SIAM J Matrix Anal Appl 34(3):946–977
37. Zhou P, Canyi L, Lin Z, Zhang C (2018) Tensor factorization for low-rank tensor completion. IEEE Transa Image Process 27(3):1152–1163
38. Liu X-Y, Aeron S, Aggarwal V, Wang X (2016) Low-tubal-rank tensor completion using alternating minimization (2016). CoRR arXiv:1610.01690
39. Srebro N, Rennie J, Jaakkola TS (2005) Maximum-margin matrix factorization. In: Advances in neural information processing systems, pp 1329–1336
40. Gu S, Zhang L, Zuo W, Feng X (2014) Weighted nuclear norm minimization with application to image denoising. In: Proceedings of the IEEE conference on computer vision and pattern recognition, pp 2862–2869
41. Du Y, Han G, Quan Y, Yu Z, Wong H-S, Philip Chen CL, Zhang J (2018) Exploiting global low-rank structure and local sparsity nature for tensor completion. IEEE Trans Cybern (99):1–13
42. Boyd S, Parikh N, Chu E, Peleato B, Eckstein J et al (2011) Distributed optimization and statistical learning via the alternating direction method of multipliers. Found Trends® Mach Learn 3(1):1–122
43. Wang B (2016) An efficient POCS interpolation method in the frequency-space domain. IEEE Geosci Remote Sens Lett 13(9):1384–1387
44. Sidiropoulos ND, De Lathauwer L, Fu X, Huang K, Papalexakis EE, Faloutsos C (2017) Tensor decomposition for signal processing and machine learning. IEEE Trans Signal Process 65(13):3551–3582
45. Rabanser S, Shchur O, Günnemann S (2017) Introduction to tensor decompositions and their applications in machine learning. arXiv preprint arXiv:1711.10781
46. Sørensen M, De Lathauwer L (2019) Fiber sampling approach to canonical polyadic decomposition and application to tensor completion. SIAM J Matrix Anal Appl 40(3):888–917

47. Kanatsoulis CI, Sidiropoulos ND, Akçakaya M, Fu X (2019) Regular sampling of tensor signals: theory and application to FMRI. In: ICASSP 2019—2019 IEEE international conference on acoustics, speech and signal processing (ICASSP). IEEE, pp 2932–2936
48. Kanatsoulis CI, Fu X, Sidiropoulos ND, Akçakaya M (2019) Tensor completion from regular sub-nyquist samples (2019). arXiv preprint arXiv:1903.00435
49. Sørensen M, Sidiropoulos ND, De Lathauwer L (2019) Canonical polyadic decomposition of a tensor that has missing fibers: a monomial factorization approach. In: ICASSP 2019—2019 IEEE international conference on acoustics, speech and signal processing (ICASSP). IEEE, pp 7490–7494
50. Fehmers GC, Höcker CFW (2003) Fast structural interpretation with structure-oriented filtering. Geophysics 68(4):1286–1293
51. Chen Y, Bai M, Guan Z, Zhang Q, Zhang M, Wang H (2019) Five-dimensional seismic data reconstruction using the optimally damped rank-reduction method. Geophys J Int 218(1):224–246
52. Chen Y, Bai M, Chen Y (2019) Obtaining free usarray data by multi-dimensional seismic reconstruction. Nat Commun 10(1):1–13

# Chapter 4
# Tensor Low-Rank Approximation for Seismic Footprint Suppression

## 4.1 Introduction

Acquisition footprints refer to periodic amplitude anomalies in 3D seismic data, typically caused by inconsistencies in amplitude due to the survey geometry, rather than actual subsurface geological features [1, 2]. These artifacts present significant challenges for interpreting seismic data, especially when methods like amplitude variation with offset (AVO) and attribute extraction are used [3, 4]. Moreover, certain attribute analysis methods, such as coherence and edge detection, can amplify the impact of these footprints, potentially obscuring true geological structures or even misinterpreting them as small faults or fractures. Real 3D seismic data often contain both random noise and footprint artifacts, which further complicate data processing [5, 6]. To address this issue, this chapter introduces an innovative denoising approach that combines an enhanced low-rank tensor approximation (LRTA) technique, designed to simultaneously mitigate both random and footprint noise, thereby improving the quality and interpretability of seismic data.

To effectively remove acquisition footprint noise, it is crucial to identify a suitable transform space where the signal and noise are distinguishable, facilitating more efficient suppression of unwanted artifacts [3]. In recent years, several transform-domain techniques have been proposed for seismic data processing, which can be broadly categorized into three main approaches: frequency-domain methods, time-frequency-domain methods, and dictionary learning (DcL) methods. The first category considers footprints as periodic spatial noise, which can be reduced using different filtering methods in the frequency domain. Common methods include wavenumber filtering [7, 8], wavenumber notch filtering [9], adaptive wavenumber filtering [10], two-pass frequency-wavenumber filtering [1], and spectral notch filtering [11]. The second category utilizes time-frequency transforms to analyze footprints in both temporal and spectral domains, reducing the overlap between the signal and footprint, which makes filtering more effective. Common time-frequency techniques include wavelet transforms [12], 3-D complex wavelet transforms [13], curvelet

 
F. Qian et al., *Tensor Computation for Seismic Data Processing*,
Earth Systems Data and Models 6, https://doi.org/10.1007/978-3-031-78900-7_4

transforms [14], basis pursuit transforms [15], and stationary wavelet transforms (SWT) [16, 17]. While these methods help reduce the impact of footprint noise, they still face the challenge of effectively removing noise while maintaining the fidelity of the underlying seismic data.

In contrast to traditional Fourier transforms and time-frequency transforms, the third category of methods utilizes dictionary learning (DL) techniques, where atoms chosen from a trained dictionary (rather than traditional sparse coefficients) are used to represent clean images and footprint patterns in seismic data [5, 18]. Once the atoms encoding the footprint features are identified through the dictionary learning process, Boolean pattern classification [5, 18, 19] or filtering methods [6] can be applied to reconstruct the footprint component using these learned atoms. This allows for the removal of footprint noise from the data, yielding a more accurate denoised output. However, despite the significant advancements made using DL-based methods for footprint removal, there remains a limitation. Specifically, most conventional DL techniques require 3D seismic data to be flattened into 2D images for processing, which inherently disrupts the 3D structure of the data, thereby compromising the effectiveness of the denoising process.

In the field of geophysical research, significant attention has been devoted to the suppression of random noise, with associated techniques typically grouped into three main strategies: spatial filtering, sparse coding in transform spaces, and low-rank decomposition. Spatial domain filtering methods typically remove noise by using weighted sums of filter coefficients combined with neighboring pixels, which effectively attenuate noise. Common techniques include median filtering, mean filtering, structure-oriented filtering, and morphological filtering, all of which rely on local pixel relationships to suppress noise. The second approach leverages sparse representation in transform domains, where signals are represented as sparse coefficients in specific domains, enabling effective noise separation. Typical sparse domains include wavelet transforms, curvelet transforms, and seismic wavelet transforms. These methods are effective at separating signal from noise, as noise tends to be dense, whereas the signal is sparse in these transform domains. The third method focuses on low-rank approximation, which aims to recover the true signal by reducing the rank of noisy seismic data matrices or tensors. Methods like matrix-based low-rank approximation (LRA), Hankel-matrix low-rank techniques, and tensor singular value decomposition (t-SVD) are commonly used. These methods not only help suppress noise but also retain the essential structural information of the signal. Furthermore, methods like structure-oriented filtering and singular value decomposition (SVD) have also been successfully applied to footprint noise. However, despite their effectiveness in random noise suppression, these existing techniques mainly focus on random noise rather than addressing the challenge of footprint removal, thereby limiting their ability to fully mitigate mixed noise or reconstruct high-quality seismic images.

Unlike conventional matrix-based denoising approaches, this chapter introduces an innovative tensor model aimed at simultaneously suppressing footprint and random noise, along with the integration of advanced optimization techniques, which serve as the primary innovations of this study.

(1) This chapter integrates unidirectional total variation (UTV) regularization into the traditional low-rank tensor approximation (LRTA) framework, extending its application from low-rank models for random noise to the simultaneous suppression of both random and footprint noise. As a result, the suppression of both random and footprint noise is reformulated as a unified sparse and low-rank tensor optimization framework, named UTV-LRTA, which overcomes the drawbacks of matrix-based approaches by eliminating the need for flattening or matricization [5, 18, 20–22]. Based on available information, this is the first study to employ a low-rank tensor decomposition approach to address both random and footprint noise in 3-D seismic data.
(2) A novel optimization technique is introduced, employing the Split Bregman iteration (SBI) to solve the UTV-LRTA model. This method retains the advantages of straightforward implementation and lower computational overhead, while also offering efficiency gains associated with SBI.
(3) Our UTV-LRTA model presents an innovative approach that goes beyond traditional matrix-based methods for footprint removal. It enables the simultaneous removal of both random noise and footprints within a 3-D spatial context, while preserving the integrity of high-dimensional image structures. In summary, this research not only advances the theoretical framework for footprint removal but also provides deeper insights into the relationship between footprint patterns and high-dimensional image structures.

Experiments conducted on synthetic and real-world datasets reveal that the proposed UTV-LRTA model outperforms three state-of-the-art (SOTA) methods, as demonstrated by both quantitative and qualitative assessments.

## 4.2 Seismic Footprint Suppression and Its Modelling

### *4.2.1 Problem Statement*

As outlined in [5, 6, 18], it is considered that the footprint noise $\mathcal{F} \in \mathbb{R}^{n_1 \times n_2 \times n_3}$ and random noise $\mathcal{N} \in \mathbb{R}^{n_1 \times n_2 \times n_3}$ are additive in nature, rather than multiplicative. Therefore, the model for the degradation caused by mixed noise can be formulated as follows:

$$\mathcal{Y} = \mathcal{X} + \mathcal{F} + \mathcal{N}, \tag{4.1}$$

The objective of denoising is to recover the original tensor $\mathcal{X}$, representing the true data, from its noisy counterpart $\mathcal{Y}$, which is affected by a mixture of random noise $\mathcal{F}$ and footprint noise $\mathcal{N}$ (Some other related notations about tensors are detailed in Table 4.1). This process is achieved by optimizing the following expression.

$$\min_{\mathcal{X},\mathcal{F}} \frac{1}{2}\|\mathcal{X}+\mathcal{F}-\mathcal{Y}\|_F^2, \tag{4.2}$$

As seen in Eq. (5.10), the problem outlined in (4.1) exemplifies a classic ill-posed inverse problem. To address this, the denoising process in (5.10) can be transformed into an unconstrained optimization problem by embedding regularization terms that reflect the properties of the image tensor and footprint characteristics within the model parameters. Consequently, minimizing the adjusted objective function, which combines data fidelity with regularization terms, becomes the natural approach, as expressed below:

$$\min_{\mathcal{X},\mathcal{F}} \frac{1}{2}\|\mathcal{X}+\mathcal{F}-\mathcal{Y}\|_F^2 + \tau P(\mathcal{X}) + \lambda P(\mathcal{F}), \tag{4.3}$$

The cost function defined in (4.3) comprises a data fidelity term, which enforces alignment between the reconstructed data and the observed noisy tensor to ensure accurate recovery. The second term, $P(\mathcal{X})$, serves as a prior for the image tensor, steering the reconstruction towards realistic outcomes while suppressing unwanted artifacts. The third term, $P(\mathcal{F})$, provides a distinct form of regularization that leverages prior information specific to the footprint $\mathcal{F}$. The parameters $\tau > 0$ and $\lambda > 0$ control the trade-off between the data fidelity term and the two regularization terms, ensuring a balanced reconstruction process.

**Table 4.1** Tensor and matrix notations

| Notation | Description |
|---|---|
| $\mathcal{X}$ | A tensor |
| $X$ | A matrix |
| $\widetilde{\mathcal{X}}$ | The Fourier domain representation of $\mathcal{X}$: $\widetilde{\mathcal{X}} = \text{fft}(\mathcal{X}, [\,], 3)$ |
| $\mathcal{X}^{(k)}$ | The $k$th frontal slice of $\mathcal{X}$; i.e., $\mathcal{X}^{(k)} \equiv \mathcal{X}(:, :, k)$ |
| fold($\mathcal{X}$) | The tensor folding operation of $\mathcal{X}$ |
| unfold($\mathcal{X}$) | The tensor unfolding operation of $\mathcal{X}$ |
| $\|X\|_*$ | The matrix nuclear norm of $X$ |
| $\|\mathcal{X}\|_1$ | The tensor $L_1$ norm of $\mathcal{X}$: $\|\mathcal{X}\|_1 = \sum_{i,j,t} \lvert \mathcal{X}_{i,j,t} \rvert$ |
| $\|\mathcal{X}\|_F$ | The tensor Frobenius norm of $\mathcal{X}$: $\|\mathcal{X}\|_F = \sqrt{\sum_{ijk} x_{ijk}^2}$ |
| $\|\mathcal{X}\|_{\circledast}$ | The TNN of $\mathcal{X}$ |
| $\langle \mathcal{X}, \mathcal{Y} \rangle$ | The inner product of $\mathcal{X}$ and $\mathcal{Y}$ |
| $X^\top$ | The matrix transpose of $X$ |
| $\mathcal{X}^\dagger$ | The tensor transpose of $\mathcal{X}$: $\mathcal{X}^{\top(1)} = \mathcal{X}^{(1)\top}, \mathcal{X}^{\top(n3+2-k)} = \mathcal{X}^{(\ell)\top}, 2 \leq k \leq n_3$ |

### 4.2.2 *Problem Modeling*

This section introduces a methodology that employs the TNN and UTV as prior constraints, capturing the characteristics of the image (image prior) and the footprint (footprint prior), respectively, to enhance the regularization of the data fidelity component. Building on this framework, the UTV-TLRA—Tensor Low-Rank Approximation) model is developed to effectively mitigate both random noise and footprint noise. The detailed formulation and implementation procedures are described in the following sections.

#### 4.2.2.1 A Low-Tubal-Rank Tensor Model for a Clear Image Tensor

Random noise in seismic images can be effectively mitigated using low-rank approximation (LRA) techniques, as extensively demonstrated in prior studies (e.g., [22–28]). In this study, the low-rank characteristics of seismic data are preserved by employing tensor nuclear norm (TNN) minimization. This approach effectively facilitates the reduction of random noise and ensures the restoration of high-quality seismic images. The detailed methodology is outlined as follows:

$$P(\mathcal{X}) = \|\mathcal{X}\|_{\circledast}. \tag{4.4}$$

It is well-established that tensor nuclear norm (TNN) minimization-based low-rank approximation (LRA) techniques have achieved superior performance in reducing noise for seismic image denoising tasks [25].

#### 4.2.2.2 UTV Model for Footprint Removal

Seismic image footprint noise typically manifests as distinct striped or grid-like patterns, exhibiting significant directional and structural properties. As highlighted in previous studies [1, 4, 18], the UTV model offers an effective approach for addressing such directional noise. To illustrate this, Fig. 4.1 demonstrates the derivatives of a tensor's frontal slice in two orthogonal directions. Specifically, Fig. 4.1b and c show the horizontal and vertical derivatives of the noisy slice depicted in Fig. 4.1a. It is evident from Fig. 4.1b that the footprint's directional characteristics have a notable effect on the horizontal derivative, whereas the vertical derivative, as displayed in Fig. 4.1c, remains relatively unaffected. In this regard, the UTV model serves two main functions: first, it enhances smoothness along the footprint direction, thereby highlighting or isolating the underlying footprint noise; second, it promotes smoothness in the direction perpendicular to the footprint, effectively aiding in the removal of footprint noise. The definition of this process is as follows:

$$P(\mathcal{F}) = \lambda_1 \left\|\nabla_y \mathcal{X}\right\|_1 + \lambda_2 \left\|\nabla_x (\mathcal{X} - \mathcal{Y})\right\|_1 . \tag{4.5}$$

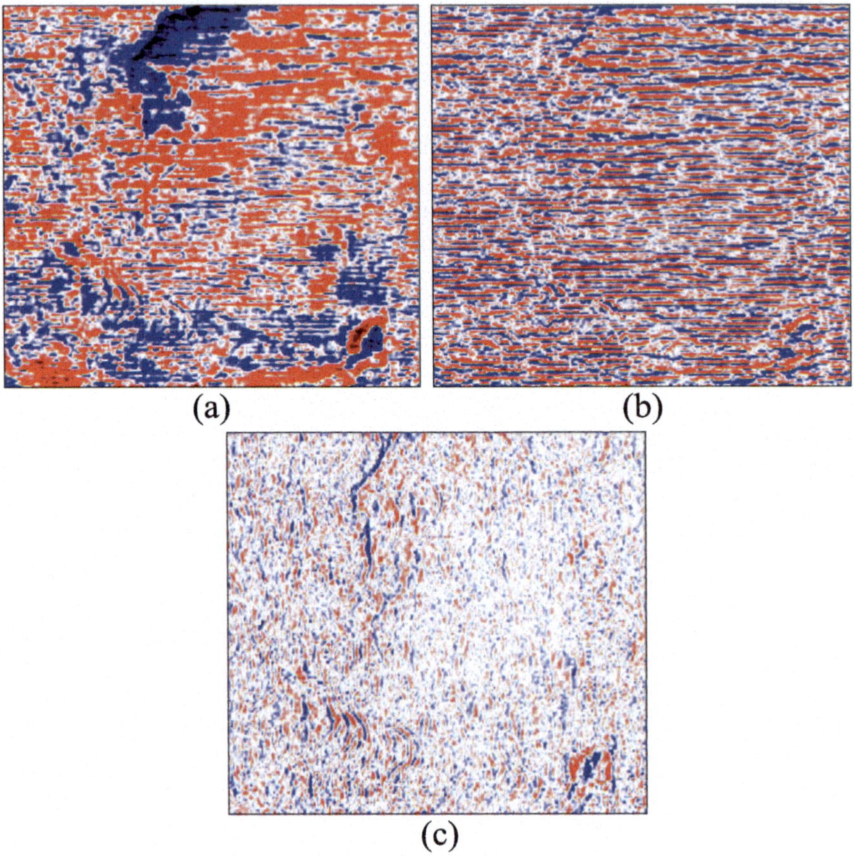


**Fig. 4.1** The directional characteristics of the seismic footprint are illustrated as follows: **a** A seismic image affected by both random noise and footprint noise, particularly at a fixed time slice. **b** The horizontal gradient of the noisy seismic image. **c** The vertical gradient of the noisy seismic image

where $\nabla_x$ and $\nabla_y$ denote the first-order derivatives with respect to $x$ and $y$, respectively.

In the UTV model defined by Eq. (4.5), seismic images generally exhibit notable spatial correlations, reflected in the form of local smoothness in the pixel domain or sparsity in the discrete gradient domain. The first term leverages the $\ell_1$-norm to enforce sparsity in the gradient of the image across footprint patterns, thereby enhancing the local smoothness of $\mathcal{X}$ and effectively eliminating the unwanted footprint artifacts. Simultaneously, considering that acquisition footprints tend to change slowly in one direction while varying rapidly in another [6, 18], the gradient of the footprint residual $(\mathcal{X} - \mathcal{Y})$ is sparse along its primary direction. This property enables effective footprint separation through the $\ell_1$ norm constraints imposed by

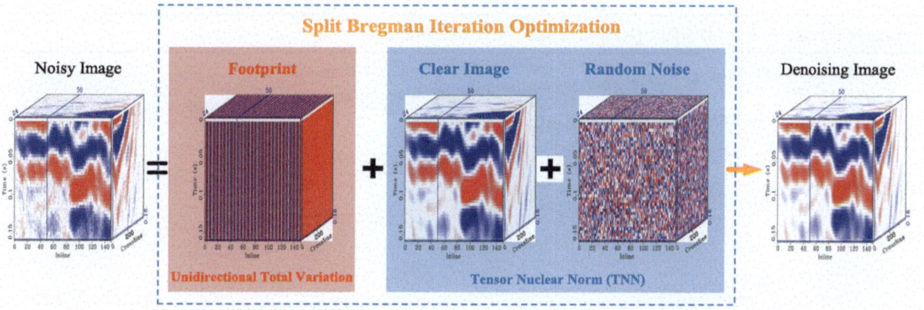


**Fig. 4.2** Diagram illustrating our Unidirectional Total Variation (UTV)—Tensor Low-Rank Approximation (TLRA) model

the second term. Consequently, UTV can be regarded as a specialized variant of directional total variation (DTV) [29, 30]. To align the footprint along the y-axis or another preferred direction (e.g., diagonal), similar to DTV, the image may need to be rotated, ensuring proper alignment for the successful application of the UTV model.

#### 4.2.2.3 Proposed UTV-TLRA Model

By incorporating the low-rank tensor model as a replacement for $P(\mathcal{X})$ and employing UTV regularization in place of $P(\mathcal{F})$, and in accordance with the recommendation in [31] to eliminate the symbol $\mathcal{F}$ from the data fidelity term in (4.3), the objective function pertaining to $\mathcal{X}$ can be re-expressed as follows:

$$\min_{\mathcal{X}} \frac{1}{2}\|\mathcal{X}-\mathcal{Y}\|_F^2 + \tau\|\mathcal{X}\|_{\circledast} + \lambda_1 \left\|\nabla_y \mathcal{X}\right\|_1 + \lambda_2 \left\|\nabla_x (\mathcal{X}-\mathcal{Y})\right\|_1 . \tag{4.6}$$

where $\lambda_1$, $\lambda_2$, and $\tau$ denote the regularization parameters that control the contribution of each component, ensuring a proper balance within the model.

## 4.3 Unidirectional Total Variation Regularized Low-Rank Tensor Approximation (UTV-TLRA)

To effectively address the issue in problem (4.6), this section transforms the unconstrained optimization problem into its equivalent constrained form, as described in Sect. 4.3.1. Subsequently, an efficient algorithm is proposed, leveraging the split Bregman iteration technique to solve the constrained optimization problem, with the detailed algorithm provided in Sect. 4.3.2 and Fig. 4.2.

**Algorithm 4.1:** The proposed TLRA method

**Input**: $\mathcal{Y}, \epsilon, \tau, \lambda_1, \lambda_2, \alpha, \beta, \gamma$.
**Output**: $\mathcal{X}_k$.

Initialize: $\mathcal{B}^0 and \mathcal{B}_x^0$, and $\mathcal{B}_y^0$ is initialized to **0**
$\mathcal{X}^0 = \mathcal{Y}, \mathcal{Z}^0 = \mathcal{Y}, \ell = 0$;
**while** *(converged=false)* **do**
    **for** $k = 1, 2, \ldots, n_3$ **do**
        Update $\mathcal{X}^{(k)\ell}$ using (4.12);
    **end**
    Update $\mathcal{Z}^\ell$ using (4.17);
    Update $\mathcal{D}_x^{\ell+1}$ using (4.20);
    Update $\mathcal{D}_y^{\ell+1}$ using (4.22);
    Update $\mathcal{B}^{\ell+1}$, $\mathcal{B}_x^{\ell+1}$, and $\mathcal{B}_y^{\ell+1}$ using (4.23);
    Check for convergence:
    $\left(\|\mathcal{X}^{\ell+1} - \mathcal{X}^\ell\|_F^2 / \|\mathcal{X}^\ell\|_F^2\right) \leq \epsilon$
**end**

### *4.3.1 Objective Function*

The primary challenge in solving the under-determined equation in (4.6) for $\mathcal{X}$ lies in the non-differentiability of the $\ell_1$-norm terms, which prevents the use of gradient-based optimization methods. The Split Bregman iteration (SBI) technique, combined with the Total Variation (TV) model, has emerged as an effective approach for solving minimization problems that involve $\ell_1$-norm regularization and has gained significant attention due to its fast solving properties [31–33]. Thus, the Split Bregman Iteration (SBI) method is naturally tailored to the proposed UTV-TLRA framework in (4.6) for reconstructing the clean third-order data tensor $\mathcal{X}$. Drawing inspiration from [31, 33], the key concept involves transforming the unconstrained optimization problem for $\mathcal{X}$ in (4.6) into an equivalent constrained form via an auxiliary variable technique. In this reformulation, $\mathcal{Z} = \mathcal{X}$, $\mathcal{D}x = \nabla x(\mathcal{X} - \mathcal{Y})$, and $\mathcal{D}y = \nabla y\mathcal{X}$. The resulting constrained optimization problem is then solved efficiently using the SBI approach, as detailed below.

$$
\begin{aligned}
&\min_{\mathcal{X}} \frac{1}{2}\|\mathcal{X} - \mathcal{Y}\|_F^2 + \tau\|\mathcal{Z}\|_{\circledast} + \lambda_1 \|\mathcal{D}_x\|_1 + \lambda_2 \left\|\mathcal{D}_y\right\|_1 \\
&\text{s.t. } \mathcal{Z} = \mathcal{X},\ \mathcal{D}_x = \nabla_x(\mathcal{X} - \mathcal{Y}),\ \mathcal{D}_y = \nabla_y\mathcal{X}.
\end{aligned} \tag{4.7}
$$

Subsequently, the constrained problem (8.8) described earlier can be transformed into an unconstrained minimization problem through the systematic application of the Bregman iteration.

$$
\begin{aligned}
\min_{\mathcal{X},\mathcal{Z},\mathcal{D}_x,\mathcal{D}_y} & \frac{1}{2}\|\mathcal{X}-\mathcal{Y}\|_F^2 + \tau\|\mathcal{Z}\|_{\circledast} + \lambda_1 \|\mathcal{D}_x\|_1 + \lambda_2 \left\|\mathcal{D}_y\right\|_1 \\
& + \frac{\alpha}{2}\|\mathcal{Z}-\mathcal{X}-\mathcal{B}\|_F^2 + \frac{\beta}{2}\|\mathcal{D}_x - \nabla_x(\mathcal{X}-\mathcal{Y}) - \mathcal{B}_x\|_F^2 \\
& + \frac{\gamma}{2}\left\|\mathcal{D}_y - \nabla_y\mathcal{X} - \mathcal{B}_y\right\|_F^2,
\end{aligned}
\tag{4.8}
$$

where $\alpha, \beta$, and $\gamma$ denote the Bregman penalization parameters, while $\mathcal{B}_x \in \mathbb{R}^{n_1\times n_2\times n_3}$, $\mathcal{B}_y \in \mathbb{R}^{n_1\times n_2\times n_3}$, and $\mathcal{B} \in \mathbb{R}^{n_1\times n_2\times n_3}$ are the Bregman tensors.

### 4.3.2 Solving Algorithm

This section employs an alternating optimization approach to tackle the problem in (4.8), where each variable is updated sequentially while the remaining variables are held constant. Consequently, the original problem (4.8) is divided into four more manageable sub-problems, each of which can be addressed independently.

#### 4.3.2.1 Updating $\mathcal{X}^{\ell+1}$

The subproblem associated with $\mathcal{X}$ is outlined as follows:

$$
\begin{aligned}
\min_{\mathcal{X}} & \frac{1}{2}\|\mathcal{X}-\mathcal{Y}\|_F^2 + \frac{\alpha}{2}\|\mathcal{Z}-\mathcal{X}-\mathcal{B}\|_F^2 + \frac{\beta}{2}\|\mathcal{D}_x - \nabla_x(\mathcal{X}-\mathcal{Y}) - \mathcal{B}_x\|_F^2 \\
& + \frac{\gamma}{2}\left\|\mathcal{D}_y - \nabla_y\mathcal{X} - \mathcal{B}_y\right\|_F^2,
\end{aligned}
\tag{4.9}
$$

Problem (4.9) can be solved by resolving each frontal slice $\mathcal{X}^{(k)}$ of $\mathcal{X}$, $1 \le k \le n_3$ separately.

$$
\begin{aligned}
\min_{\mathcal{X}^{(k)}} & \frac{1}{2}\|\mathcal{X}^{(k)}-\mathcal{Y}^{(k)}\|_F^2 + \frac{\alpha}{2}\|\mathcal{Z}^{(k)}-\mathcal{X}^{(k)}-\mathcal{B}^{(k)}\|_F^2 \\
& + \frac{\beta}{2}\|\mathcal{D}_x^{(k)} - \nabla_x(\mathcal{X}^{(k)}-\mathcal{Y}^{(k)}) - \mathcal{B}_x^{(k)}\|_F^2 \\
& + \frac{\gamma}{2}\|\mathcal{D}_y^{(k)} - \nabla_y\mathcal{X}^{(k)} - \mathcal{B}_y^{(k)}\|_F^2,
\end{aligned}
\tag{4.10}
$$

It is worth highlighting that problem (4.10) represents a convex function with respect to the real variable $\mathcal{X}^{(k)}$, which simplifies the process of finding the optimal solution, expressed as follows:

$$\begin{aligned}&\left(\alpha + \beta \nabla_x^\top \nabla_x + \gamma \nabla_y^\top \nabla_y + \mathcal{I}^{(k)}\right) \mathcal{X}^{(k)\ell+1} \\ &= \alpha \left(\mathcal{Z}^{(k)\ell} - \mathcal{B}^{(k)\ell}\right) + \beta \nabla_x^\top \left(\mathcal{D}_x^{(k)\ell} + \nabla_x \mathcal{Y}^{(k)} - \mathcal{B}_x^{(k)\ell}\right) \\ &+ \gamma \nabla_y^\top \left(\mathcal{D}_y^{(k)} - \mathcal{B}_y^{(k)}\right) + \mathcal{Y}^{(k)}.\end{aligned} \tag{4.11}$$

where $(\cdot)^\top$ indicates the standard transpose operation applied to the 2-D frontal slices of the tensors, while $\ell$ denotes the iteration count. Building upon the methodologies outlined in [33, 34], a two-dimensional Fourier transform is utilized to obtain a closed-form solution for the aforementioned problem, as expressed below:

$$\mathcal{X}^{(k)\ell+1} = \mathbb{F}^{-1}\left(\frac{\mathcal{A}_1^{(k)}}{\mathcal{A}_2^{(k)}}\right). \tag{4.12}$$

In the above equation,

$$\begin{aligned}\mathcal{A}_1^{(k)} &= \alpha \mathbb{F}\left(\mathcal{Z}^{(k)\ell} - \mathcal{B}^{(k)\ell}\right) + \beta \mathbb{F}\left(\nabla_x\right)^* \circ \mathbb{F}\left(\mathcal{D}_x^{(k)\ell} + \nabla_x \mathcal{Y}^{(k)} - \mathcal{B}_x^{(k)\ell}\right) \\ &+ \gamma \mathbb{F}\left(\nabla_y\right)^* \circ \mathbb{F}\left(\mathcal{D}_y^{(k)\ell} - \mathcal{B}_y^{(k)\ell}\right) + \mathbb{F}\left(\mathcal{Y}^{(k)}\right),\end{aligned} \tag{4.13}$$

and

$$\begin{aligned}\mathcal{A}_2^{(k)} &= \alpha + \beta \mathbb{F}\left(\nabla_x\right)^* \circ \mathbb{F}\left(\nabla_x\right) \\ &+ \gamma \mathbb{F}\left(\nabla_y\right)^* \circ \mathbb{F}\left(\nabla_y\right) + \mathbb{F}\left(\mathcal{I}^{(k)}\right),\end{aligned} \tag{4.14}$$

Here, $\mathbb{F}(\cdot)$ and $\mathbb{F}^{-1}(\cdot)$ denote the 2D Fourier transform and its inverse, respectively. The symbol "o" represents the element-wise product between two matrices, while $*$ indicates the conjugate of a matrix.

#### 4.3.2.2 Updating $\mathcal{Z}^{\ell+1}$

The subproblem associated with $\mathcal{Z}$ is formulated as follows:

$$\min_{\mathcal{Z}} \tau \|\mathcal{Z}\|_{\circledast} + \frac{\alpha}{2} \|\mathcal{Z} - \mathcal{X} - \mathcal{B}\|_F^2, \tag{4.15}$$

It is clear that the subproblem in (4.15) must be solved using tensor singular value decomposition (t-SVD), which is conceptually similar to the conventional matrix-based singular value decomposition (SVD).

$$[\mathcal{U}, \Theta, \mathcal{V}] = \text{tSVD}(\mathcal{X}^{\ell+1} + \mathcal{B}^{\ell}). \tag{4.16}$$

Next, the closed-form expression for $\mathcal{Z}^{\ell+1}$ is derived by applying the soft-thresholding operator with a threshold value of $\tau/\alpha$ [35], as demonstrated below.

$$\mathcal{Z}^{\ell+1} = \mathcal{U} * \text{shrink}\left(\Theta, \frac{\tau}{\alpha}\right) * \mathcal{V}^{\dagger}, \tag{4.17}$$

where

$$\text{shrink}\left(\Theta, \frac{\tau}{\alpha}\right) = \frac{\Theta}{|\Theta|} * \max\left(\Theta - \frac{\tau}{\alpha}, 0\right). \tag{4.18}$$

It is worth emphasizing that the implementation of this shrinkage operation is straightforward. Each iteration only requires performing tensor singular value decomposition (t-SVD) and basic tensor operations.

#### 4.3.2.3 Updating $\mathcal{D}_x^{\ell+1}$

The subproblem associated with $\mathcal{D}_x$ is described as follows:

$$\min_{\mathcal{D}_x} \lambda_1 \|\mathcal{D}_x\|_1 + \frac{\beta}{2} \|\mathcal{D}_x - \nabla_x(\mathcal{X} - \mathcal{Y}) - \mathcal{B}_x\|_F^2 . \tag{4.19}$$

The subproblem in (4.19) can also be addressed using the soft shrinkage operator presented in (4.18):

$$\mathcal{D}_x^{\ell+1} = \text{shrink}\left(\nabla_x(\mathcal{X}^{\ell+1} - \mathcal{Y}) + \mathcal{B}_x^{\ell}, \frac{\lambda_1}{\beta}\right). \tag{4.20}$$

#### 4.3.2.4 Update $\mathcal{D}_y^{\ell+1}$

The subproblem associated with $\mathcal{D}_y$ is defined as follows:

$$\min_{\mathcal{D}_y} \lambda_2 \left\|\mathcal{D}_y\right\|_1 + \frac{\gamma}{2} \left\|\mathcal{D}_y - \nabla_y \mathcal{X} - \mathcal{B}_y\right\|_F^2 . \tag{4.21}$$

Similar to the subproblem related to $\mathcal{D}_x$, the analytical solution for subproblem (4.21) is:

$$\mathcal{D}_y^{\ell+1} = \text{shrink}\left(\nabla_y \mathcal{X}^{\ell+1} + \mathcal{B}_y^{\ell}, \frac{\lambda_2}{\gamma}\right). \tag{4.22}$$

#### 4.3.2.5 Updating the Bregman Tensors $\mathcal{B}^{\ell+1}$, $\mathcal{B}_x^{\ell+1}$, and $\mathcal{B}_y^{\ell+1}$

Finally, $\mathcal{B}^{\ell+1}$, $\mathcal{B}_x^{\ell+1}$, and $\mathcal{B}_y^{\ell+1}$ represent the Bregman tensors and are updated according to the following equations:

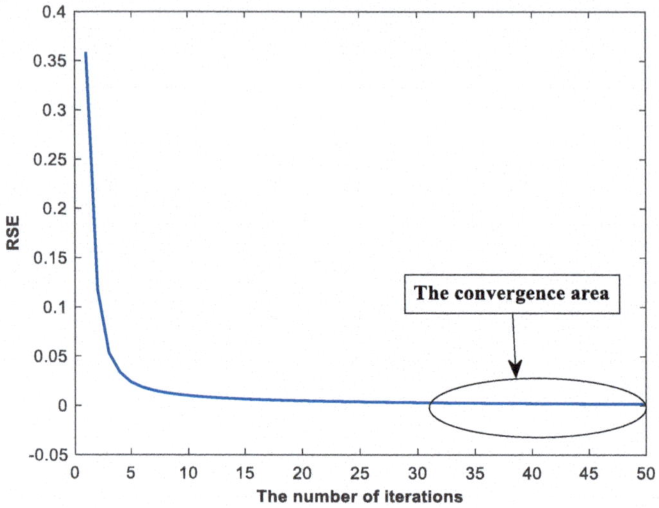


**Fig. 4.3** Convergence curve of the Split Bregman Iteration (SBI) algorithm as applied to the synthetic dataset

$$\begin{cases} \mathcal{B}^{\ell+1} = \mathcal{B}^k + \left(\mathcal{X}^{\ell+1} - \mathcal{Z}^{\ell+1}\right) \\ \mathcal{B}_x^{\ell+1} = \mathcal{B}_x^\ell + \left(\nabla_x\left(\mathcal{X}^{\ell+1} - \mathcal{Y}\right) - \mathcal{D}_x^{\ell+1}\right) \\ \mathcal{B}_y^{\ell+1} = \mathcal{B}_y^\ell + \left(\nabla_y \mathcal{X}^{\ell+1} - \mathcal{D}_y^{\ell+1}\right). \end{cases} \tag{4.23}$$

The approach for solving the proposed UTV-TLRA model is outlined in Algorithm 4.1. It is essential to note that the stopping criterion in Algorithm 4.1 is defined as $\left(\left|\mathcal{X}^{\ell+1} - \mathcal{X}^\ell\right| F^2 / \left|\mathcal{X}^\ell\right| F^2\right) \leq \epsilon$, and the iterations will cease once this condition is met. Furthermore, the Split Bregman iteration (SBI) method [32] guarantees the convergence of Algorithm 4.1 in solving (4.8), with the convergence process depicted in Fig. 4.3.

### 4.3.3 Computational Complexity Analysis

The computational complexity of an algorithm is generally dictated by its most resource-intensive operations [36]. In Algorithm 4.1, the dominant computational expense arises from the tSVD procedure in line 7, as outlined in Eq. (4.16). Based on Eqs. (2.2) and (2.4), tSVD involves performing $n_3$ independent SVD computations in the Fourier domain. Each SVD computation has a complexity of $O\left(n_1 n_2^2\right)$. Consequently, disregarding the computational cost of the Fourier transform, the overall complexity of the tSVD process, for each iteration of the algorithm, can be expressed as $O\left(n_1 n_2^2 n_3\right)$. Additionally, the SBI method effectively enhances the convergence speed of the proposed UTV-TLRA model, as evidenced in Fig. 4.3.

## 4.4 Field Examples of UTV-TLRA

To evaluate the effectiveness of the proposed UTV-TLRA model, a comprehensive set of preliminary experiments was conducted on both synthetic and real-world datasets. The experimental framework includes a thorough explanation of the performance metrics, the baseline models considered for comparison, and the parameter settings, as outlined in Sects. 4.4.1.1 and 4.4.1.4. Furthermore, detailed experimental analyses and results for synthetic and real-world data are provided in Sects. 8.4.2 and 8.4.3, respectively.

### *4.4.1 Experimental Setup*

#### 4.4.1.1 Performance Metrics

To evaluate the denoising performance of the UTV-TLRA model when ground truth data is accessible, quantitative metrics are utilized to complement visual assessments. Consistent with methodologies from prior research, the signal-to-noise ratio (SNR) is adopted as a key quantitative metric, defined as follows.

$$\mathrm{SNR[dB]} = 10\log_{10}\frac{\|\mathcal{X}\|_2^2}{\|\mathcal{X}-\widehat{\mathcal{X}}\|_2^2}, \tag{4.24}$$

where $\mathcal{X}$ and $\widehat{\mathcal{X}}$ denote the clean data tensor and the estimated data tensor, respectively.

#### 4.4.1.2 Baseline Algorithms

The developed UTV-TLRA model incorporates both TNN and UTV elements. To evaluate the impact of integrating UTV, a comparison is conducted with conventional TLRA approaches, such as the DRR method. Additionally, the UTV-TLRA model is compared to the FR-Net technique. Although both models utilize the UTV term, they employ different strategies for modeling the useful signal, allowing for an examination of the effects of these differences. This analysis underscores the benefits of tensor-based techniques over matrix-based methods for footprint removal, demonstrating the enhanced performance made possible by tensor operations. Detailed descriptions of the compared methods are provided below.

1. *Footprint removal network (FR-Net)* [37]: FR-Net aims to decompose seismic images contaminated by footprints into separate image and footprint components without supervision. This is achieved using a deep autoencoder network that incorporates UTV regularization.

2. *Statistics-guided residual dictionary learning (SGRDL)* [5][1]: SGRDL serves as a framework for identifying dictionary atoms in seismic images and distinguishing footprint characteristics from these learned atoms. Subsequently, SGRDL employs a statistical analysis approach to determine whether the dictionary atoms are influenced by footprints or are free from such effects.
3. *Fxy Filtering (Fxy)* [38][1]: In the Fxy domain, random and footprint noise are treated as outliers or undesirable transient events. To tackle this, the Fxy filtering technique employs a 2-D smoothing filter applied selectively to specific frequency slices, leaving the remaining slices unaltered. This method effectively isolates noise within a targeted frequency range while preserving the integrity of the rest of the data.

The Fxy, SGRDL, and our UTV-TLRA models were implemented using MATLAB on a Windows 10 system equipped with an Intel Core i5-6500 CPU and 32 GB RAM. Conversely, the FR-Net model was executed on the Google Colab platform utilizing TensorFlow.

#### 4.4.1.3 Parameter Settings of UTV-TLRA

Determining suitable parameters is a critical aspect of the UTV-TLRA model. To address this, an additional empirical tuning approach is applied to identify the optimal parameter values. Specifically, six key parameters—$\alpha$, $\beta$, $\gamma$, $\tau$, $\lambda_1$, $\lambda_2$—are closely linked to the model's noise reduction effectiveness. These parameters can be categorized into two groups: $\alpha$ and $\tau$ primarily regulate the handling of footprint noise, while the remaining parameters focus on controlling random noise. Based on the noise characteristics of the input tensors in our experiments, the parameters are configured as $\tau = 0.1, \lambda_1 = 50, \lambda_2 = 10, \alpha = 1, \beta = 200$, and $\gamma = 50$ for synthetic data and $\tau = 0.1, \lambda_1 = 1, \lambda_2 = 1, \alpha = 0.1, \beta = 50$, and $\gamma = 50$ for the Penobscot-3D dataset. Further details on parameter selection are provided in Sect. 4.5.2.

#### 4.4.1.4 Parameter Settings of the Baseline Methods

Experiments on synthetic data are conducted to assess the SNR performance of baseline methods, leveraging the availability of ground truth data during the suppression of random plus footprint noise. A grid search strategy is utilized to determine the optimal parameters that maximize the SNR and enhance model performance. For the real Penobscot-3D dataset, which serves as a benchmark, the SGRDL and FR-Net methods are evaluated using parameter settings directly adopted from [5, 37]. In the case of Fxy filtering, the filtering window size in the Fxy domain is fine-tuned to achieve optimal denoising results [39].

---

[1] https://github.com/chenyk1990/reproducible_research/tree/master/foot.

### 4.4.2 Validation on Synthetic Data

To assess the performance and applicability of the proposed UTV-TLRA model, it is tested on a synthetic physical model dataset illustrated in Fig. 4.5a. This dataset closely emulates real field data, providing a more realistic benchmark than conventional seismic forward modeling data. Random and footprint noise are then artificially introduced into the clean dataset, creating noisy seismic images that serve as inputs for the UTV-TLRA model, as shown in Fig. 4.5d. The added random noise is designed to replicate the characteristics of actual seismic noise encountered in field conditions. The clean dataset is used as a reference to compute the SNR for each method, facilitating a quantitative comparison of SOTA algorithms with the proposed model. As presented in Table 4.2, the FR-Net, Fxy, and SGRDL methods yield SNR values of 23.07, 17.59, and 22.22 dB, respectively, while the UTV-TLRA model outperforms them, achieving an SNR of 25.51 dB.

For a more thorough evaluation, Fig. 4.5 demonstrates the denoising effectiveness of the UTV-TLRA model in comparison with SOTA methods under conditions of random noise and footprint noise, where the initial SNR is 14.66 dB. This visualization was produced using the Madagascar software package.[2] To gain deeper insights into the denoising performance, the data volumes in Fig. 4.5 are examined

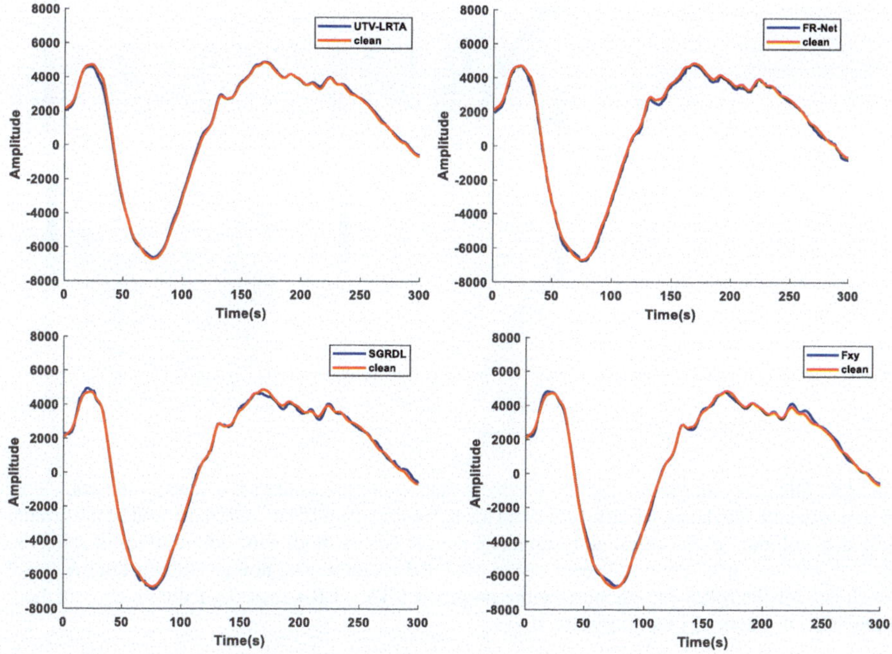


**Fig. 4.4** Comparison of denoising results for a single trace on the synthetic dataset

[2] https://reproducibility.org/wiki/Main_Page.

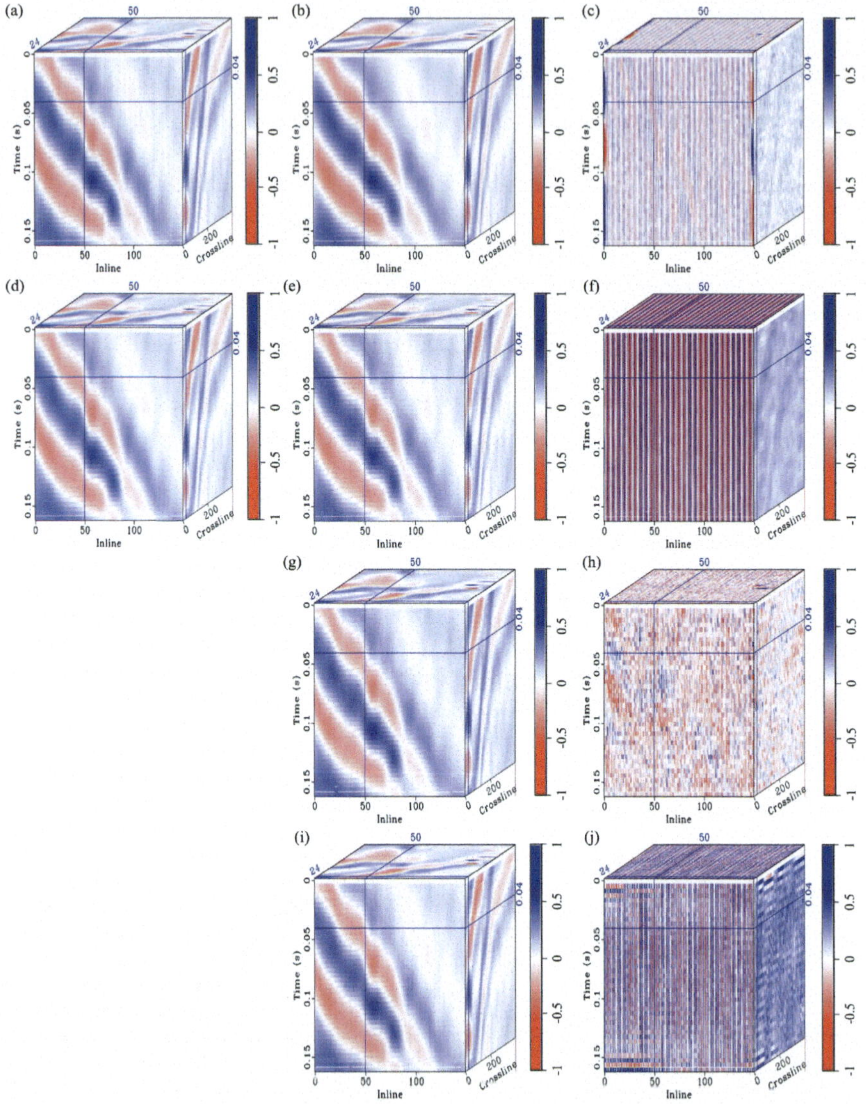


**Fig. 4.5** The denoising results on the synthetic dataset are presented as follows: **a** the volume of the noisy data; **b** the denoised result obtained using the proposed UTV-TLRA method; **d** the ground truth data volume; (e) the denoised result using the FR-Net method; **g** the denoised result using the SGRDL method; **i** the denoised result using the DRR method. The noise residuals after removal are shown for the following methods: **c** the proposed UTV-TLRA method; **f** the FR-Net method; **h** the SGRDL method; **j** the Fxy method

through slices along the time, inline, and xline directions, with corresponding results displayed in Figs. 4.6, 4.7, and 4.8. These slices reveal that the output of the UTV-TLRA model closely matches the ground truth, whereas the SOTA methods display

**Table 4.2** Comparative evaluation of random plus footprint noise removal between state-of-the-art and proposed UTV-TLRA methods on synthetic datasets

| Methods | Synthetic dataset (dB) |
|---|---|
| Fxy | 17.59 |
| SGRDL | 22.22 |
| FR-Net | 23.07 |
| Proposed UTV-TLRA | **25.51** |

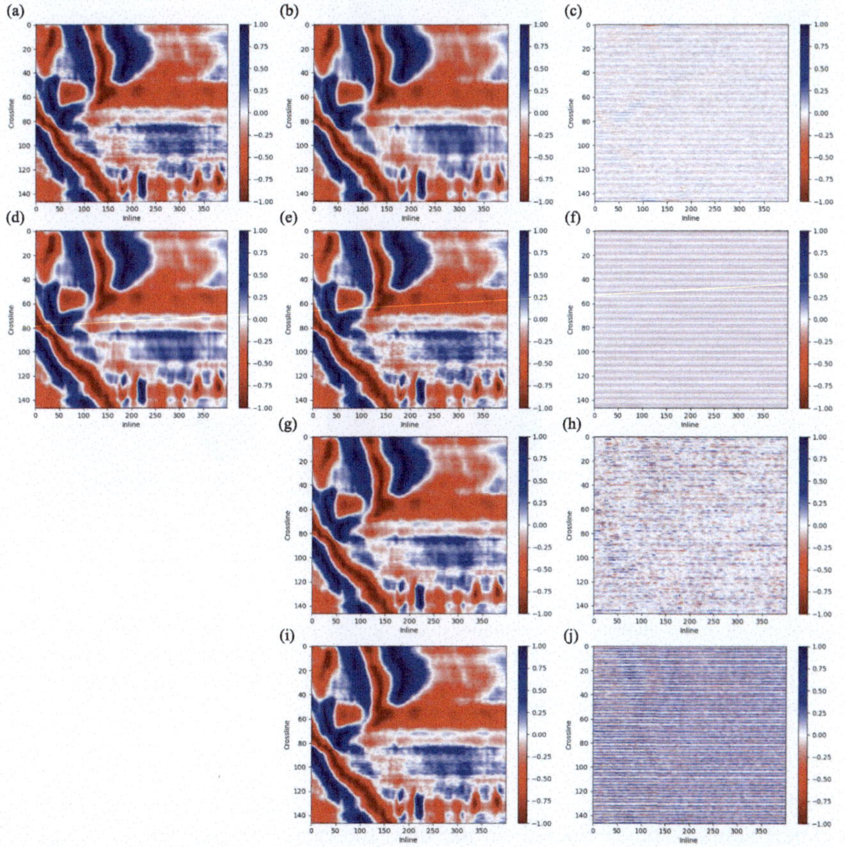

**Fig. 4.6** Denoising results on the synthetic dataset in terms of the constant time slice comparison. **a** Original 27th time section. The denoising outcomes are generated by (**b**) the proposed UTV-TLRA method, **d** the ground truth data, **e** the FR-Net method, **g** the SGRDL method, and **i** the DRR method. The residuals (noise) are removed by (**c**) the proposed UTV-TLRA method, **f** the FR-Net method, **h** the SGRDL method, and (**j**) the Fxy method

noticeable discrepancies. This conclusion is further substantiated by the single-trace comparison shown in Fig. 4.4.

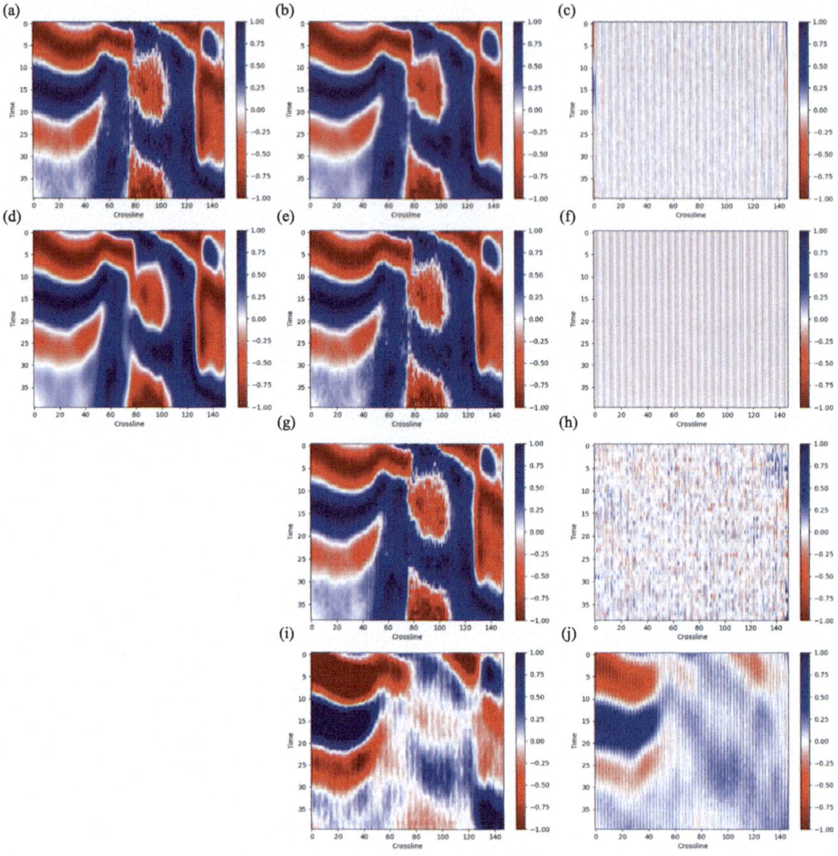


**Fig. 4.7** Denoising results on the synthetic dataset in terms of the inline section comparison. **a** Original 300th inline section. The denoising outcomes are generated by (**b**) the proposed UTV-TLRA method, **d** the ground truth data, **e** the FR-Net method, **g** the SGRDL method, and **i** the DRR method. The residuals (noise) are removed by (**c**) the proposed UTV-TLRA method, **f** the FR-Net method, **h** the SGRDL method, and **j** the Fxy method

To analyze the UTV-TLRA model's robustness across different noise levels, two sub-experiments were conducted. In the first, the random noise intensity was held constant while the footprint noise was varied. In the second, the random noise was adjusted while keeping the footprint noise fixed. The outcomes of these experiments, depicted in Figs. 4.14 and 4.13, illustrate that the UTV-TLRA model consistently achieves higher SNR compared to SOTA methods across various noise conditions. Considering both visual and numerical evaluations, it is evident that the UTV-TLRA model demonstrates superior performance on synthetic datasets.

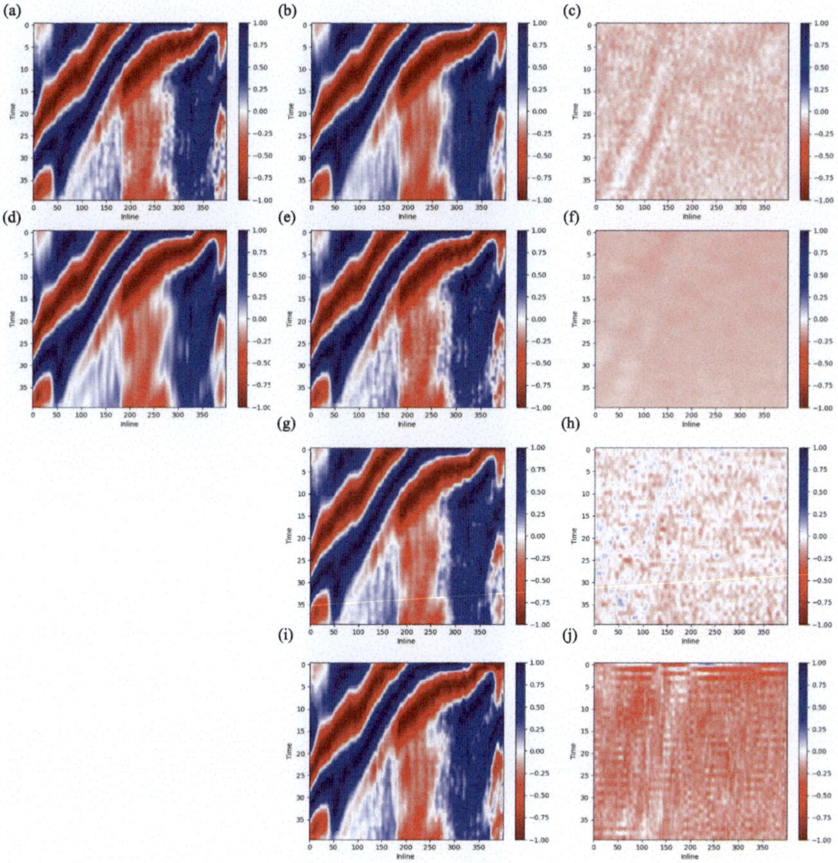

**Fig. 4.8** Denoising results on the synthetic dataset in terms of the crossline section comparison. **a** Original 70th crossline section. The denoising outcomes are generated by (**b**) the proposed UTV-TLRA method, **d** the ground truth data, the FR-Net method, **g** the SGRDL method, and **i** the DRR method. The residuals (noise) are removed by (**c**) the proposed UTV-TLRA method, (f) the FR-Net method, **h** the SGRDL method, and **j** the Fxy method

### *4.4.3 Validation on Real Field Data*

In this section, we use the publicly available Penobscot-3D dataset from dGB Earth Sciences[3] to evaluate the proposed methods. The dataset is affected by both random noise and footprint artifacts in the shallow time slices, as shown in Figs. 4.9a, 4.10a, 4.11a, and 4.12a. The footprint artifacts manifest as a series of stripes of varying lengths, oriented in the crossline direction. This dataset is well-suited for benchmarking the performance of various removal techniques, and has been used in previous studies, including [5, 6].

[3] https://wiki.seg.org/wiki/Penobscot_3D.

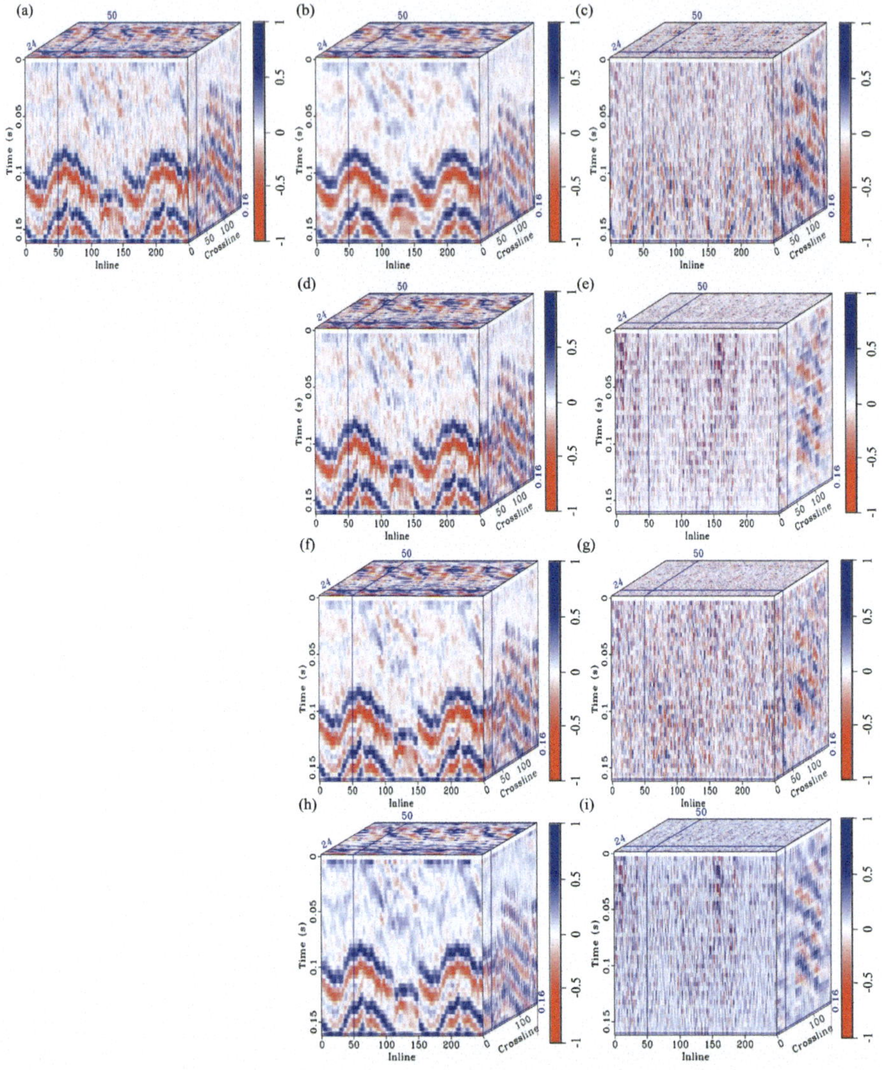

**Fig. 4.9** Denoising results on the Penobscot-3D dataset. **a** Noisy data volume. The denoising outcomes are generated by (**b**) the proposed UTV-TLRA method, **d** the FR-Net method, **f** the SGRDL method, and **h** the DRR method. The residuals (noise) are removed by (**c**) the proposed UTV-TLRA method, **e** the FR-Net method, **g** the SGRDL method, and **i** the Fxy method

Figures 4.9, 4.10, 4.11, and 4.12 demonstrate the effectiveness of different methods in suppressing mixed noise. Specifically, Fig. 4.9 provides a comprehensive comparison of the noise reduction performance of the proposed UTV-TLRA approach against state-of-the-art (SOTA) techniques across the entire dataset.

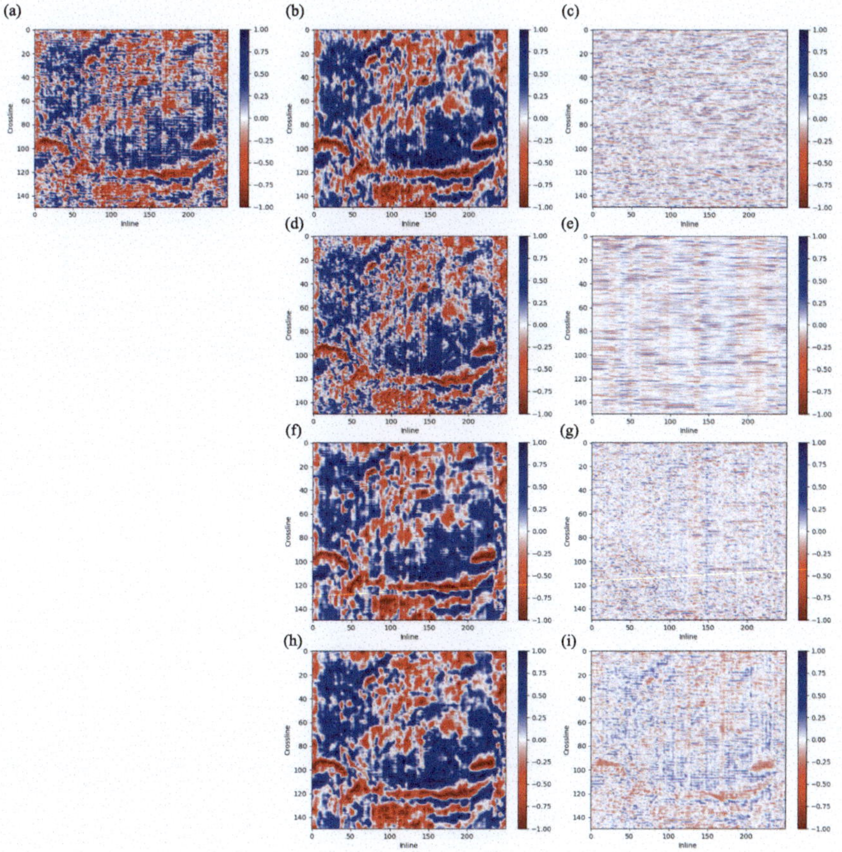

**Fig. 4.10** Denoising results on the Penobscot-3D dataset in terms of the constant time slice comparison. **a** Original 1st time section. The denoising outcomes are generated by (**b**) the proposed UTV-TLRA method, **d** the FR-Net method, **f** the SGRDL method, and **h** the DRR method. The residuals (noise) are removed by (**c**) the proposed UTV-TLRA method, **e** the FR-Net method, **g** the SGRDL method, and **i** the Fxy method

Figures 4.10, 4.11, and 4.12 further illustrate the denoising results for specific slices: constant time, xline, and inline sections, respectively. The experimental results show that the proposed UTV-TLRA model exhibits excellent denoising performance, especially in handling complex noise consisting of both random and footprint noise components. While the FR-Net demonstrates strong capabilities in removing footprint artifacts, it is less effective in suppressing random noise. In contrast, the Fxy filtering method struggles with removing footprint noise, as shown in Fig. 4.10h

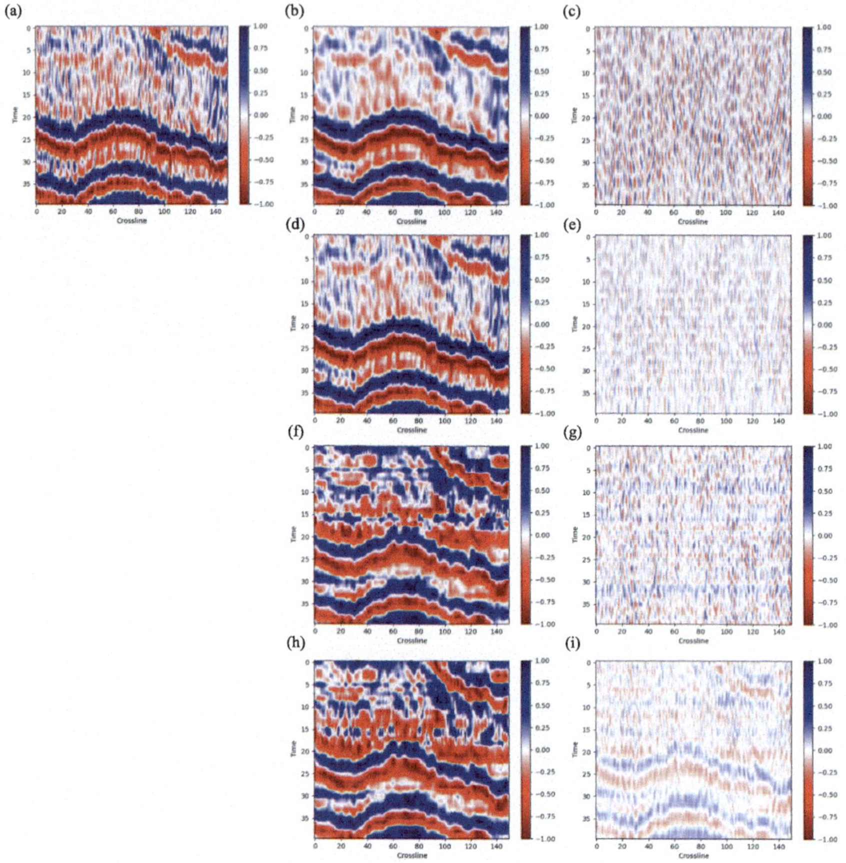

**Fig. 4.11** Denoising results on the Penobscot-3D dataset in terms of the inline section comparison. **a** Original 10th inline section. The denoising outcomes are generated by (**b**) the proposed UTV-TLRA method, **d** the FR-Net method, **f** the SGRDL method, and **h** the DRR method. The residuals (noise) are removed by (**c**) the proposed UTV-TLRA method, **e** the FR-Net method, **g** the SGRDL method, and **i** the Fxy method

and i. SGRDL achieves partial suppression of both noise types, but significant footprint noise remains in the results shown in Fig. 4.10f and g. Overall, the UTV-TLRA model effectively reduces mixed noise while preserving crucial local features and the overall structural integrity of seismic images, as demonstrated in Figs. 4.9, 4.10, 4.11, and 4.12.

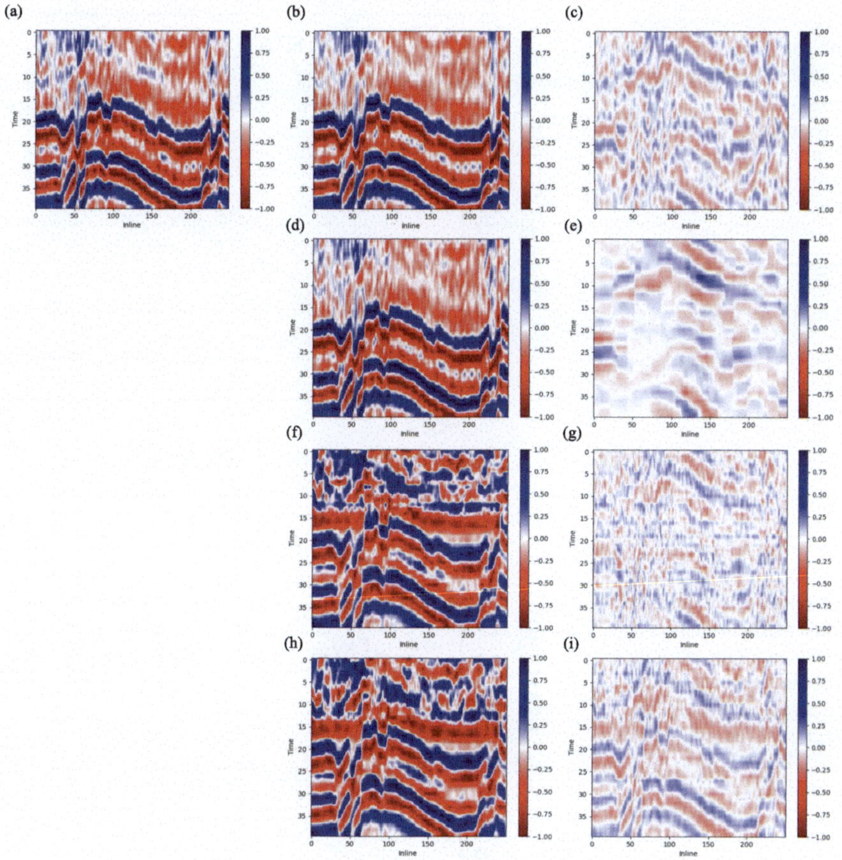

**Fig. 4.12** Denoising results on the Penobscot-3D dataset in terms of the crossline section comparison. **a** Original 70th crossline section. The denoising outcomes are generated by (**b**) the proposed UTV-TLRA method, **d** the FR-Net method, **f** the SGRDL method, and **h** the DRR method. The residuals (noise) are removed by (**c**) the proposed UTV-TLRA method, **e** the FR-Net method, **g** the SGRDL method, and **i** the Fxy method

## 4.5 The Limitations and Strengths of UTV-TLRA

### *4.5.1 Advantages and Disadvantages*

#### 4.5.1.1 Advantages

The key benefit of the UTV-TLRA model over conventional approaches is its capability to effectively manage mixed noise. It efficiently mitigates both random and footprint noise in both synthetic and real-world data, as shown in Figs. 4.4, 4.12. The improved performance of the UTV-TLRA model can be attributed to two main

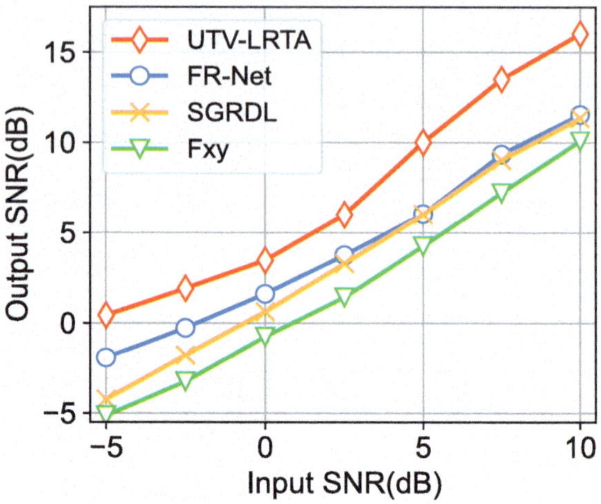


**Fig. 4.13** SNR comparison for different levels of random noise intensity among the proposed method and existing approaches, while maintaining a constant footprint noise level. Symbols "○, ◊, ∇, and ×" represent FR-Net, the proposed UTV-TLRA approach, Fxy filtering, and SGRDL, respectively

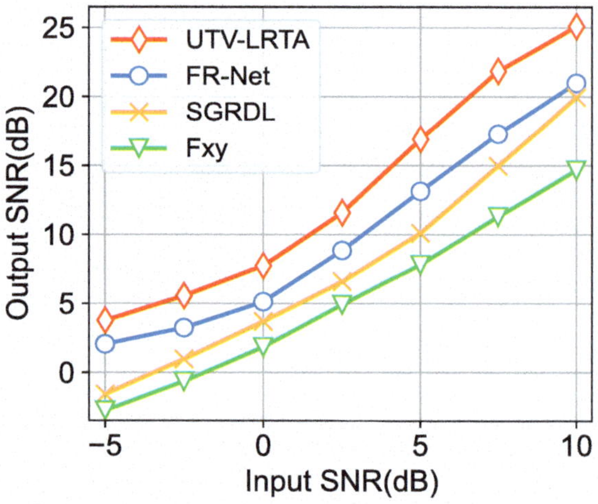


**Fig. 4.14** SNR (Signal-to-Noise Ratio) comparison across different noise intensities for the proposed methods and existing techniques, with the level of random noise fixed. The markers ("○, ◊, ∇, and ×") represent the following methods: FR-Net, the proposed UTV-TLRA method, Fxy filtering, and SGRDL

factors: first, the precise modeling and elimination of footprint noise using UTV, and second, the effective suppression of random noise achieved by the TNN. Moreover, a notable advantage is that all computations are performed within the tensor domain, utilizing tensor operations, which helps maintain the structural integrity of high-dimensional seismic data.

The UTV-TLRA model demonstrates remarkable computational efficiency, mainly attributed to the use of the SBI algorithm. As presented in Table 4.3, a comparison of the computational complexities between the proposed UTV-TLRA method and several state-of-the-art (SOTA) techniques, By evaluating both synthetic and real seismic datasets, the comparison highlights that the UTV-TLRA model provides faster processing times. This contrasts with FR-Net, which shows a higher computational cost due to its large number of parameters for training. However, it is worth

**Table 4.3** Comparison of computation time between the proposed UTV-TLRA and state-of-the-art methods

| Method | Synthetic dataset (s) | Penobscot-3D (s) |
|---|---|---|
| Fxy | 6.79 | 2.98 |
| SGRDL | 1173.51 | 358.11 |
| FR-Net | 161.79 | 87.12 |
| Proposed UTV-TLRA | **40.71** | **21.26** |

noting that this comparison may not be entirely conclusive, as platform differences, as mentioned in Sect. 4.4.1.2, could potentially affect the observed computational efficiency of the UTV-TLRA model.

#### 4.5.1.2 Disadvantages

The effectiveness of the UTV-TLRA model is contingent upon specific model assumptions, including low-rank and directional characteristics. However, the assumption of global low-rank may not always be valid, particularly in cases involving seismic data with complex geological formations [40, 41].

This limitation can have a notable impact on the effectiveness of noise reduction. Similarly, when footprint noise lacks a clear directional pattern, particularly when it doesn't exhibit a consistent stripe structure with a uniform wavelength and direction, the UTV regularization becomes less efficient. This reduces the overall performance of the UTV-TLRA model. However, even in such situations, the UTV-TLRA model can still address both random and footprint noise by leveraging the low-rank assumption found in TNN, though its effectiveness largely relies on the accuracy of the approximations made by TNN. These challenges are common limitations associated with model-based denoising approaches.

Like other low-rank models [42–44], the UTV-TLRA model, when combined with the SBI algorithm, depends on an optimization process that is sensitive to the choice of parameters. This process seeks to achieve optimal reduction of mixed noise through precise tuning of these parameters. The model's parameters can be categorized into two sets: the first includes four key parameters—$\lambda_1$, $\lambda_2$, $\beta$, and $\gamma$—which play a pivotal role in the removal of footprint noise. The second set comprises two parameters, $\tau$ and $\alpha$, which are essential for the model's ability to reduce random noise. If these parameters are set too low, residual noise may persist, while overly high values can lead to important signals being misclassified as noise. More detailed information on parameter selection and its effects is discussed in the following section.

### *4.5.2 Analysis of the Parameters*

The UTV-TLRA model (4.8) involves six parameters: $\tau, \lambda_1, \lambda_2, \alpha, \beta, \gamma$. Selecting appropriate values for these parameters presents a typical challenge in many optimization problems, as their effectiveness depends on the specific levels of random and footprint noise. The ideal approach for determining the optimal parameters is to maximize the SNR within a six-dimensional parameter space. However, this approach can be computationally demanding and time-consuming, making it difficult to achieve the global optimum.

To tackle this issue, two main approaches are utilized to optimize the parameters, inspired by previous works in [31, 33]. First, a robustness evaluation is performed on the three regularization parameters, $\tau, \lambda_1, \lambda_2$, which aids in reducing the number of parameters that require precise tuning. Next, a grid search method is employed to identify the best values for the Bregman parameters, $\alpha, \beta, \gamma$, in the UTV-TLRA model, ensuring optimal noise reduction. Although this search technique may occasionally converge to a local optimum, it still delivers notable performance in noise suppression, as shown in the experimental results above.

#### 4.5.2.1 Robust Analysis of $\tau$, $\lambda_1$, and $\lambda_2$

To evaluate the consistency of the selected regularization parameters as baseline values, we perform a detailed analysis of the three regularization parameters under varying levels of footprint noise (i.e., case 1 and case 2). The robustness evaluation demonstrates that the Signal-to-Noise Ratio (SNR) changes with different values of the parameters $\tau$, $\lambda_1$, and $\lambda_2$, as shown in Fig. 4.15. In Figs. 4.15a–c, it is clear that the effectiveness of suppressing random and footprint noise remains consistent, even when these parameters are adjusted. This indicates that there are optimal ranges for $\tau$, $\lambda_1$, and $\lambda_2$ that depend on the level of noise. For instance, the optimal SNR can be empirically determined when $\lambda_1 \in [50, 500]$, as seen in Fig. 4.15b. Additionally, experiments using real-field data suggest that these parameter ranges tend to be applicable across various datasets.

#### 4.5.2.2 Grid Search of $\alpha$, $\beta$, and $\gamma$

As outlined in previous works [31–33], parameter selection through empirical approaches is commonly used to identify suitable Bregman parameters, ensuring optimal solutions tailored to specific tasks. In this study, we utilize a grid search technique to determine the ideal values for the three Bregman parameters: $\alpha$, $\beta$, and $\gamma$. The goal is to enhance the signal-to-noise ratio (SNR) for synthetic data or achieve the best possible denoising results for real-world field data. To simplify the process, we empirically limit the search range for these parameters, setting $\alpha \in [0.01, 1]$, $\beta \in [50, 200]$, and $\gamma \in [50, 100]$.

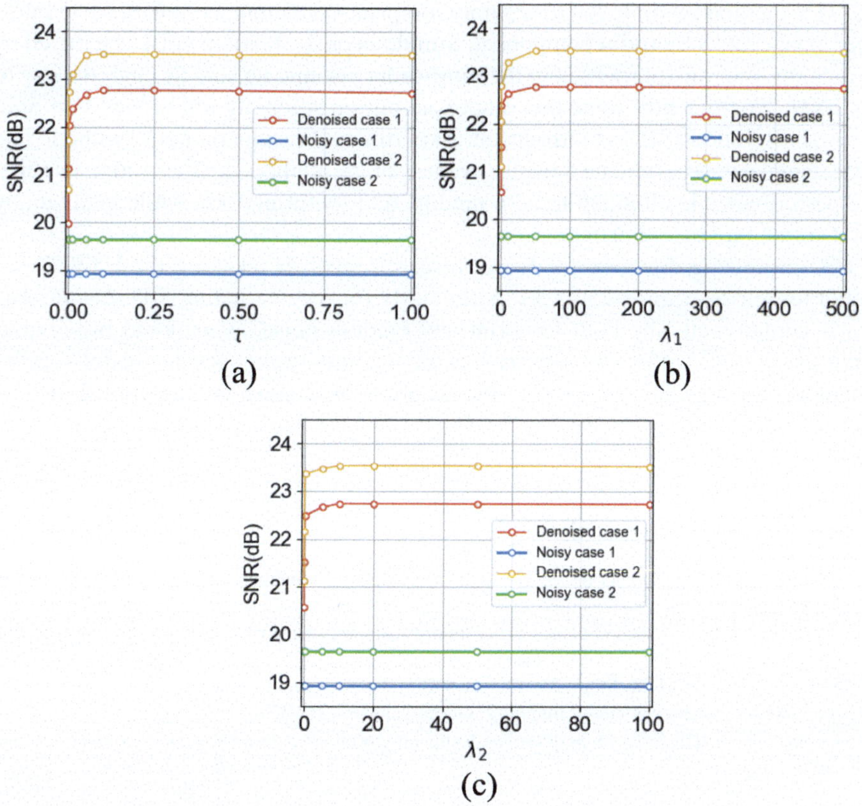


**Fig. 4.15** The sensitivity analysis of the regularization parameters $\tau$, $\lambda_1$, and $\lambda_2$ is presented as follows: **a** The relationship between the Signal-to-Noise Ratio (SNR) values and the parameter$\tau$ is shown, with fixed values of $\lambda_1 = 50$, $\lambda_2 = 10$. **b** The variation of SNR values with respect to the parameter $\lambda_1$ is illustrated, maintaining $\tau = 0.1$, $\lambda_2 = 10$. **c** Lastly, the change in SNR values concerning the parameter $\lambda_2$ is depicted, with $\lambda_1 = 50$, $\tau = 0.1$ held constant

Ultimately, through a series of comprehensive and computationally intensive experiments, our goal is to identify the best-fitting UTV-TLRA model parameters tailored to each of the two datasets.

## 4.6 Summary

This chapter introduces the UTV-TLRA model, a novel approach designed to address 3-D seismic noise caused by random disturbances and footprint interference. The proposed method tackles the problem of mixed noise reduction within a unified framework, combining low-rank and sparse tensor decomposition techniques. Specifically, the UTV method is employed to separate footprint-contaminated seismic data into a clean image and a footprint component. Simultaneously, random noise is reduced by leveraging low-rank properties in the third-order seismic tensors through the use of the TNN. To efficiently solve this model, an optimization algorithm based on Split Bregman Iteration (SBI) is introduced. Experimental results on both synthetic and real seismic datasets validate the effectiveness of the method, demonstrating its superior performance in eliminating both random and footprint noise while maintaining good computational efficiency.

One promising direction for future research involves adapting the UTV-TLRA model for tensor completion tasks, particularly for reconstructing 3-D seismic data that is contaminated by both footprint and random noise, as explored in previous studies such as [28, 45]. This approach could potentially enhance the model's ability to handle missing data and improve the accuracy of seismic data interpolation.

## References

1. Chopra S, Larsen G (2000) Acquisition footprint-its detection and removal. CSEG Rec 25(8):16–20
2. Schuster GT, Liu Z (2001) Seismic array theorem and rapid calculation of acquisition footprint noise. Geophysics 66(6):1843–1849
3. Marfurt KJ, Scheet RM, Sharp JA, Harper MG (1998) Suppression of the acquisition footprint for seismic sequence attribute mapping. Geophysics 63(3):1024–1035
4. Cooper JK (2010) Seismic acquisition footprint: modelling and mitigation. Master's thesis, Graduate Studies
5. Chen W, Saad OM, Wang H, Chen Y (2021) Statistics-guided residual dictionary learning for footprint noise removal. IEEE Trans Geosci Remote Sens 60:1–11
6. Gómez JL, Velis DR (2020) Footprint removal from seismic data with residual dictionary learning. Geophysics 85(4):V355–V365
7. Gulunay N, Martin F, Martinez R (1994) 3d data acquisition artifacts removal-spot editing in the spatial-temporal frequency domain. In: 56th EAEG meeting. European Association of Geoscientists & Engineers, p cp–47
8. Soubaras R (2002) Attenuation of acquisition footprint for non-orthogonal 3d geometries. In: SEG technical program expanded abstracts 2002. Society of Exploration Geophysicists, pp 2142–2145
9. Drummond J(J)M, Budd AJL, Ryan JW (2000) Adapting to noisy 3d data-attenuating the acquisition footprint. In: SEG technical program expanded abstracts 2000. Society of Exploration Geophysicists, pp 9–12
10. Falconer S, Marfurt KJ (2008) Attribute-driven footprint suppression. In: 2008 SEG annual meeting. OnePetro
11. Gülünay N (1999) Acquisition geometry footprints removal. In: SEG technical program expanded abstracts 1999. Society of Exploration Geophysicists, pp 637–640
12. Alali A, Machado G, Marfurt KJ (2018) Attribute-assisted footprint suppression using a 2d continuous wavelet transform. Interpretation 6(2):T457–T470
13. Zhou Yu, Abma R, Etgen J, Sullivan C (2017) Attenuation of noise and simultaneous source interference using wavelet denoising. Geophysics 82(3):V179–V190
14. Górszczyk A, Malinowski M, Bellefleur G (2015) Enhancing 3d post-stack seismic data acquired in hardrock environment using 2d curvelet transform. Geophys Prospect 63(4-Hard Rock Seismic imaging):903–918

15. Zhang R (2009) Footprint suppression with basis pursuit denoising. In: 2009 SEG annual meeting. OnePetro
16. Chen X, Yang W, He Z, Zhong W (2012) Adaptive acquisition footprint suppression based on a 3d stationary wavelet transform: a case study from china. J Appl Geophys 77:1–6
17. Cvetkovic M, Falconer S, Marfurt KJ, Chávez-Pérez S (2007) 2d stationary-wavelet transform-based acquisition footprint suppression. In: SEG technical program expanded abstracts 2007. Society of Exploration Geophysicists, pp 2590–2594
18. Liu D, Gao L, Wang X, Chen W (2021) A dictionary learning method with atom splitting for seismic footprint suppression. Geophysics 86(6):V509–V523
19. Gao L, Xiaokai W, Chen W, Chao W (2016) A local morphological-component analysis-based acquisition footprint suppression method. In: 2016 SEG international exposition and annual meeting. OnePetro
20. Al-Bannagi MS, Fang K, Kelamis PG, Douglass GS (2004) Acquisition footprint suppression via the truncated SVD technique. In: SEG technical program expanded abstracts 2004. Society of Exploration Geophysicists, pp 1957–1960
21. Al-Bannagi MS, Fang K, Kelamis PG, Douglass GS (2005) Acquisition footprint suppression via the truncated SVD technique: case studies from Saudi Arabia. Lead Edge 24(8):832–834
22. Chen Y, Zhang D, Jin Z, Chen X, Shaohuan Z, Huang W, Gan S (2016) Simultaneous denoising and reconstruction of 5-d seismic data via damped rank-reduction method. Geophys J Int 206(3):1695–1717
23. Chen Y, Zhou Y, Chen W, Shaohuan Z, Huang W, Zhang D (2017) Empirical low-rank approximation for seismic noise attenuation. IEEE Trans Geosci Remote Sens 55(8):4696–4711
24. Feng J, Liu X, Li X, Wenxi X, Liu B (2021) Low-rank tensor minimization method for seismic denoising based on variational mode decomposition. IEEE Geosci Remote Sens Lett 19:1–5
25. Qian F, Zhang C, Feng L, Cai L, Zhang G, Guangmin H (2020) Tubal-sampling: bridging tensor and matrix completion in 3-d seismic data reconstruction. IEEE Trans Geosci Remote Sens 59(1):854–870
26. Feng J, Li X, Liu X, Chen C, Chen H (2021) Seismic data denoising based on tensor decomposition with total variation. IEEE Geosci Remote Sensing Lett 18(7)
27. Anvari R, Kahoo AR, Mohammadi M, Khan NA, Chen Y (2019) Seismic random noise attenuation using sparse low-rank estimation of the signal in the time–frequency domain. IEEE J Sel Top Appl Earth Observations Remote Sens 12(5):1612–1618
28. Gao J, Cheng J, Sacchi MD (2016) Five-dimensional seismic reconstruction using parallel square matrix factorization. IEEE Trans Geosci Remote Sens 55(4):2124–2135
29. Wang D, Gao J, Liu N, Jiang X (2020) Structure-oriented DTGV regularization for random noise attenuation in seismic data. IEEE Trans Geosci Remote Sens 59(2):1757–1771
30. Bayram I, Kamasak ME (2012) A directional total variation. In: 2012 proceedings of the 20th European signal processing conference (EUSIPCO). IEEE, pp 265–269
31. Chang Y, Fang H, Yan L, Liu H (2013) Robust destriping method with unidirectional total variation and framelet regularization. Opt Express 21(20):23307–23323
32. Goldstein T, Osher S (2009) The split Bregman method for l1-regularized problems. SIAM J Imaging Sci 2(2):323–343
33. Chang Y, Yan L, Fang H, Luo C (2015) Anisotropic spectral-spatial total variation model for multispectral remote sensing image destriping. IEEE Trans Image Process 24(6):1852–1866
34. Ng MK, Chan RH, Tang W-C (1999) A fast algorithm for deblurring models with Neumann boundary conditions. SIAM J Sci Comput 21(3):851–866
35. Donoho DL (1995) De-noising by soft-thresholding. IEEE Trans Inform Theory 41(3):613–627
36. Arora S, Barak B (2009) Computational complexity: a modern approach. Cambridge University Press
37. Qian F, Guo W, Liu Z, Hongtao Yu, Zhang G, Guangmin H (2022) Unsupervised erratic seismic noise attenuation with robust deep convolutional autoencoders. IEEE Trans Geosci Remote Sens 60:1–16
38. Hodgson L, Whitcombe D, Lancaster S, Lecocq P (2002) Frequency slice filtering-a novel method of seismic noise attenuation. In: SEG technical program expanded abstracts 2002. Society of Exploration Geophysicists, pp 2214–2216

39. Liu Y, Zheng Z (2021) Noniterative f-xy streaming prediction filtering for random noise attenuation on seismic data. IEEE Trans Geosci Remote Sens 60:1–9
40. Qian F, Liu Z, Wang Y, Zhou Y, Guangmin H (2022) Ground truth-free 3-d seismic random noise attenuation via deep tensor convolutional neural networks in the time-frequency domain. IEEE Trans Geosci Remote Sens 60:1–17
41. Qian F, Wang Y, Zheng B, Liu Z, Zhou Y, Guangmin H (2022) Multidimensional seismic data denoising using framelet-based order-p tensor deep learning. IEEE Trans Geosci Remote Sens 60:1–18
42. Wang C, Zhu Z, Hanming G, Xinming W, Liu S (2018) Hankel low-rank approximation for seismic noise attenuation. IEEE Trans Geosci Remote Sens 57(1):561–573
43. Chen Y, Huang W, Zhang D, Chen W (2016) An open-source matlab code package for improved rank-reduction 3d seismic data denoising and reconstruction. Comput Geosci 95:59–66
44. Bai M, Huang G, Wang H, Chen Y (2020) Seismic signal enhancement based on the low-rank methods. Geophys Prospect 68(9):2783–2807
45. Qian F, Liu Z, Wang Y, Liao S, Pan S, Guangmin H (2021) DTAE: deep tensor autoencoder for 3-d seismic data interpolation. IEEE Trans Geosci Remote Sens 60:1–19

# Chapter 5
# Tensor Deep Learning for Seismic Data Reconstruction

## 5.1 Introduction

Compared to the STC model presented in Chap. 3, we demonstrate that the DTAE model offers a more effective approach for 3-D seismic data reconstruction, which is the main focus of this chapter. A significant limitation of the STC model, like other tensor-based methods, is its reliance on the strict low-rank assumption when interpolating real-world data. In practice, the pursuit of a perfectly low-rank tensor can be unrealistic, potentially compromising the recovery accuracy. This issue is addressed through the optimization process based on tensor backpropagation (TBP). This chapter makes the following key contributions:

(1) *Model*: Building on the t-NN theory, we introduce a foundational DTAE model and two variants (as explained in the subsequent introduction to the DTAE framework). We frame the reconstruction of 3D seismic data as a nonlinear regression problem. By fully exploiting the potential of DTAE, we can accurately predict the missing traces from a corrupted 3D tensor. To the best of our knowledge, this is the first application of tensor deep learning techniques for the reconstruction of missing traces in 3D seismic data.
(2) *Solution*: To improve model parameter optimization, we propose an enhanced version of the traditional backpropagation algorithm, called the TBP algorithm, which is specifically designed to optimize the weight tensor. This advancement allows the DTAE model to produce a reconstructed tensor that effectively recovers the corrupted input data, reducing the gap between the reconstructed output and the target seismic tensor.
(3) *Framework*: To support the deployment of the DTAE model and the TBP algorithm, we derive a theoretical link between tensor and matrix autoencoders (AEs) by utilizing the characteristics of the t-product (refer to Definition 5.2 in Appendix 5.8). This theoretical basis underpins the practical application of DTAE, including the estimation of weight tensors through a denoising autoencoder (DA), along with two extensions—stacked sparse denoising autoencoder

 
F. Qian et al., *Tensor Computation for Seismic Data Processing*,
Earth Systems Data and Models 6, https://doi.org/10.1007/978-3-031-78900-7_5

(SSDA) and deep convolutional autoencoder (DCAE)—designed for processing individual frequency slices.

By assessing a simulated model and two datasets from real-world surveys, the experimental results demonstrate that the proposed method exhibits a significant performance improvement compared to three advanced existing techniques.

3D deep learning and the DTAE model share similar advantages in several aspects, particularly in capturing complex multidimensional relationships, which both outperform traditional matrix-based deep learning methods [1, 2]. Given their similarities in framework, it is reasonable to compare the DTAE model with common 3D deep learning models, especially 3D convolutional neural networks (CNNs) [3] and 3D autoencoders (AEs) [4]. Since DTAE follows the t-NN framework [5, 6], it has a more compact parameter space, enabling faster training compared to conventional 3D deep learning models. Furthermore, another significant advantage of DTAE is its ability to generate a larger number of training samples from the same seismic dataset by utilizing spectral images. This capability is particularly valuable in scenarios with limited data. Typically, in settings without overlapping windows, 3D deep learning methods can only generate one training sample per data window [3], whereas DTAE can effectively expand the training dataset by generating multiple spectral images (see Sect. 5.5.1.3). This advantage is especially important in seismic surveys, where the available data is often limited.

## 5.2 Seismic Data Reconstruction and Its Formulation

As mentioned earlier, existing tensor-based methods face several key limitations that hinder their potential for performance optimization. To address this bottleneck, this section introduces an innovative DTAE model. First, in Sect. 5.2.1, we clearly define the reconstruction problem. Then, in Sect. 5.2.2, by incorporating a deep tensor neural network (t-NN), we reformulate the reconstruction problem as a nonlinear regression task for more efficient solving. The foundational concepts and theories of tensor algebra related to this discussion are provided in detail in Appendix 5.8.

### *5.2.1 Problem Statement*

3D data reconstruction aims to recover missing seismic traces from the partially collected dataset $\mathcal{T}_\Omega \in \mathbb{R}^{N_1 \times N_2 \times N_3}$, where specific traces are absent compared to the reference seismic tensor $\mathcal{T} \in \mathbb{R}^{N_1 \times N_2 \times N_3}$. Specifically, let the collected traces $\mathcal{T}_\Omega$ be represented by the set $\Omega$. Assuming that the missing traces are randomly distributed, the observation or sampling process can be expressed as:

$$\mathcal{T}_\Omega = \mathcal{P}_\Omega(\mathcal{T}). \tag{5.1}$$

where $\mathcal{P}_\Omega(\cdot)$ is a sampling operator that maps the tensor $\mathcal{T}$ onto the set $\Omega$, with the specific expression given as follows:

$$\mathcal{P}_\Omega(\mathcal{T}) = \begin{cases} \mathcal{T}(i, j, :), & \text{if } (i, j) \in \Omega, \\ 0, & \text{otherwise}, \end{cases} \tag{5.2}$$

The $(i, j)$th element of $\mathcal{P}_\Omega(\mathcal{T})$ corresponds to $\mathcal{T}(i, j, :)$ when $(i, j) \in \Omega$, and is set to zero otherwise. Since the number of observed measurements greatly exceeds the count of missing elements in $\mathcal{T}$, the reconstruction task is severely ill-posed, which makes direct solving unfeasible without additional constraints on $\mathcal{T}$.

To estimate the complete tensor $\mathcal{T}$ from partial observations $\mathcal{T}_\Omega$, traditional non-learning-based methods typically rely on the low-rank assumption [7–9], attempting to overcome the challenges of the inference process by exploiting this assumption. These methods commonly provide an effective solution to the inference problem (8.1) through low-rank approximation (LRA). However, in certain cases, particularly when surveys are conducted in geologically complex regions, the seismic reflection data may not satisfy the strict low-rank condition. If the missing data does not exhibit low-rank characteristics, traditional non-learning-based methods will struggle to effectively recover the missing seismic traces.

### 5.2.2 Problem Formulation

To address the aforementioned challenges, we propose an innovative approach utilizing deep neural networks (DNNs) to solve the inverse problem (8.1) without relying on traditional prior assumptions. Unlike existing methods [10, 11], our approach leverages high-dimensional techniques by employing a specifically pretrained deep tensor autoencoder (DTAE) model (for further details, refer to Sect. 5.3) to assist in the reconstruction of 3D seismic data. Our approach follows a typical supervised learning framework, which includes two main stages: training and prediction. During the training phase, when a sufficient number of training samples (e.g., field data pairs consisting of corrupted data $\mathcal{T}_\Omega$ and complete data $\mathcal{T}$) are available, we model the corruption process of 3D seismic data as

$$\mathcal{T} = \eta(\mathcal{T}_\Omega), \tag{5.3}$$

where $\eta(\cdot) : \mathbb{R}^{N_1 \times N_2 \times N_3} \to \mathbb{R}^{N_1 \times N_2 \times N_3}$ represents a stochastic corruption process that corrupts the observed seismic tensor $\mathcal{T}_\Omega$; consequently, $\eta(\cdot) = \mathcal{P}_\Omega^{-1}(\cdot)$. Therefore, the learning objective for the reconstruction task becomes

$$\mathcal{L}_1(\mathcal{T}, \mathcal{T}_\Omega; \Theta) = \underset{\Theta}{\operatorname{argmin}}\ \mathbb{E}_{\mathcal{T}_\Omega} \| f(\mathcal{T}, \Theta) - \mathcal{T}_\Omega \|_F^2. \tag{5.4}$$

Based on this formulation, the primary objective is to train a DTAE (Deep Tensor Autoencoder) model to learn a function $f(\cdot)$, which depends on the parameters $\Theta$, and effectively approximates $\eta^{-1}(\cdot)$ or $\mathcal{P}_\Omega(\cdot)$. Simply put, the reconstruction of 3D seismic data can be approached through a unified framework by modeling the problem appropriately and selecting the suitable mapping function $\eta(\cdot)$ for each specific case.

During the prediction phase, after determining the function $f(\cdot)$, the reconstructed result is obtained by merging the complete tensor from the input with the predicted output. The key to recovering the missing traces is to effectively leverage the available known data. During this process, it is crucial to ensure that the known traces, $\mathcal{T}_\Omega$, remain unchanged. Thus, the final interpolated solution, $\widehat{\mathcal{T}}$, can be formulated in the following way:

$$\widehat{\mathcal{T}} = \eta(\mathcal{T}_\Omega) \odot (1 - \Omega) + \mathcal{T}_\Omega. \tag{5.5}$$

where the symbol $\odot$ denotes the element-wise tensor product, with the detailed definition provided in Definition 5.4 of Appendix 5.8.

## 5.3 Basic Deep Tensor Auto-Encoder (DTAE)

To address the problem (7.4), we propose a novel approach in Sect. 5.3.1 that employs a DTAE for 3D seismic data reconstruction (Table 5.1). By incorporating the tensor product (t-product) operation into the standard backpropagation (BP) algorithm, we further develop the tensor Backpropagation (TBP) algorithm to effectively optimize the model parameters, as detailed in Sect. 5.3.2. Additionally, to facilitate the implementation of the DTAE model, the necessary background on tensor neural networks (t-NNs) and matrix autoencoders (AEs) is provided "Tensor Neural Networks (t-NNs)' and "Deep Matrix Autoencoders (DMAEs)" in Appendix 5.8, respectively.

**Table 5.1** Performance comparison of the proposed approach against existing methods

| Dataset | PMF (dB) | TNN (dB) | MSSA (dB) | DTAE (dB) |
|---|---|---|---|---|
| Synthetic VSP | 15.53 | 15.20 | 15.26 | **19.51** |
| F3 Netherlands | 14.41 | 15.92 | 16.12 | **21.26** |
| Parihaka-3D | 6.53 | 9.10 | 5.85 | **19.11** |

### 5.3.1 Basic DTAE Model

The objective is to train a DTAE model that learns from input pairs formed by the observed seismic data tensor $\mathcal{T}_\Omega$ and the corresponding ideal target data tensor $\mathcal{T}$. In this way, the model can generate a fully sampled seismic data tensor $\widehat{\mathcal{T}}$ when provided with new observed data $\mathcal{T}_\Omega$. To construct the training set, we extract multiple local 3D data cubes $\{X_n\}_{n=1}^N$ from the complete seismic dataset $\mathcal{T}$, which serve as the ground truth for the data reconstruction task. Subsequently, by randomly removing traces from these complete cubes $\{\mathcal{X}n\}_{n=1}^N$, we generate the corresponding observed cubes or labeled data $\{\mathcal{Y}n\}_{n=1}^N$. Each cube $\mathcal{X}_n$ and $\mathcal{Y}_n$ has dimensions of $n_1 \times n_2 \times n_3$, where $N$ represents the total number of cubes or samples.

For each training pair $(\mathcal{X}_n, \mathcal{Y}_n)$, a new DTAE model is constructed, with its architecture shown in Fig. 5.1. DTAE is a tensor neural network (t-NN), which extends the structure of the matrix autoencoder (AE) described in Appendix 5.8 to higher-dimensional data representations. The model consists of two main components: encoding and decoding layers. During the encoding process, each layer of the DTAE applies an affine transformation to the input data using a set of tensor projection matrices $\{\mathcal{W}_\ell\}_{\ell=1}^K$ and bias tensors $\mathcal{B}_\ell$, progressively extracting deeper features of the data.

In this model, $\mathcal{W}_{\ell-1} \in \mathbb{R}^{n_\ell \times n_{\ell-1} \times n_3}$ and $\mathcal{B}_\ell \in \mathbb{R}^{n_\ell \times 1 \times n_3}$, where $K$ represents the total number of layers. Each encoding layer employs the Rectified Linear Unit (ReLU) as the activation function $\sigma(\cdot)$, which introduces non-linearity to the input data. Specifically, the output of the $\ell$th layer is computed by applying the projection matrices and bias tensors, followed by the ReLU activation function.

$$\mathcal{X}_{n,\ell} = \mathbf{h}\left(\mathcal{X}_{n,\ell-1}\right) = \sigma\left(\mathcal{W}_{\ell-1} * \mathcal{X}_{n,\ell-1} + \overrightarrow{\mathcal{B}}_\ell\right), \quad \ell \in [K], \tag{5.6}$$

The symbol $*$ denotes the tensor product (t-product) operation, as illustrated in Fig. 2.4c, and $\mathcal{X}_{n,\ell} \in \mathbb{R}^{n_\ell \times n_2 \times n_3}$ represents the output tensor of the $\ell$th layer, where $\ell$ indicates the layer index. During the decoding phase, due to the symmetric design of

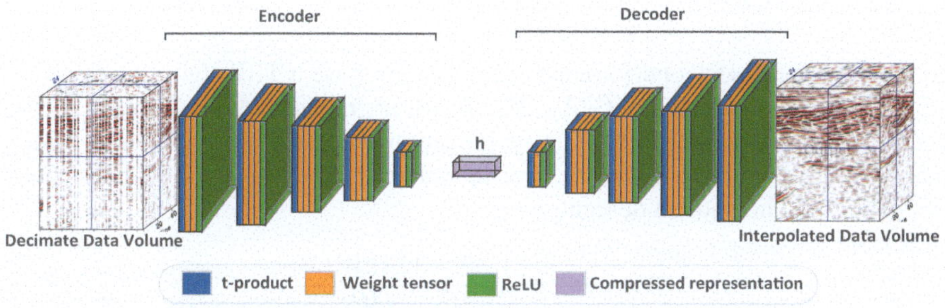


**Fig. 5.1** The schematic diagram illustrates the architecture of the DTAE applied to 3-D seismic data reconstruction. It is important to note that the design of the network architecture is based on the tensor-tensor product (t-product) method

the DTAE, the tensor dictionaries or projection matrices from the encoding layers are reused in the corresponding decoding layers. Thus, each decoding layer first applies an affine transformation to the input $\mathcal{X}_{n,\ell}$ and then follows with a nonlinear ReLU activation function.

$$\mathbf{g}\left(\mathcal{X}_{n,\ell-1}\right) = \sigma\left(\mathcal{W}_{\ell-1}^{\dagger} * \mathbf{h}\left(\mathcal{X}_{n,\ell-1}\right) + \vec{\mathcal{B}}_{\ell-1}\right), \quad \ell \in [K] \tag{5.7}$$

In our supervised learning framework, the goal of training the DTAE model is to adjust the parameters $\Theta = \{\mathcal{W}_\ell, \mathcal{B}_\ell\}_{\ell=1}^{K}$ such that the reconstruction error between the model's output tensor values $\left\{\mathcal{X}_{n,\ell}\right\}_{n=1}^{N}$ and the corresponding target tensor values $\{\mathcal{Y}_n\}_{n=1}^{N}$ is minimized.

$$\mathcal{L}_2(\mathcal{X}_n, \mathcal{Y}_n; \Theta) = \arg\min_{\Theta} \frac{1}{N} \sum_{n=1}^{N} \frac{1}{2} \|\mathcal{X}_{n,\ell} - \mathcal{Y}_n\|_{\mathrm{F}}^2. \tag{5.8}$$

Similar to traditional autoencoders, the parameters $\Theta$ of the DTAE model can generally be optimized using the backpropagation (BP) algorithm. However, since DTAE operates as a tensor-based neural network (t-NN), its optimization requires a tensor variant of the BP algorithm, known as tensor backpropagation (TBP), rather than the conventional BP method. Various optimization algorithms based on t-NN and TBP have been developed to handle the different tensor operators and their respective decompositions. Specifically, these methods include t-NNs based on the t-product [5, 6], t-NNs using Tucker decomposition [12], t-NNs based on tensor-train decomposition [13], and t-NNs utilizing Kronecker product [14]. Additionally, there has been significant progress in applying tensor networks in deep learning, as demonstrated in [15]. In this work, we develop a novel TBP algorithm within the t-product-based t-NN framework, specifically designed to optimize the parameters of the DTAE model.

### *5.3.2 TBP Algorithm*

To improve the model's performance (5.8), we propose a TBP algorithm aimed at minimizing the loss function $\mathcal{L}_3(\mathcal{X}_n, \mathcal{Y}_n; \Theta)$ using gradient descent, thereby adjusting the network parameters $\Theta$. The optimization process relies on layer evolution rules, with the first step being the determination of how to update the parameters $\mathcal{W}_{\ell-1}$ and $\mathcal{B}_\ell$ via gradient descent.

$$\begin{aligned} \mathcal{W}_{\ell-1} &\leftarrow \mathcal{W}_{\ell-1} - \alpha \frac{\partial \mathcal{L}_3(\mathcal{W}, \mathcal{B})}{\partial \mathcal{W}_{\ell-1}}, \\ \mathcal{B}_\ell &\leftarrow \mathcal{B}_\ell - \alpha \frac{\partial \mathcal{L}_3(\mathcal{W}, \mathcal{B})}{\partial \mathcal{B}_\ell}, \end{aligned} \tag{5.9}$$

where $\alpha$ denotes the learning rate, an important adjustable parameter in the optimization process that determines the magnitude of each update during iteration. These two formulations provide us with a more comprehensive insight into the core concepts of gradient descent. The first principle is implemented by calculating the partial derivative $\partial \mathcal{L}_3 / \partial \mathcal{W}_{\ell-1}$, with the detailed mathematical expression provided as follows:

$$\delta \mathcal{W}_{\ell-1} = \frac{\partial \mathcal{L}_3}{\partial \mathcal{W}_{\ell-1}} = \frac{\partial \mathcal{L}_3}{\partial \mathcal{X}_{n,\ell}} \frac{\partial \mathcal{X}_{n,\ell}}{\partial \mathcal{W}_{\ell-1}}. \tag{5.10}$$

We first focus on $\partial \mathcal{L}_3 / \partial \mathcal{X}_{n,\ell}$, which can be obtained by differentiating the equation (5.8). The detailed derivation steps are as follows:

$$\frac{\partial \mathcal{L}_3}{\partial \mathcal{X}_{n,\ell}} = \frac{1}{N} \sum_{n=1}^{N} \|\mathcal{Y}_n - \mathcal{X}_{n,\ell}\|_{\mathrm{F}}. \tag{5.11}$$

We will discuss the term $\delta \mathcal{W}_{\ell-1}$ in more detail in the subsequent section. The next critical step is to compute $\partial \mathcal{L}_3 / \partial \mathcal{B}_\ell$. Similar to the treatment of $\delta \mathcal{W}_\ell$, this term can be reorganized and expressed in a different form, as shown below:

$$\delta \mathcal{B}_\ell = \frac{\partial \mathcal{L}_3}{\partial \mathcal{B}_\ell} = \frac{\partial \mathcal{L}_3}{\partial \mathcal{X}_{n,\ell}} \frac{\partial \mathcal{X}_{n,\ell}}{\partial \mathcal{B}_\ell}. \tag{5.12}$$

The next step is to further derive the partial derivatives $\partial \mathcal{X}_{n,\ell} / \partial \mathcal{W}_{\ell-1}$ and $\partial \mathcal{X}_{n,\ell} / \partial \mathcal{B}_\ell$, with the detailed calculations presented as follows:

#### 5.3.2.1 Update $\partial \mathcal{X}_{n,\ell} / \partial \mathcal{W}_{\ell-1}$

With Definition 5.6 and Eq. (5.6), $\partial \mathcal{X}_{n,\ell} / \partial \mathcal{W}_{\ell-1}$ is defined as follows:

$$\begin{aligned} \frac{\partial \mathcal{X}_{n,\ell}}{\partial \mathcal{W}_{\ell-1}} &= \frac{\partial \sigma \left( \mathcal{W}_{\ell-1} * \mathcal{X}_{n,\ell-1} + \vec{\mathcal{B}}_\ell \right)}{\partial \mathcal{W}_{\ell-1}} \\ &= \sigma'(\mathcal{Z}_{n,\ell}) * \mathcal{X}^{\dagger}_{n,\ell-1}, \end{aligned} \tag{5.13}$$

we define $\mathcal{Z}_{n,\ell} = \mathcal{W}_{\ell-1} * \mathcal{X}_{n,\ell-1} + \vec{\mathcal{B}}_\ell$, Here, $\sigma'(\cdot)$ denotes the derivative of the activation function. Using this expression, we can reformulate (5.10) in tensor notation as follows:

$$\begin{aligned} \frac{\partial \mathcal{L}_3}{\partial \mathcal{W}_{\ell-1}} &= \frac{\partial \mathcal{L}_3}{\partial \mathcal{X}_{n,\ell}} \frac{\partial \mathcal{X}_{n,\ell}}{\partial \mathcal{W}_{\ell-1}} \\ &= \frac{1}{N} \sum_{n=1}^{N} \|\mathcal{Y}_n - \mathcal{X}_{n,\ell}\|_{\mathrm{F}} (\sigma'(\mathcal{Z}_{n,\ell}) * \mathcal{X}^{\dagger}_{n,\ell-1}). \end{aligned} \tag{5.14}$$

#### 5.3.2.2 Update $\partial \mathcal{X}_{n,\ell}/\partial \mathcal{B}_\ell$

By applying the backpropagation (BP) algorithm, we can further derive and compute the following partial derivatives:

$$\frac{\partial \mathcal{X}_{n,\ell}}{\partial \mathcal{B}_\ell} = \text{sum}\left(\sigma'\left(\mathcal{Z}_{n,\ell}\right), 2\right), \tag{5.15}$$

where sum$(\cdot, 2)$ denotes the summation along the second dimension, i.e., the summation of the lateral slices. By using this relation and appropriately rearranging the terms, we can express equation (8.8) in the following form:

$$\begin{aligned}\frac{\partial \mathcal{L}_3}{\partial \mathcal{B}_\ell} &= \frac{\partial \mathcal{L}_3}{\partial \mathcal{X}_{n,\ell}} \frac{\partial \mathcal{X}_{n,\ell}}{\partial \mathcal{B}_\ell} \\ &= \frac{1}{N}\sum_{n=1}^{N} \|\mathcal{Y}_n - \mathcal{X}_{n,\ell}\|_{\text{F}}\, \text{sum}\left(\sigma'\left(\mathcal{Z}_{n,\ell}\right), 2\right).\end{aligned} \tag{5.16}$$

Within the TBP algorithm, the efficient management of t-product-based tensor operations and decompositions remains unresolved, even though this approach could establish a robust theoretical basis for Eqs. (5.14) and (5.16). Ideally, t-product-based tensor operations would be smoothly incorporated into TensorFlow's backpropagation (BP) framework [16]; but this presents several practical challenges. To address this issue, similar to the approaches proposed in [5, 6, 14], we can transform tensor computations into equivalent matrix representations and use standard BP methods to learn the corresponding matrix weights. The detailed implementation will be discussed in the following section.

## 5.4 Adapting Matrix AE Methods for the Tensor AE Scenario

To tackle the challenges in implementing DTAE, we establish a mathematical connection between tensor and matrix autoencoders (AEs), utilizing the properties described in Definition 5.2 in Appendix 5.8. Building on this relationship, we introduce a theoretical framework that decomposes the DTAE method into multiple independent deep matrix autoencoder (DMAE) tasks, each corresponding to a frontal slice in the DCT domain. While this framework is based on a traditional matrix autoencoder, it is extended to incorporate the SSDA method discussed in Sect. 5.4.2, and further generalized in Sect. 5.4.3 to cover the DCAE approach.

### 5.4.1 The Connection to the DMAE Method

The DTAE framework processes each frontal slice independently by applying a matrix autoencoder (AE) in the DCT domain. To provide a clearer understanding, the relationship between the DTAE and DMAE models will be analyzed from two distinct perspectives.

#### 5.4.1.1 The Connection to the DMAE Network

Equations (5.6) and (5.7) in the DTAE model require the computation of the t-product, as previously discussed. Based on the properties in Eq. (5.30), we can reformulate these two equations as follows:

$$\begin{aligned} \mathbf{h}\left(\widetilde{\mathcal{X}}_{n,\ell-1}^{(k)}\right) &= \sigma\left(\widetilde{\mathcal{W}}_{\ell-1}^{(k)}\widetilde{\mathcal{X}}_{n,\ell-1}^{(k)} + \widetilde{\mathcal{B}}_{\ell}^{(k)}\right), \\ \mathbf{g}\left(\widetilde{\mathcal{X}}_{n,\ell-1}^{(k)}\right) &= \sigma\left(\widetilde{\mathcal{W}}_{\ell-1}^{(k)\top}\mathbf{h}\left(\widetilde{\mathcal{X}}_{n,\ell-1}^{(k)}\right) + \widetilde{\mathcal{B}}_{\ell-1}^{(k)}\right). \end{aligned} \tag{5.17}$$

In this scenario, the computational process of the DTAE model closely mirrors that of a matrix autoencoder operating in the DCT domain, as illustrated in equation (5.33).

#### 5.4.1.2 The Connection to the BP algorithm

During training, the effectiveness and simplicity of the TBP algorithm are based on the derivative computations in Eq. (5.14). By fully leveraging the properties of the t-product, we can reformulate Eq. (5.14) into the following expression:

$$\frac{\partial \widetilde{\mathcal{L}}_3^{(k)}}{\partial \widetilde{\mathcal{W}}_{\ell-1}^{(k)}} = \frac{1}{N}\sum_{n=1}^{N} \|\mathcal{Y}_n^{(k)} - \mathcal{X}_{n,\ell}^{(k)}\|_{\mathrm{F}}(\sigma'(\mathcal{Z}_{n,\ell}^{(k)})\mathcal{X}_{n,\ell-1}^{(k)\top}). \tag{5.18}$$

Building on the derivation of the previous equation and following the analysis outlined earlier, the TBP algorithm for training the DTAE model can be interpreted as applying an independent autoencoder to each frontal slice within the DCT domain. By consolidating the insights from points (1) and (2), we can extract the key elements of the DTAE model and its training method, leading to the following implementation guidelines:

$$f_\theta(\cdot) \leftarrow \mathrm{DMAE}(\widetilde{\mathcal{X}}_n^{(k)},\ \widetilde{\mathcal{Y}}_n^{(k)},\ \alpha,\ N_{\mathrm{bs}},\ P_{AE},\ \mathrm{type}). \tag{5.19}$$

The main concept is to concentrate on a specific frequency range of the seismic signals and selectively process a subset of the frontal slices. This strategy prevents

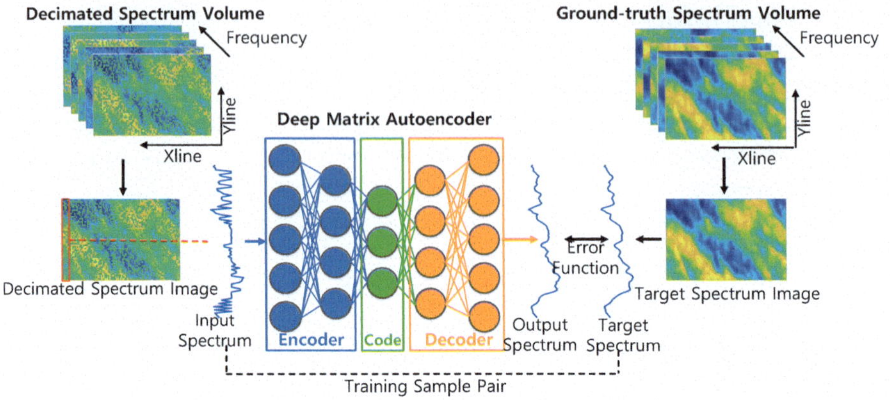


**Fig. 5.2** The schematic diagram shows the detailed implementation of the DTAE model depicted in Fig. 5.1. The weight parameters of the DTAE are determined by using a matrix autoencoder for each individual slice in the DCT domain

---

**Algorithm 5.1:** TBP Algorithm

---

**Input**: $\mathcal{X}_n$, $\mathcal{Y}_n$, $\alpha$, $N_{\text{bs}}$, $N_r$, $N_t$, $\mathcal{P}_\Omega$, $P_{AE}$, Type.
**Output**: $\widehat{\mathcal{X}}_n$.

```
// Training of the DTAE model
```

$\widetilde{\mathcal{X}}_n \leftarrow \text{dct}(\mathcal{X}_n, [\,], 3)$,
$\mathcal{X}^0 = \mathcal{Y}, \mathcal{Z}^0 = \mathcal{Y}, \ell = 0$;
**for** $k = 1, \cdots, N_3$ **do**
  $f_\theta(\cdot) \leftarrow \text{DMAE}(\widetilde{\mathcal{X}}_n^{(k)}, \widetilde{\mathcal{Y}}_n^{(k)}, \alpha, N_{\text{bs}}, P_{AE}, \text{type}), n \in [N_r]$.
**end**

```
// Testing of the DTAE model
```

**for** $k = 1, \cdots, N_3$ **do**
  $\widetilde{\mathcal{P}}_n^{(k)} \leftarrow f_\theta(\widetilde{\mathcal{X}}_n^{(k)}), n \in [N_t]$.
**end**
$\widehat{\mathcal{X}}_n \leftarrow \text{idct}(\widetilde{\mathcal{P}}_n, [\,], 3), n \in [N_t]$.
$\widehat{\mathcal{X}}_n = \eta(\widehat{\mathcal{X}}_n) \odot (1 - \mathcal{P}_\Omega) + \widehat{\mathcal{X}}_n$.

---

unnecessary computational resources from being allocated to irrelevant regions due to the band-limited nature of the signals. In practice, we train a matrix autoencoder (AE) using the backpropagation (BP) algorithm only on the chosen frequency slices, as shown in Fig. 5.2. After training, the matrix AE is capable of effectively reconstructing the missing parts of the decimated spectral image. While it is theoretically possible to train a separate matrix AE for each frequency, this approach is generally not advisable, as it would drastically reduce the available training data.

The process is outlined in Algorithm 5.1, which involves two primary computational phases. Initially, a training set is formed from pairs of spectral images ($\widetilde{\mathcal{X}}(n)$, $\widetilde{\mathcal{Y}}n$), where $n \in [N_t]$. The network architecture DMAE$(\cdot)$ must first be defined, followed by the estimation of the parameters $\theta$, with $N_r$ representing the total number of training samples. Upon completion of training, the learned mapping function $f_\theta(\cdot)$ is used to predict the missing traces $\widetilde{\mathcal{Y}}n$ for new decimated spectral images $\widetilde{\mathcal{X}}n$, where $n \in [N_t]$ and $N_t$ denotes the number of test samples. Moreover, $P_{AE}$ refers to the private parameter set associated with each autoencoder (AE), and $N_{bs}$ indicates the batch size, which defines the number of samples processed by the DCAE in each iteration.

## 5.4.2 Theoretical Expansion to Include SSDAs

We have found that integrating a more advanced autoencoder (AE) approach into the DTAE model results in notable performance gains when compared to using a traditional AE. This suggests that the principles of stacked sparse denoising auto-encoder (SSDA) [17] naturally align with the DTAE model. The enhanced DTAE model, which replaces the conventional denoising autoencoder (DA) applied to the frontal slices in the DCT domain with SSDA, is termed SSDA-DTAE. This model effectively leverages the advantages of tensor sparse coding (TSC) and neural networks, particularly through the use of SSDA. The formal definition of the SSDA-DTAE model is as follows:

$$\begin{aligned}\mathcal{L}_4(\mathcal{X}, \mathcal{Y}; \Theta) &= \frac{1}{N}\sum_{n=1}^{N}\frac{1}{2}\|\mathcal{Y}_n - \mathbf{g}(\mathcal{X}_n)\|_{\mathrm{F}}^2 + \beta\mathbb{KL}(\hat{\rho}\|\rho)\\ &+\frac{\lambda}{2}\sum_{k=1}^{K}\left(\|\mathcal{W}_\ell\|_{\mathrm{F}}^2 + \|\widehat{\mathcal{W}}_\ell\|_{\mathrm{F}}^2\right),\end{aligned} \tag{5.20}$$

In this framework, $\beta > 0$ and $\lambda > 0$ serve as tuning parameters to control the level of uncertainty in the observed constraints. Meanwhile, $\mathbb{KL}(\cdot|\cdot)$ introduces sparsity regularization, encouraging sparsity in the hidden units. Its specific definition is as follows:

$$\mathbb{KL}(\widehat{\rho}\|\rho) = \sum_{j=1}^{|\widehat{\rho}|}\rho\log\frac{\rho}{\widehat{\rho}_j} + (1-\rho)\log\frac{(1-\rho)}{1-\widehat{\rho}_j}. \tag{5.21}$$

where $\hat{\rho}$ denotes the mean activation of the hidden layer, which is defined as follows:

$$\widehat{\rho} = \frac{1}{N}\sum_{n=1}^{N}\mathbf{h}(\mathcal{X}_n), \tag{5.22}$$

where $\mathbf{h}(\cdot)$ and $\mathbf{g}(\cdot)$ are defined in Eqs. (5.6) and (5.7), respectively. After applying the discrete cosine transform (DCT), we decompose the problem into several independent subproblems, following the established tensor-matrix autoencoder relationship, as outlined below:

$$
\begin{aligned}
\mathcal{L}_3(\widetilde{\mathcal{X}}^{(k)}, \widetilde{\mathcal{Y}}^{(k)}; \theta) = {} & \frac{1}{N} \sum_{n=1}^{N} \frac{1}{2} \left\| \widetilde{\mathcal{Y}}_n^{(k)} - \mathbf{g}\left(\widetilde{\mathcal{X}}_n^{(k)}\right) \right\|_{\mathrm{F}}^2 + \beta \mathbb{KL}(\hat{\rho} \| \rho) \\
& + \frac{\lambda}{2} \sum_{\ell=1}^{K} \left( \left\| \widetilde{\mathcal{W}}_\ell^{(k)} \right\|_{\mathrm{F}}^2 + \left\| \widehat{\mathcal{W}}_\ell^{(k)} \right\|_{\mathrm{F}}^2 \right).
\end{aligned}
\tag{5.23}
$$

Problem (5.23) has effectively been transformed into a form suitable for solving with SSDAs.

### 5.4.3 *Expanding the Approach to Convolutional AEs*

It has been previously demonstrated that the DTAE provides a flexible and efficient solution for the problem of 3D seismic data reconstruction. The main concept behind DTAE is to bypass the direct computation of the tensor product (t-product) in the time domain, which is often complex and computationally demanding. Instead, by leveraging the TensorFlow optimization framework and the properties discussed in Definition 2.6, we have designed a more streamlined computational approach. With this approach, we are able to efficiently perform the tensor product on the DCT-domain frontal slices, significantly enhancing the computational efficiency of DTAE.

Considering the advantage of convolutional autoencoders (DCAE) in reconstruction accuracy [17, 18], replacing traditional autoencoders (AE) with convolutional autoencoders may enhance the performance of deep tensor autoencoders (DTAE). The model variant is named DCAE-DTAE. In this approach, convolutional operations replace matrix multiplication, which prevents the computation process from being decomposed into a sequence of DCAE operations on frequency slices, as seen in SSDA-DTAE. Although a formal theoretical foundation is lacking, by leveraging insights from existing studies [19, 20], and t-SVD techniques, the DTAE framework can be empirically adapted for the DCAE-DTAE model, leading to enhanced performance.

$$
f_\theta(\cdot) \leftarrow \mathrm{DMAE}(\widetilde{\mathcal{X}}_n^{(k)},\ \widetilde{\mathcal{Y}}_n^{(k)},\ \alpha,\ N_{\mathrm{bs}},\ P_{DCAE}, \mathrm{DCAE}), \tag{5.24}
$$

where $f_\theta(\cdot)$ denotes the mapping function in the convolutional autoencoder (AE), which is defined as follows:

$$
\begin{aligned}
\mathcal{F}_\ell &= \Phi_\ell\left(\widetilde{\mathcal{X}}_{n,\ell}^{(k)}\right) = \sigma\left(\widetilde{\mathcal{X}}_{n,\ell}^{(k)} \circledast \mathcal{W}_\ell + B_\ell\right), \\
\widehat{\mathcal{X}}_{n,\ell}^{(k)} &= \sigma\left(\mathcal{F}_\ell \circledast \tilde{\mathcal{W}}_\ell + D_\ell\right).
\end{aligned}
\tag{5.25}
$$

In this work, the symbol $\circledast$ is used to indicate the convolution operation, $\mathcal{W}_\ell$ represents a set of filters, $B_\ell$ stands for the bias applied uniformly across the entire map, $D_\ell$ corresponds to the bias for each input channel, and $\tilde{\mathcal{W}}_\ell$ denotes the inverse operation (deconvolution) applied across both weight dimensions. It should be noted that although this framework could potentially be extended to other deep learning (DL) models, exploring such possibilities is outside the scope of this study.

## 5.5 Field Examples of DTAE

To assess the performance of the proposed DTAE framework for 3D seismic data reconstruction, we conduct three experiments using both synthetic and real field data. These include the synthetic Vertical Seismic Profile (VSP) data experiment (Sect. 8.4.2), the F3 Netherlands dataset experiment (Sect. 5.5.3), and the Parihaka-3D dataset experiment (Sect. 5.5.4). Before presenting the experiments, we provide details on four key preparatory elements: datasets (Sect. 5.5.1.1), experimental protocol (Sect. 5.5.1.2), training data (Sect. 5.5.1.3), and network parameters (Sect. 5.5.1.4). In each experiment, the reconstruction performance of the proposed DTAE model is compared both quantitatively and visually with that of three state-of-the-art (SOTA) reconstruction techniques. The DTAE model, a deep learning-based approach, is implemented using TensorFlow [16], while the SOTA methods are developed using MATLAB. All experiments are executed on a system with a 4-core Intel i5-6500 CPU, 32 GB of RAM, and an NVIDIA TITAN X (Pascal) GPU.

### *5.5.1 Experimental Setup*

#### 5.5.1.1 Datasets

Our objective is to validate the reconstruction performance of the proposed DTAE framework through its application to one synthetic dataset and two real field datasets. All datasets used in this chapter are publicly accessible and freely available, with detailed descriptions provided in the following sections.

1) *3-D 9C synthetic vertical seismic profiling (VSP) data*[1]: These four 3D, 9C datasets are derived from an anisotropic model and an intricate synthetic model originating from the RCP Postle Field project, which exhibits horizontal transverse isotropy (HTI). The choice of these synthetic datasets is also motivated by their ability to provide sufficient training data, enabling the DTAE framework to achieve accurate data reconstruction.
2) *F3 Netherlands*[2]: This dataset is a well-known and representative 3D poststack seismic dataset widely used for testing in seismic data processing [21, 22]. It

[1] Data available here: https://wiki.seg.org/wiki/3D_9C_synthetic_VSP_data.

[2] Data available here: https://terranubis.com/datainfo/F3-Demo-2020.

covers an area of approximately 384 $km^2$ in the offshore region of the Dutch Central Graben basin.

3) *Parihaka-3D*[3]: The Parihaka-3D dataset presents intricate geological features, including several generations of faults, winding channel systems, and significant unconformities, which frequently challenge the low-rank assumption.

In conclusion, these datasets span a broad spectrum from straightforward to complex concerning the low-rank assumption, making them suitable benchmarks for comparing reconstruction algorithms that effectively demonstrate the features of DTAE.

#### 5.5.1.2 Protocols

As the full dataset is accessible, the signal-to-noise ratio (SNR), a widely used quantitative measure, is chosen to assess the reconstruction effectiveness of both the DTAE framework and the SOTA methods. Generally, a higher SNR indicates better data reconstruction quality, while a lower SNR suggests the opposite. The specific definition of SNR in the context of 3D reconstruction is as follows:

$$\text{SNR [dB]} = 20 \log_{10} \frac{1}{\text{RSE}}, \tag{5.26}$$

where RSE represents the relative square error (RSE) between the interpolated results and the clean data, which is defined in detail as follows:

$$\text{RSE} = \frac{\|\mathcal{T} - \widehat{\mathcal{T}}\|_{\text{F}}}{\|\mathcal{T}\|_{\text{F}}}. \tag{5.27}$$

In this study, $\mathcal{T}$ represents the fully sampled ground truth dataset, while $\widehat{\mathcal{T}}$ denotes the interpolated data derived from the incomplete dataset. In addition to the conventional Signal-to-Noise Ratio (SNR) metric, we introduce local similarity to assess reconstruction quality, following the methods outlined in [20, 23]. Compared to the SNR metric, local similarity provides a more accurate measure of reconstruction quality, as it captures structural and detail information within local regions, offering a more sensitive evaluation than global SNR. The detailed advantages and application of this method can be found in [23]. The local similarity score varies between 0 and 1, where higher values correspond to improved reconstruction quality.

It is essential to assess the effectiveness of the proposed DTAE 3-D data reconstruction method by comparing it with traditional techniques under identical data conditions. As discussed in Sect. 5.1, three well-established methods—MSSA, TNN, and PMF—were chosen for this comparison due to their proven success in reconstruction tasks. These methods have been validated in numerous studies as capable of

[3] Data available here: https://wiki.seg.org/wiki/Parihaka-3D.

providing stable and efficient reconstruction results, thus serving as reliable benchmarks for the comparative analysis in this study.

(1) *MSSA*[4] [24]: In this approach, seismic data are arranged into a Hankel block matrix according to specific time and frequency parameters. The occurrence of missing samples results in an increased rank of the Hankel block matrix, which prompts the use of MSSA rank reduction as a method for reconstructing the missing traces.
(2) *TNN*[5] [25, 26]: This approach employs a low-rank tensor approximation of seismic data using t-SVD, along with a complexity-penalized algorithm to recover the missing traces.
(3) *PMF* [7, 27]: This approach performs low-rank matrix factorizations for multiple unfolded versions of tensors in various configurations, allowing for the reconstruction of a tensor in matrix form.

To ensure the fairness of the comparisons, each benchmark method undergoes a detailed parameter tuning process [19, 20, 25–27]. For instance, to determine the best tensor rank that enhances reconstruction quality, we adjust the low-rank value within the range of 2–25, increasing it in steps of 2, to explore how different rank values influence the reconstruction performance.

#### 5.5.1.3 Training and Test Data

This section explains the process of preparing training and test datasets for the DTAE model. In practical seismic surveys, the data is split into training, test, and prediction blocks according to the distribution of missing traces in the area. Specifically, the seismic data with missing traces is assigned to the prediction block, while the remaining data is used for training and testing. Since both the training and prediction blocks typically contain large amounts of seismic data that cannot be directly fed into the DTAE model, they are divided into smaller cubes. When the number of cubes is sufficiently large, non-overlapping 3D windows of size $n_1 \times n_2 \times n_3$ voxels can be used.

Once the cubes are obtained, we generate input-label pairs through the following three steps: (1) Following the approach of [1, 28], all cubes from the training block are considered as label data, denoted by $\{\mathcal{Y}_n\}_{n=1}^N$; (2) The corresponding input data is generated by actively randomly decimating the complete cube set using a fixed sampling ratio, denoted by $\{\mathcal{X}_n\}_{n=1}^N$; (3) As described in Algorithm 5.1, the final input to the DMAE consists of decimated spectral images and complete spectral image pairs in the DCT domain, $\left(\widetilde{\mathcal{X}}_n^{(k)}, \widetilde{\mathcal{Y}}_n^{(k)}\right)$, where $n \in [N_t]$ and $k \in [n_3]$. Therefore, we

[4] Code available here: https://github.com/chenyk1990/reproducible_research/tree/master/nonMada/usarray.

[5] Code available here: https://github.com/jamiezeminzhang/Tensor_Completion_and_Tensor_RPCA.

perform a discrete cosine transform (DCT) along the time dimension for all cubes and select the frontal slices within a predefined frequency range as spectral images.

The spectral images employed for prediction are produced following the same three steps outlined previously. however, there is no need to generate the spectral images that serve as labels, since during the prediction phase, the label data are not input to the DTAE. Instead, they are used solely as ground truth for evaluating the model's reconstruction accuracy. In practical measurement scenarios, such label data are typically unavailable during testing.

#### 5.5.1.4 Network Parameters

The parameter tuning process requires a series of extensive and computationally intensive experiments. In theory, a comprehensive grid search over the entire potential parameter space is needed, which includes network structure parameters (such as activation functions, layer types, number of neurons, and layer depth) and optimizer parameters (type and hyperparameters). Searching such a high-dimensional parameter space is often infeasible. Therefore, a greedy fashion is typically employed to adjust the DTAE network, modifying one or a few components at a time, with the goal of finding near-optimal parameters for the three different datasets.

The experiments using the synthetic VSP dataset will demonstrate that the DCAE model within the DTAE framework achieves better performance in spectral image reconstruction compared to both DA and SSDA models. This finding leads to the decision to evaluate only the performance of the DCAE model on the F3 and Parihaka-3D real-world datasets. Specifically, a U-Net-style convolutional autoencoder [29] is used across all three datasets. Therefore, all experiments include the network parameters for DA and SSDA models in the DTAE framework from the synthetic data experiments, as well as the network parameters for DCAE in both synthetic and real-world data experiments. The specific settings are as follows.

*Parameter configurations for DTAE framework based on DA and SSD:* In both the DA and SSDA networks, the encoder and decoder consist of three layers, each of size ($60 \times 60$), with the ReLU activation function applied at each layer. The training data comprises 60,000 synthetic VSP spectral images ($64 \times 32$), which are reshaped into one-dimensional vectors of size $2048 \times 1$ before being input into the DA and SSDA networks. The Adam optimizer is employed during training, with the learning rate initialized at 0.0001, and adjustments made according to the number of iterations. Weight updates are performed using the standard backpropagation (BP) algorithm, with a batch size of 50. For the SSDA network, specific hyperparameters are set as $\lambda = 1e-4$, $\beta = 1e-2$, and $\rho = 0.05$, based on the recommendations from [17].

The DCAE model, built on convolutional neural networks, consists of 18 convolutional layers, each using a kernel size of $3 \times 3$. A ReLU activation function is applied after each convolutional layer to introduce non-linearity. The architecture is divided into an encoder and a decoder, with a max-pooling layer placed after the last convolutional layer in the encoder to downsample the feature maps and capture more significant features. In the decoder, a concatenation layer is added after the

**Table 5.2** Parameters of the DTAE model

| No. | Type | #kernels | No. | Type | #kernels |
|---|---|---|---|---|---|
| 1 | Conv1+Conv2 | $3 \times 3 \times 48$ | 10 | Concatenate1 | $2 \times 2$ |
| 2 | MaxPooling1 | $2 \times 2$ | 11 | Conv11+Conv12 | $3 \times 3 \times 384$ |
| 3 | Conv3+Conv4 | $3 \times 3 \times 96$ | 12 | Concatenate2 | $2 \times 2$ |
| 4 | MaxPooling2 | $2 \times 2$ | 13 | Conv13+Conv14 | $3 \times 3 \times 192$ |
| 5 | Conv5+Conv6 | $3 \times 3 \times 192$ | 14 | Concatenate3 | $2 \times 2$ |
| 6 | MaxPooling3 | $2 \times 2$ | 15 | Conv15+Conv16 | $3 \times 3 \times 96$ |
| 7 | Conv7+Conv8 | $3 \times 3 \times 384$ | 16 | Concatenate4 | $2 \times 2$ |
| 8 | MaxPooling4 | $2 \times 2$ | 17 | Conv17+Conv18 | $3 \times 3 \times 48$ |
| 9 | Conv9+Conv10 | $3 \times 3 \times 768$ | | | |

final convolutional layer to restore the image's spatial resolution. Both the pooling and concatenation layers have a size of $2 \times 2$. The entire network is trained for 20 epochs using the Adam optimizer. The number of neurons in each convolutional layer is described by the following expression:

$$n_{\ell+1} = \begin{cases} 2n_\ell, & \text{if } n_{\ell+1} \leq l_t, \\ l_t, & \text{otherwise}, \end{cases} \tag{5.28}$$

In this model, $n_\ell$ denotes the number of neurons in layer $\ell$, while $n_{\ell+1}$ indicates the number of neurons in the next layer, $\ell + 1$. As the depth of the network increases, the number of neurons in each layer progressively increases until it reaches the set maximum value, $l_t$. Specifically, for the VSP dataset, the number of neurons in the first layer is set to $n_1 = 32$, with the maximum number of neurons set to $l_t = 512$. For the F3 and Parihaka-3D datasets, the number of neurons in the first layer, $n_1$, is set to 48, with the maximum number of neurons adjusted to $l_t = 768$. Additionally, for training the DCAE model, we set the batch size for each dataset: 48 for the F3 and Parihaka-3D datasets, and 32 for the VSP dataset. A detailed list of the network parameters can be found in Table 5.2.

### 5.5.2 Evaluation on Synthetic Data

To validate the correctness of the proposed DTAE framework, we conduct data reconstruction experiments on the synthetic 3-D 9C VSP dataset. Specifically, we select the htins_vsp01_zz data from the 3-D 9C VSP dataset, which contains 42,025 shot gathers, each recorded at 31 downhole locations with 500 time samples per trace, providing sufficient training data for DTAE. The process of generating input-label pairs for DTAE involves the following three steps: (1) The simulated shot gathers shown in Fig. 5.5a are treated as input data, denoted as $\mathcal{T}$. Then, 50% of the traces in each shot gather are randomly removed to generate the observed data $\mathcal{T}_\Omega$ (see Fig. 5.5d). (2) The input-label pairs $\mathcal{T}$ and $\mathcal{T}_\Omega$ are further divided into non-overlapping 3-D cubes of size $48 \times 64 \times 450$. (3) A Discrete cosine transform (DCT) is applied along the time axis to both $\mathcal{T}$ and $\mathcal{T}_\Omega$, generating spectral images as the final input to DTAE. Due to the band-limited characteristics of seismic signals, we focus only on the spectral image data within the frequency range of 5 –200 data points, while setting other frequencies to zero. In the synthetic experiments, we randomly choose 90% of the damaged 3-D cubes along with their corresponding complete cubes for the training set, reserving the remaining 10% for testing.

Since the DTAE framework ultimately performs reconstruction through an autoencoder, it is essential to evaluate the impact of choosing different types of autoencoders (AEs) on reconstruction performance. In this experiment, we compare three typical models: DA, SSDA, and DCAE, for spectrogram image reconstruction, with the results shown in Fig. 5.4. In particular, the first and second columns of Fig. 5.4 represent the actual spectrogram images and the corresponding spectrogram images with missing data, respectively. The third, fourth, and fifth columns correspond to the reconstructed spectrogram images generated by the DA, SSDA, and DCAE models, respectively. A visual comparison clearly demonstrates that the data reconstruction performance of the DCAE model outperforms that of the DA and SSDA autoencoders, which is also confirmed by the survey of autoencoders in [30]. Therefore, in all subsequent experiments, we naturally choose to embed the DCAE model into the DTAE framework, rather than selecting the DA or SSDA models.

The factor that most significantly influences the reconstruction performance of algorithms is undoubtedly the amount of known trace information, meaning that the sampling scheme, at different rates, typically has a considerable impact on the effectiveness of reconstruction methods. To effectively assess this factor, we perform reconstruction performance comparisons between DTAE and other state-of-the-art (SOTA) algorithms at various sampling rates ranging from 10 to 90%, with the results shown in Fig. 5.3. From Fig. 5.3, it is evident that, under decimation levels between 10 and 60%, DTAE demonstrates significant performance improvements, outperforming existing SOTA methods. This is due to DTAE's relaxation of the low-rank assumption, as low signal-to-noise ratios severely disrupt low-rank structures. However, when faced with higher levels of missing data, such as 80 or 90%, DTAE sometimes performs worse than the existing SOTA methods.

We chose to compare the reconstruction performance of various methods under 50% traces missing conditions, as shown in Figs. 5.6, 5.5, and Table 5.1. It is worth noting that these volume displays were generated using the Madagascar software package.[6] From a qualitative comparison perspective, the ground truth data is shown in Figs. 5.5a and 5.6a, while the undersampled data is shown in Figs. 5.5d and 5.6f. The reconstruction results for PMF, TNN, MSSA, and the proposed DTAE model are shown in Figs. 5.5b–f and 5.6b–e. From these, it can be observed that our DTAE model provides reconstruction results that are closer to the ground truth data compared to

[6] http://www.ahay.org.

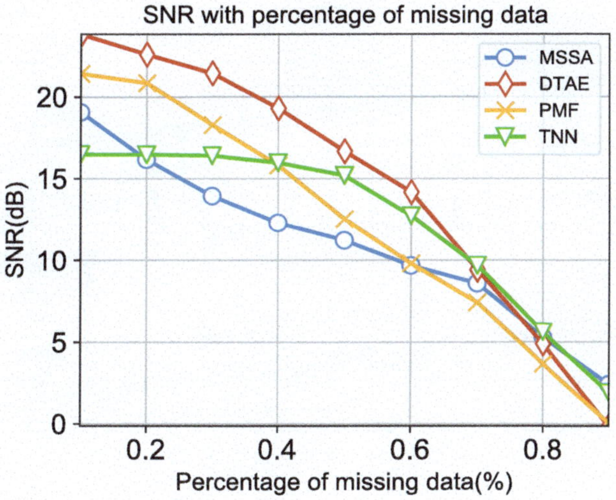


**Fig. 5.3** In the comparison of Signal-to-Noise Ratio (SNR) with respect to the percentage of missing data, the symbols $\diamondsuit$, $\times$, $\bigtriangledown$, and $\bigcirc$ represent the DTAE model, PMF, TNN, and MSSA methods, respectively

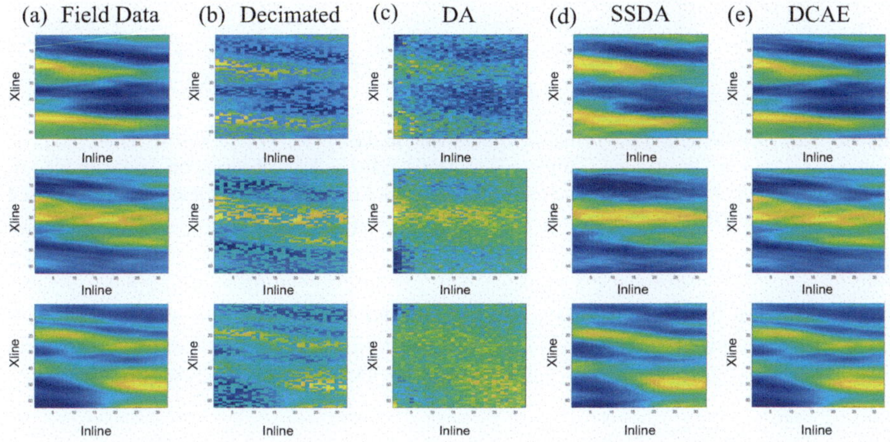


**Fig. 5.4** A comparison of the reconstruction results across different frequency slices in a 3-D synthetic example is presented. **a** The original noise-free data. **b** The data after decimation, and **c–e** The reconstruction errors obtained using the following model variants: **c** DA model, **d** SSDA model, and **e** DCAE model

other state-of-the-art algorithms. This conclusion is further supported by quantitative metrics such as local similarity and signal-to-noise ratio (SNR). Figure 5.6g–j provide a comparison of local similarity between the ground truth and the reconstruction results, aiming to quantify the reconstruction errors in fine details. From the local similarity results shown in Fig. 5.6g–j, it is evident that the larger orange regions,

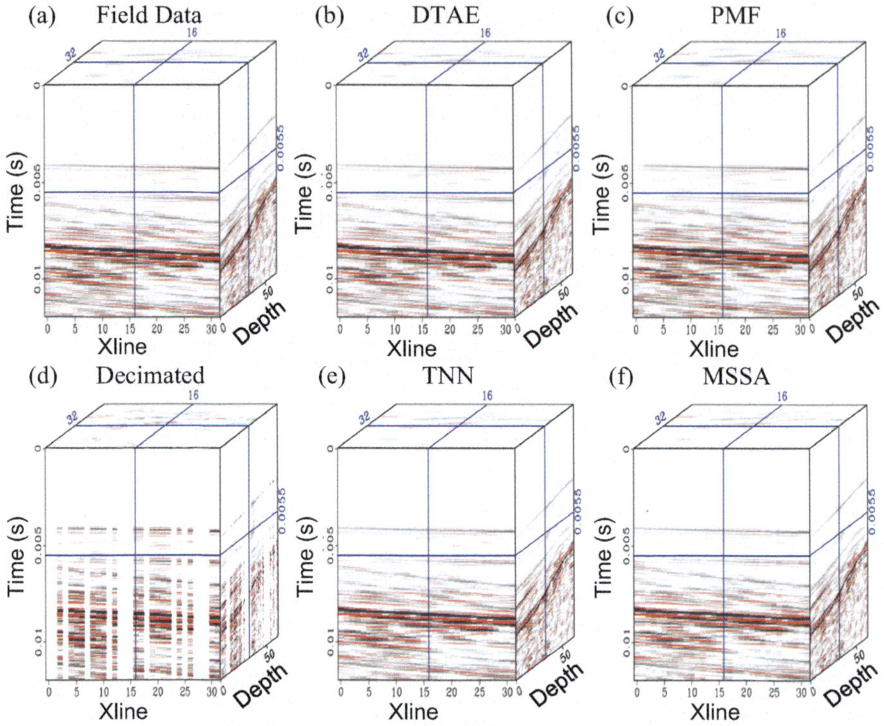


**Fig. 5.5** The reconstruction performance comparison in the 3D 9C synthetic VSP dataset experiment is as follows: **a** The original complete 3D VSP data; **b** The VSP observation data with 50% of the traces randomly removed; **c–f** The reconstruction results using the TNN method, the proposed DTAE model, the PMF method, and the MSSA method, respectively

which represent higher similarity, and the smaller yellow regions, which represent lower similarity, suggest that the DTAE reconstruction results are closer to the ground truth. This conclusion is further validated by the SNR metric, as shown in Table 5.1.

### *5.5.3 Evaluation on Real F3 Netherlands Data*

In this section, the DTAE model is used on real field data from the F3 Netherlands site to verify that the proposed method demonstrates superior data reconstruction performance in practical applications compared to state-of-the-art (SOTA) methods. Since our research primarily focuses on data reconstruction rather than denoising, we use the Dip_steered_median_filter data from the F3 dataset as the clean reference data $\mathcal{T}$, as shown in Fig. 5.7a. Subsequently, we randomly discard 50% of the traces to create the observed data $\mathcal{T}_\Omega$, as illustrated in Fig. 5.7d. Finally, following the procedure described in the theoretical section, $\mathcal{T}_\Omega$ is divided into independent 3D data

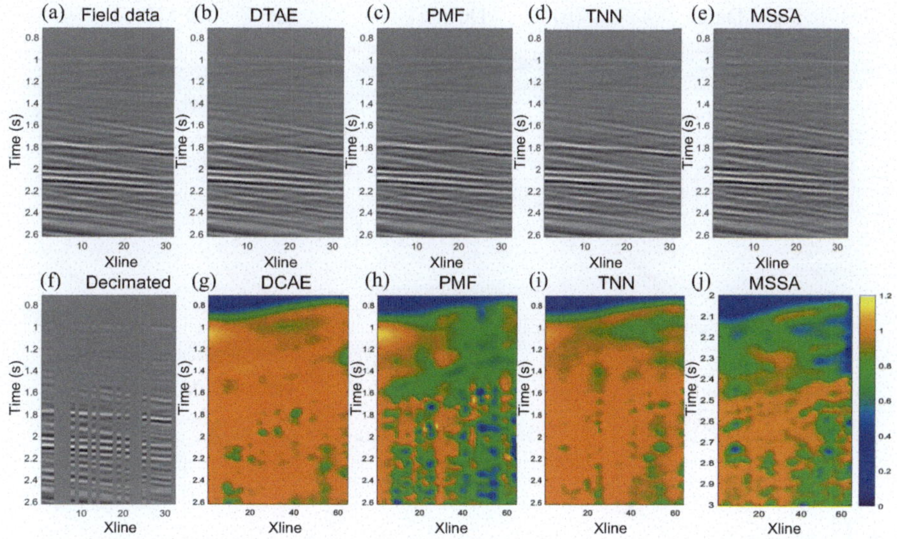


**Fig. 5.6** Reconstruction comparison on the second depth section of the 3D 9C synthetic VSP dataset, using different interpolation methods: **a** Original unprocessed data; **b–e** Data interpolated using the DTAE model, TNN, PMF, and MSSA; **f** Decimated data. Additionally, **g–j** Local similarity assessments between the reconstructed data and the original data after interpolation with DTAE, PMF, TNN, and MSSA, respectively

blocks of size $48 \times 48 \times 450$, completing the preparation of DTAE training samples. Considering that the actual input to DTAE consists of discrete cosine transform (DCT) domain spectral images, we further perform a DCT transformation along the time dimension of the 3D data blocks, selecting the frequency range from 5 to 300 data points to generate spectral images of size $48 \times 48$. It is important to highlight that two adjacent data blocks are used as the test set, with the remaining blocks reserved for training.

The results presented in Table 5.1 show that the DTAE method outperforms the three state-of-the-art benchmark methods by a significant margin in terms of Signal-to-Noise Ratio (SNR). Moreover, the reconstruction results produced by DTAE closely match the ground truth, whereas the benchmark methods display noticeable discrepancies, as shown in the visualizations in Figs. 5.7 and 5.8b–e. Additionally, Fig. 5.8f–j highlights the local similarity between the true data and the outputs from each method. In contrast to the benchmark methods, the local similarity map for DTAE is predominantly orange, indicating a similarity close to 90%, with other colors appearing less frequently. This significant difference clearly demonstrates that the DTAE method provides the closest reconstruction to the true field data, exhibiting the best reconstruction performance.

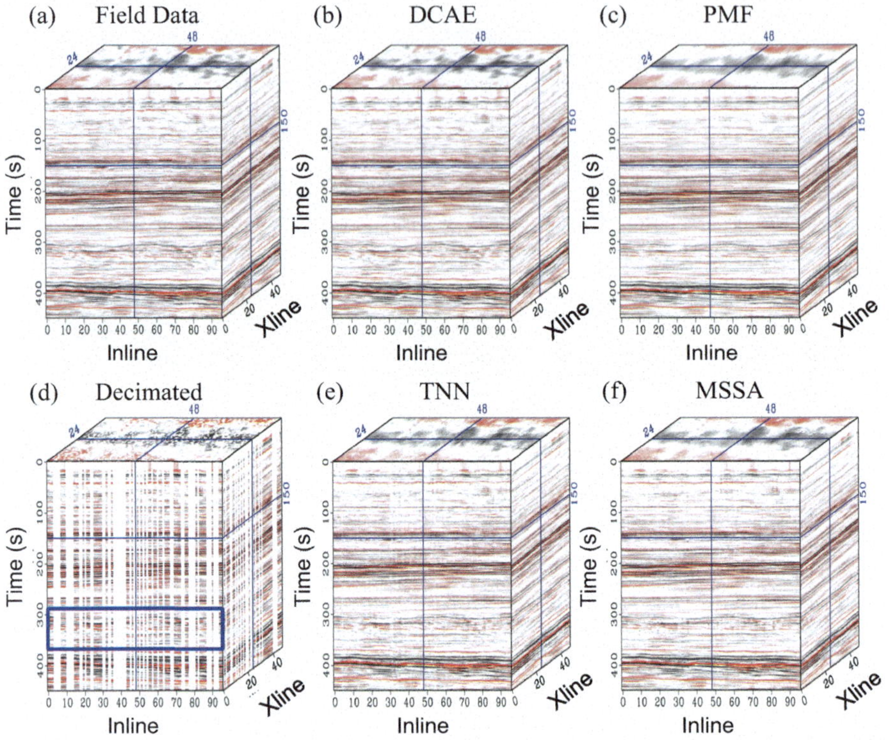


**Fig. 5.7** In the comparison of reconstruction methods on the 3D F3 Netherlands dataset: **a** The original 3D field data; **b** The data interpolated using the DTAE model; **c** The data interpolated using the PMF method; **d** The decimated 3D field data with 50% of the traces missing; **e** The data interpolated using the TNN method; (f) The data interpolated using the MSSA method

### *5.5.4 Evaluation on the Real Parihaka-3D Data*

We selected the Parihaka-3D seismic exploration data to further validate the stability and robustness of the DTAE method in practical scenarios. The reason for choosing this dataset is its more complex geological structure compared to the F3 dataset, which is beneficial for testing the significant improvement in data recovery performance of DTAE when the low-rank assumption does not hold. As shown in Fig. 5.9a, considering that the actual Parihaka-3D data has a high signal-to-noise ratio, we use the original 3D volume data as the clean sample $\mathcal{T}$ and directly generate the actual observed data through random fiber sampling. The specific configuration parameters for the training and testing datasets, including block size, spectrograms, and the number of training samples, are consistent with those of the real F3 dataset, and these settings are not sensitive to different real-world data. The only difference is

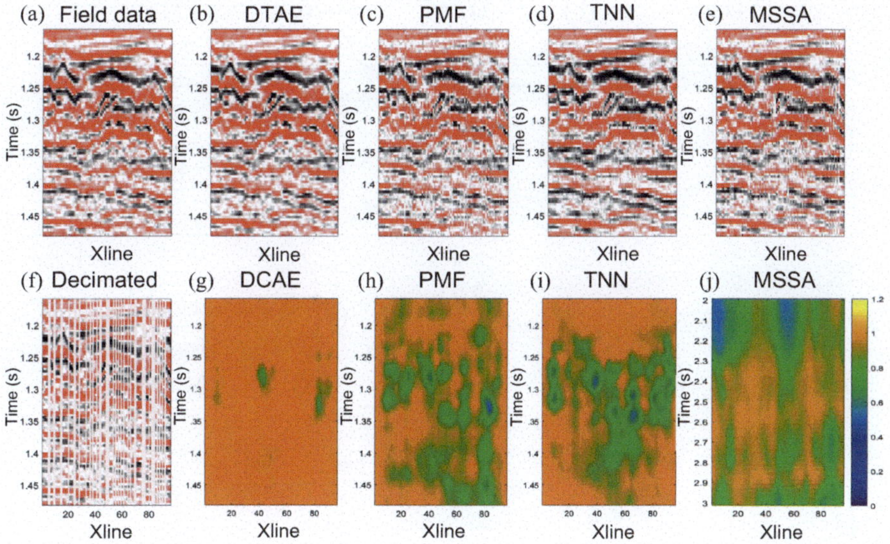


**Fig. 5.8** A reconstruction comparison is conducted on the first $x$-line section of the 3-D F3 Netherlands dataset using different interpolation methods: **a** The original data; **b**–**e** The interpolated data using the DTAE model, TNN, PMF, and MSSA methods, respectively; **f** The decimated original data. Subsequently, local similarity assessments between the original and reconstructed data are performed using **g** DTAE, **h** PMF, **i** TNN, and **j** MSSA methods

that the frequency band range varies due to differences in acquisition equipment. For the Parihaka-3D seismic exploration data, the DCT domain is adjusted to 5–165 data points.

Experimental results have demonstrated that different sampling rates significantly affect data recovery performance. We select a typical 50% sampling rate for visual comparison between the DTAE and baseline methods, as shown in Fig. 5.9. The quantitative comparison is divided into two parts: the signal-to-noise ratio (SNR) comparison, as shown in Table 5.1. The recovery result of the DTAE network achieves the highest SNR of 19.11 dB, while the TNN, PMF, and MSSA methods yield SNR values of 6.53, 9.10, and 5.85 dB, respectively. Another quantitative metric, the local similarity comparison, is shown in Fig. 5.10g–j, where the similarity between the recovery results of each method and the actual field data is computed. It is clear that the local similarity map of DTAE shows significantly more orange than the other state-of-the-art methods, indicating a better match with the true field data. Notably, in areas with missing traces, the DTAE map is predominantly green, while the SOTA maps show mostly blue, reflecting lower similarity. These findings suggest that the three SOTA methods, particularly MSSA, encounter challenges in reconstructing missing traces. As theoretically expected, this confirms the importance of the low-rank assumption, which does not strictly apply to the Parihaka-3D data, influencing the performance of these SOTA methods.

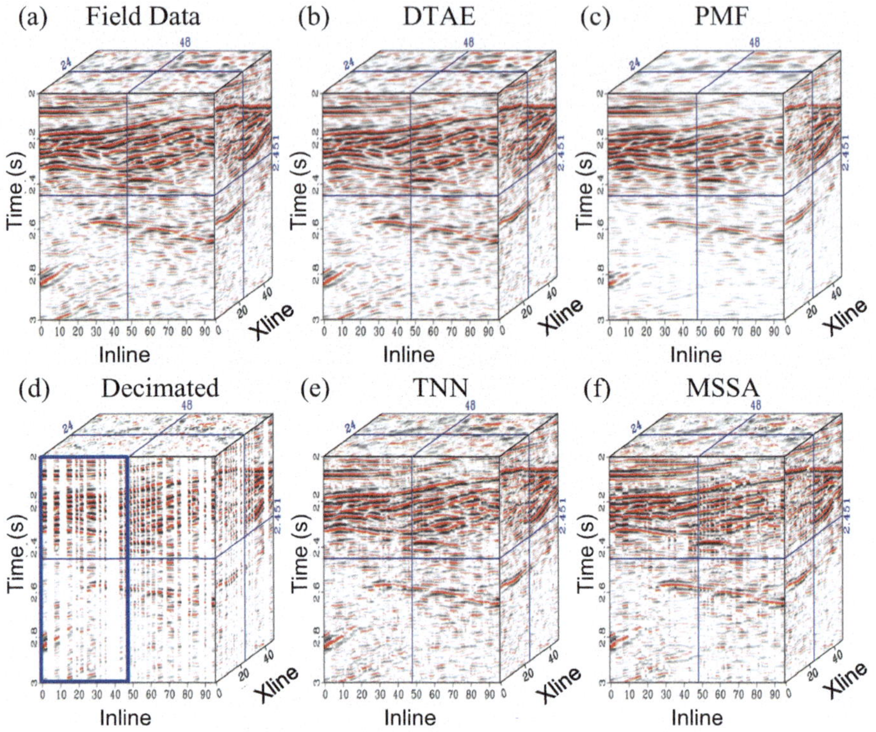


**Fig. 5.9** In the reconstruction comparison on the Parihaka-3D dataset, the results are as follows: **a** the original complete 3D field data; **b** the data interpolated using the DTAE model; **c** the data reconstructed using PMF; **d** the decimated data with 50% of traces missing; **e** the data interpolated using TNN; and **f** the data interpolated using MSSA

## 5.6 The Limitations of DTAE

The aforementioned experiments have verified that the proposed DTAE effectively performs 3D seismic data reconstruction under high noise conditions. However, due to the absence of noise consideration during the construction of training samples, its 3D data reconstruction capability is limited in low signal-to-noise ratio (SNR) scenarios, which differentiates it from the methods in [1, 28]. Theoretically, DTAE possesses some noise suppression capabilities against random noise, and the key lies in how to obtain clean training samples from noisy seismic data. The most conventional approach is to apply traditional denoising methods, such as structure-guided filtering, MSSA, etc., to remove noise from the actual data to generate clean training samples. These clean samples can then be used, along with the incomplete seismic data, to train the DTAE model, achieving both 3D data reconstruction and random noise suppression simultaneously. However, this approach introduces a prac-

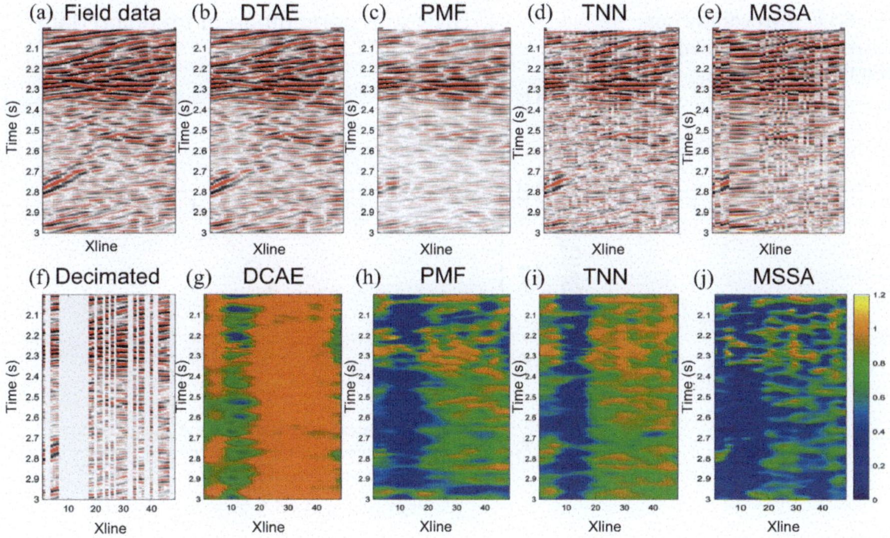


**Fig. 5.10** Comparison of reconstruction effects for the second $x$-line profile data in the Parihaka-3D experiment. **a** Complete data. **b–e** Data reconstructed using the proposed DTAE, PMF, TNN, and MSSA methods, respectively. **f** Observational data with 50% missing. **g–j** Local similarity measurements between the complete data and the data reconstructed using DTAE, PMF, TNN, and MSSA methods, respectively

tical issue: the noise suppression performance of DTAE will be constrained by the effectiveness of the traditional denoising methods in reducing random noise.

Our DTAE model, as a typical supervised learning model, requires a large number of training samples to train the frequency domain DMAE network. For instance, in the synthetic VSP experiment, 60,000 spectral images were needed for training, which represents the second limitation of our DTAE. The performance of data reconstruction in DTAE is undoubtedly influenced by the number of training samples. As shown in Figs. 5.11 and 5.12, it is evident that increasing the amount of training sample data improves the data reconstruction accuracy of DTAE. However, training with a large amount of sample data results in a time-consuming training process. Although the prediction speed is relatively fast, the comparison of execution times with other algorithms is shown in Table 5.3. When the computation time of DTAE includes both training and prediction, it requires more time than the PMF or TNN methods, and its execution time is similar to that of MSSA (note that these baseline methods are unsupervised and do not involve training time). The latter, however, incurs time delays due to the need for large-scale Hankel matrix computations.

Like all supervised learning methods, the DTAE model relies on the assumption that the predicted and training data share similar distribution characteristics, as indicated by the diagonal elements in Table 5.4, where both datasets come from the same survey area. However, when the training and test datasets originate from different survey regions, which corresponds to a transfer learning scenario, the recon-

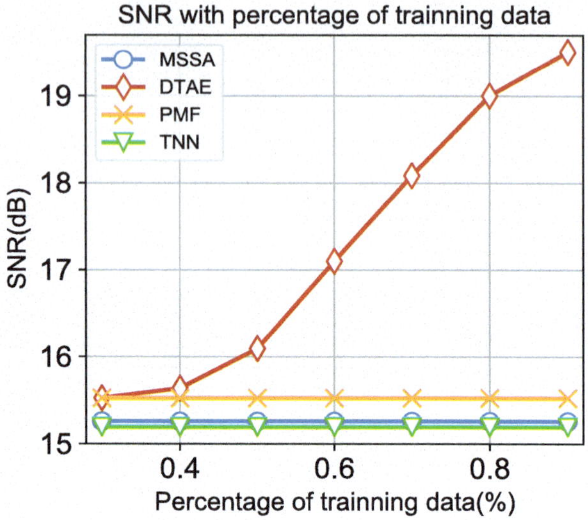


**Fig. 5.11** SNR comparison across varying percentages of training data in the synthetic VSP experiment.

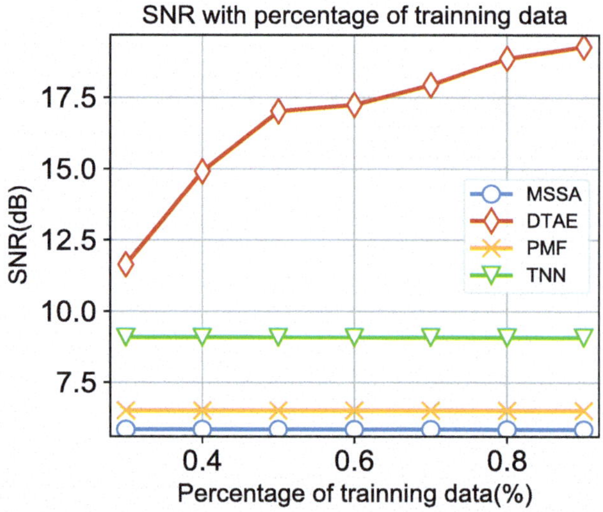


**Fig. 5.12** Comparison of data recovery performance between the proposed DTAE and baseline methods under different training data percentages in the Parihaka-3D experiment

struction performance of the DTAE model significantly degrades, as shown by the non-diagonal elements in Table 5.4. Addressing this issue will be a key challenge for future research.

**Table 5.3** Comparison of the computation time between the proposed DTAE and existing baseline methods, with DTAE trained on 90% of the available training data

| Dataset | PMF (s) | TNN (s) | MSSA (s) | DTAE (s) |
|---|---|---|---|---|
| Synthetic VSP | 7.2468 | 119.0322 | 84.5965 | 1763.9 |
| F3 Netherlands | 16.6801 | 452.4752 | 1567.2 | 1705.5 |
| Parihaka-3D | 16.0475 | 452.4752 | 1511.6 | 1693.8 |

**Table 5.4** Comparison of the proposed DTAE network's transfer learning-based data recovery capability across three synthetic and real fields datasets

| Trained model | VSP (dB) | F3 (dB) | Parihaka-3D (dB) |
|---|---|---|---|
| Synthetic VSP's model | 19.51 | 18.12 | 12.46 |
| F3's model | 17.29 | 21.26 | 17.34 |
| Parihaka-3D's model | 10.31 | 19.55 | 19.11 |

## 5.7 Summary

This chapter presents a DTAE framework, which includes a basic version and two variants, aimed at establishing a data-driven, nonlinear mapping between high-dimensional seismic unknown and known datasets for the purpose of 3D seismic data reconstruction. We introduce the tensor product (t-product) operation into deep learning to construct a novel tensor-based deep learning model, and accordingly adapt the standard BP method to TBP for training. Through the use of the convolutional characteristics of the tensor product, we create a mathematical connection between tensor autoencoders (TAEs) and matrix autoencoders (AEs), providing a theoretical foundation for the implementation of tensor autoencoders. With this theoretical connection, the training of the DTAE model can be specifically transformed into the training of matrix autoencoders for every frontal slice in the DCT domain. The experimental results, obtained from a synthetic dataset and two real-world datasets, show that the proposed DTAE framework surpasses three leading reconstruction methods, delivering superior reconstruction performance even under low signal-to-noise ratio conditions.

The future development of the DTAE framework can be directed toward the following three areas: first, integrating with more advanced deep learning models, such as those proposed in [31] or [32]; second, expanding to more complex seismic data completion tasks, such as prestack seismic data completion [33] and five-dimensional seismic reconstruction. [7]; third, adapting to seismic data completion in more complex noise environments, especially in non-Gaussian random noise scenarios (Table 5.3).

## 5.8 Appendix

### Background on Tensors

tThis section defines the following notation: a real-valued third-order seismic data tensor is denoted as $\mathcal{T} \in \mathbb{R}^{n_1 \times n_2 \times n_3}$, where $\mathcal{T}_{i,j,k}$ represents the $(i, j, k)$th element of the tensor, with $i$, $j$, and $k$ being the indices along the $x$-line, $y$-line, and time dimensions, respectively. For each slice, the horizontal slice is denoted as $\mathcal{T}(i, :, :)$, while the frontal slice and lateral slice are depicted as $\mathcal{T}(:, j, :)$ and $\mathcal{T}(:, :, k)$, which can be simplified as $\overrightarrow{\mathcal{T}}_j$ and $\mathcal{T}^{(k)}$, respectively. Additionally, scalars are represented by lowercase letters, such as $t$, vectors are denoted by bold lowercase letters, like $\mathbf{t}$, and matrices are represented by bold uppercase letters, such as $T$.

In the derivation of the DTAE model in this chapter, we introduce two important norms: the Frobenius norm for matrices and tensors. For the matrix case, the Frobenius norm is defined as $\|T\|_F = \left(\sum_{ij} |T_{ij}|^2\right)^{1/2}$, while for the tensor case, the Frobenius norm is defined as $\|\mathcal{T}\|_F = \left(\sum_{ijk} |\mathcal{T}_{ijk}|^2\right)^{1/2}$. Based on these notations, we further define the tensor product (t-product) and related concepts, which form the mathematical foundation for the development of the DTAE model. Next, we will provide detailed descriptions of these key mathematical definitions, which are crucial for the subsequent model derivation.

**Definition 5.1** (*t-product in the time domain* [34, 35]). Given $\mathcal{A} \in \mathbb{R}^{n_1 x r \times n_3}$ and $\mathcal{B} \in \mathbb{R}^{r \times n_2 \times n_3}$, the t-product is defined as

$$\mathcal{T}(i, j, :) = \sum_{s=1}^{n_2} \mathcal{A}(i, s, :)\mathcal{B}(s, j, :). \tag{5.29}$$

Building on the formula above, the tensor product (t-product) can be interpreted as a convolution-like operation performed on two separate tensors [36]. Following the principles of convolution theory, this operation can be computed efficiently in the frequency domain using a multiplicative process, as explained in the definition below.

**Definition 5.2** (*t-product in the frequency domain* [36, 37]). The t-product transforms the convolution in the time domain into a product in the frequency domain for computation, and is defined as follows:

$$\mathcal{T} = \mathcal{A} * \mathcal{B} \Longleftrightarrow \widetilde{\mathcal{T}}^{(k)} = \widetilde{\mathcal{A}}^{(k)} \cdot \widetilde{\mathcal{B}}^{(k)}, \quad k \in [n_3]. \tag{5.30}$$

To carry out this transformation, we substitute the traditional Fourier transform with the Discrete Cosine Transform (DCT), thereby avoiding complex number operations in the deep learning model, as outlined in Sect. 5.4 [38]. This decision not only streamlines the implementation but also improves the model's generalizability, since deep learning models working in the complex domain often necessitate more specialized processing. Specifically, we define $\widetilde{\mathcal{A}} = \text{dct}(\mathcal{A}, [], 3)$, where the DCT is applied along the third dimension of the tensor $\mathcal{A}$. Subsequently, we recover the original tensor $\mathcal{A}$ using the inverse DCT, denoted as $C = \text{idct}(\widetilde{\mathcal{A}}, [], 3)$. It is important to note that $[n_3]$ denotes the set of integers $1, 2, \ldots, n_3$.

**Definition 5.3** (*Tensor transpose*) Given that $\mathcal{A} \in \mathbb{R}^{n_1 \times n_2 \times n_3}$, the operation $\mathcal{A}^\dagger \in \mathbb{R}^{n_2 \times n_1 \times n_3}$ involves transposing each frontal slice and reversing the order of slices from the 2nd to the $n_3$th slice. This operation generalizes the matrix transpose to higher-order tensors, introducing a more complex structural transformation.

**Definition 5.4** (*Elementwise tensor product* [39]). The elementwise tensor product $\mathcal{T} = \mathcal{A} \odot \mathcal{B}$ of two tensors $\mathcal{A}$ and $\mathcal{B}$, both of size $\mathbb{R}^{n_1 \times n_2 \times n_3}$, results in a tensor $\mathcal{T}$ of the same dimensions. The $(i, j, k)$th element of $\mathcal{T}$ is computed as $\mathcal{T}(i, j, k) = \mathcal{A}(i, j, k)\mathcal{B}(i, j, k)$.

### Tensor Neural Networks (t-NNs)

Building upon the t-product concept introduced in Definition 5.2, Newman et al. [5] proposed the t-NN framework, which provides a novel approach to constructing neural networks capable of effectively processing tensor data. Following this framework, the matrix autoencoder (AE) can be naturally extended to a tensor autoencoder (AE). However, before proceeding with the construction of the tensor autoencoder, it is important to begin by revisiting several key concepts that are fundamental to the development of the t-NN framework.

**Definition 5.5** (*Matrix representation of a neural network* [5, 40]) Consider a feed-forward neural network (FNN) as depicted in Fig. 2.4a. Let $W_{\ell-1}$ denote the weight matrix of the $(\ell - 1)$th layer, with dimensions $n_\ell \times n_{\ell-1}$; $X_{\ell-1}$ represents the output of the $(\ell - 1)$th layer, with dimensions $n_{\ell-1} \times n_2$; $X_0$ is the input data matrix, with dimensions $n_1 \times n_2$; and $\mathbf{b}_\ell$ is the bias vector of the $\ell$th layer, with length $n_\ell$. The forward propagation process is then defined as follows.

$$X_\ell = \sigma\left(W_{\ell-1}X_{\ell-1} + \mathbf{b}_\ell\right), \quad \ell \in [K], \tag{5.31}$$

where $\sigma(\cdot)$ represents an elementwise nonlinear activation function,[7] and $K$ represents the total number of layers in the network, as shown in Fig. 2.4b. This setup naturally suggests the potential for extending the neural network into a tensor-based framework. In contrast to the conventional matrix-based method, the t-product framework substitutes the previously discussed fully connected layers with layers of a different structure, allowing for the processing of higher-dimensional tensor data.

**Definition 5.6** (*Tensor representation of a neural network* [5, 6]) Let us consider the tensors $\mathcal{X}_{\ell-1} \in \mathbb{R}^{n_{\ell-1} \times n_2 \times n_3}$, $\mathcal{W}_\ell \in \mathbb{R}^{n_\ell \times n_{\ell-1} \times n_3}$, and $\overrightarrow{\mathcal{B}}_\ell \in \mathbb{R}^{n_\ell \times 1 \times n_3}$. We then define tensor forward propagation as

[7] In this study, the ReLU function is used, defined as $\sigma(x) = \max(0, x)$.

$$\mathcal{X}_\ell = \sigma\left(\mathcal{W}_{\ell-1} * \mathcal{X}_{\ell-1} + \overrightarrow{\mathcal{B}}_\ell\right), \qquad \ell \in [K]. \tag{5.32}$$

where $\mathcal{W}_{\ell-1}$ represents the weight mapping from layer $\ell - 1$ to layer $\ell$. The operator "+" indicates the addition of the bias vector $\overrightarrow{\mathcal{B}}_\ell$ to each lateral slice of $\mathcal{W}\ell - 1 * \mathcal{X}_{\ell-1}$, as shown in Fig. 2.4c. This operation can be interpreted as shifting the result of each slice to obtain the activation values for layer $\ell$.

**Deep Matrix Autoencoders (DMAEs)**

To ensure the paper is self-contained, we provide a brief overview of Deep Matrix Autoencoders (DMAEs) [18]. An Autoencoder (AE) typically consists of two main components: the encoder, which maps the input data to a latent feature space through multiple layers of nonlinear transformations, and the decoder, which maps these encoded features back into the reconstruction space using additional nonlinear functions. For a set of input data $\{X_n, Y_n\}_{n=1}^N$, the hidden representations at each layer are determined as follows:

$$\begin{aligned} X_{n,\ell} &= \mathbf{h}\left(X_{n,\ell-1}\right) = \sigma\left(W_{\ell-1} X_{n,\ell-1} + \mathbf{b}_\ell\right), \\ \mathbf{g}\left(X_{n,\ell-1}\right) &= \sigma\left(W_{\ell-1}^\top \mathbf{h}\left(X_{n,\ell-1}\right) + \mathbf{b}_{\ell-1}\right), \quad \ell \in [K], \end{aligned} \tag{5.33}$$

here, $(\cdot)^\top$ denotes the transpose of a matrix, and $N$ represents the total number of input sample pairs. After processing $X_{n,\ell-1}$, the output $X_{n,\ell}$ is obtained by reversing the encoding process. The objective of the autoencoder (AE) is to adjust the network parameters to accurately reconstruct the input data. To achieve this, the optimization goal is to minimize the reconstruction error, and the cost function is defined as follows:

$$\mathcal{L}_4(Y_n, X_{n,\ell-1}; \theta) = \arg\min_\theta \frac{1}{N} \sum_{n=1}^N \frac{1}{2} \left\| Y_n - X_{n,\ell} \right\|_F^2, \tag{5.34}$$

where $\theta = \{W_\ell, \mathbf{b}_\ell\}_{\ell=1}^K$ represents the parameters of the AE model.

## References

1. Zhang H, Yang X, Ma J (2020) Can learning from natural image denoising be used for seismic data interpolation? Geophysics 85(4):WA115–WA136
2. Chai X, Tang G, Wang S, Peng R, Chen W, Li J (2020) Deep learning for regularly missing data reconstruction. IEEE Trans Geosci Remote Sens 58(6):4406–4423
3. Kamnitsas K, Ledig C, Newcombe VFJ, Simpson JP, Kane AD, Menon DK, Rueckert D, Glocker B (2017) Efficient multi-scale 3d CNN with fully connected CRF for accurate brain lesion segmentation. Med Image Anal 36:61–78
4. Chen Z, Zhang H (2019) Learning implicit fields for generative shape modeling. In: Proceedings of the IEEE/CVF conference on computer vision and pattern recognition, pp 5939–5948

5. Newman E, Horesh L, Avron H, Kilmer M (2018) Stable tensor neural networks for rapid deep learning. arXiv preprint arXiv:1811.06569
6. Horesh L, Newman E, Kilmer ME, Avron H (2020) Generating and managing deep tensor neural networks. US Patent App. 16/189,744, May 14 2020
7. Gao J, Cheng J, Sacchi MD (2016) Five-dimensional seismic reconstruction using parallel square matrix factorization. IEEE Trans Geosci Remote Sens 55(4):2124–2135 (2016)
8. Kreimer N, Sacchi MD (2012) A tensor higher-order singular value decomposition for prestack seismic data noise reduction and interpolation. Geophysics 77(3):V113–V122
9. Kreimer N, Stanton A, Sacchi MD (2013) Tensor completion based on nuclear norm minimization for 5d seismic data reconstruction. Geophysics 78(6):V273–V284
10. Liang J, Ma J, Zhang X (2014) Seismic data restoration via data-driven tight frame. Geophysics 79(3):V65–V74
11. Yu S, Ma J, Zhang X, Sacchi MD (2015) Interpolation and denoising of high-dimensional seismic data by learning a tight frame. Geophysics 80(5):V119–V132
12. Chien J-T, Bao Y-T (2017) Tensor-factorized neural networks. IEEE Trans Neural Networks Learn Syst 29(5):1998–2011
13. Novikov A, Podoprikhin D, Osokin A, Vetrov D (2015) Tensorizing neural networks. In: Proceedings of the 28th International Conference on Neural Information Processing Systems, vol 1, pp 442–450
14. Yu D, Deng L, Seide F (2016) Computer-implemented deep tensor neural network. US Patent 9,292,787, March 22 2016
15. Liao H-J, Liu J-G, Wang L, Xiang T (2019) Differentiable programming tensor networks. Phys Rev X 9(3):031041
16. Abadi M, Barham P, Chen J, Chen Z, Davis A, Dean J, Devin M, Ghemawat S, Irving G, Isard M et al (2016) {TensorFlow}: a system for {Large-Scale} machine learning. In: 12th USENIX symposium on operating systems design and implementation (OSDI 16), pp 265–283
17. Xie J, Xu L, Chen E (2012) Image denoising and inpainting with deep neural networks. In: Advances in neural information processing systems, pp 341–349
18. Hinton GE, Zemel R (1993) Autoencoders, minimum description length and helmholtz free energy. Adv Neural Inform Process Syst 6:3–10
19. Oropeza V, Sacchi M (2011) Simultaneous seismic data denoising and reconstruction via multichannel singular spectrum analysis. Geophysics 76(3):V25–V32
20. Zhang D, Zhou Y, Chen H, Chen W, Shaohuan Z, Chen Y (2017) Hybrid rank-sparsity constraint model for simultaneous reconstruction and denoising of 3d seismic data. Geophysics 82(5):V351–V367
21. Hao W, Zhang B, Lin T, Li F, Liu N (2019) White noise attenuation of seismic trace by integrating variational mode decomposition with convolutional neural network. Geophysics 84(5):V307–V317
22. Wu X, Shi Y, Fomel S, Liang L, Zhang Q, Yusifov AZ (2019) Faultnet3d: predicting fault probabilities, strikes, and dips with a single convolutional neural network. IEEE Trans Geosci Remote Sens 57(11):9138–9155
23. Chen Y, Zhang D, Jin Z, Chen X, Shaohuan Z, Huang W, Gan S (2016) Simultaneous denoising and reconstruction of 5-d seismic data via damped rank-reduction method. Geophys J Int 206(3):1695–1717
24. Chen Y, Bai M, Chen Y (2019) Obtaining free usarray data by multi-dimensional seismic reconstruction. Nat Commun 10(1):1–13
25. Ely G, Aeron S, Hao N, Kilmer ME (2015) 5d seismic data completion and denoising using a novel class of tensor decompositions. Geophysics 80(4):V83–V95
26. Zhang Z, Ely G, Aeron S, Hao N, Kilmer M (2014) Novel methods for multilinear data completion and de-noising based on tensor-SVD. In: Proceedings of the IEEE conference on computer vision and pattern recognition, pp 3842–3849
27. Gao J, Stanton A, Sacchi MD (2015) Parallel matrix factorization algorithm and its application to 5d seismic reconstruction and denoising. Geophysics 80(6)

28. Wang Y, Wang B, Ning T, Geng J (2020) Seismic trace interpolation for irregularly spatial sampled data using convolutional autoencoder. Geophysics 85(2):V119–V130
29. Liu G, Reda FA, Shih KJ, Wang T-C, Tao A, Catanzaro B (2018) Image inpainting for irregular holes using partial convolutions. In: Proceedings of the European conference on computer vision (ECCV), pp 85–100
30. Liu W, Wang Z, Liu X, Zeng N, Liu Y, Alsaadi FE (2017) A survey of deep neural network architectures and their applications. Neurocomputing 234:11–26
31. Saad OM, Chen Y (2021) A fully unsupervised and highly generalized deep learning approach for random noise suppression. Geophys Prospect 69(4):709–726
32. Saad OM, Chen Y (2020) Deep denoising autoencoder for seismic random noise attenuation. Geophysics 85(4):V367–V376
33. Gao J, Sacchi MD, Chen X (2013) A fast reduced-rank interpolation method for prestack seismic volumes that depend on four spatial dimensions. Geophysics 78(1):V21–V30
34. Kilmer ME, Martin CD (2011) Factorization strategies for third-order tensors. Linear Algebra Its Appl 435(3):641–658
35. Kilmer ME, Braman K, Hao N, Hoover RC (2013) Third-order tensors as operators on matrices: a theoretical and computational framework with applications in imaging. SIAM J Matrix Anal Appl 34(1):148–172
36. Kernfeld E, Kilmer M, Aeron S (2015) Tensor-tensor products with invertible linear transforms. Linear Algebra Its Appl 485:545–570
37. Lu C, Peng X, Wei Y (2019) Low-rank tensor completion with a new tensor nuclear norm induced by invertible linear transforms. In: Proceedings of the IEEE/CVF Conference on Computer Vision and Pattern Recognition, pp 5996–6004
38. Bipin Kumar Tripathi (2017) On the complex domain deep machine learning for face recognition. Appl Intell 47(2):382–396
39. Liu X-Y, Aeron S, Aggarwal V, Wang X (2019) Low-tubal-rank tensor completion using alternating minimization. IEEE Trans Inform Theory 66(3):1714–1737
40. Denil M, Shakibi B, Dinh L, Ranzato M, De Freitas N (2013) Predicting parameters in deep learning. In: Advances in neural information processing systems, pp 2148–2156

# Chapter 6
# Transform-Based Tensor Deep Learning for Seismic Random Noise Attenuation

**Abstract** In 3D seismic data processing, the key challenge of removing random noise lies in isolating the unique high-dimensional structures embedded in true signals. Deep learning (DL) has shown significant potential in this area, particularly for extracting hidden patterns from noisy data. However, the scarcity of clean seismic data in practical applications poses a challenge. Many DL-based denoising techniques, particularly those relying on matrix-based methods, face challenges in effectively managing complex high-dimensional structures, especially in unsupervised environments, which can limit their effectiveness in denoising 3D seismic data. To address these limitations, this chapter introduces a novel denoising strategy that integrates tensor convolutional neural networks (TCNN) with Stein's unbiased risk estimator (SURE), referred to as SURE-TCNN. The proposed method focuses on directly learning the intrinsic high-dimensional structures of seismic data without relying on noise-free datasets. SURE offers an approximately unbiased estimation of the mean squared error (MSE), enabling SURE-TCNN to achieve performance comparable to supervised MSE-TCNN models that require clean ground truth data.This method utilizes the characteristics of tensor-tensor products (t-product) to build a theoretical link between tensor representations and their matrix counterparts. This connection facilitates efficient parameter optimization by processing each frontal slice of the tensor in the time-frequency domain, such as the wavelet transform domain, using a matrix-based SURE-CNN. This design streamlines implementation while maintaining theoretical consistency.Experimental results on both synthetic and real-world datasets show that the proposed SURE-TCNN method significantly outperforms three leading denoising techniques in terms of its effectiveness in reducing noise. Moreover, the method exhibits strong robustness and broad applicability across various scenarios.

## 6.1 Introduction

In seismic data analysis, noise reduction plays a vital role in enhancing the precision of subsurface structure analysis. Recently, several methods have been introduced to tackle noise suppression in seismic datasets, which can be broadly categorized into

 
F. Qian et al., *Tensor Computation for Seismic Data Processing*,
Earth Systems Data and Models 6, https://doi.org/10.1007/978-3-031-78900-7_6

model-driven and data-driven strategies. The former relies on physical modeling of seismic signal properties, while the latter leverages data-driven techniques to uncover hidden patterns and features within the signals. Both approaches aim to enhance the inherent structural characteristics of true seismic signals to effectively suppress noise.

Model-driven denoising methods suppress noise in seismic data by constructing mathematical models of seismic images, using techniques such as sparse representation [1, 2] and low-rank approximation (LRA ) [3, 4]. In the field of sparse representation , the goal is to represent local features of seismic images as linear combinations of selected basis functions through various transformations. Common transformations include wavelet transform [1, 5], curvelet transform [6, 7], shearlet transform [8, 9], seislet transform [10], Radon transform [11], empirical mode decomposition [12], and learned dictionaries [2, 13]. The core goal of these methods is to reveal the underlying signal features of seismic data by selecting appropriate basis sets. In contrast, low-rank approximation (LRA ) methods focus on extracting global structures from noisy seismic data. LRA methods can be classified into approaches that rely on matrix structures and those that utilize tensor representations. Matrix-based methods treat seismic data as matrices and apply rank-reduction techniques for denoising, including unstructured methods [3, 14] and structured denoising methods based on Hankel matrices [4, 15]. Tensor-based methods handle multi-dimensional data by tensor decomposition, with typical techniques including tensor singular value decomposition (t-SVD ) [16, 17] and CANDECOMP/PARAFAC (CP) decomposition [18]. Although model-driven methods can suppress noise to some extent, they often heavily rely on manually designed priors, such as sparsity priors [1, 2] or low-rank priors [3, 4], which limits their adaptability and generalization capability when faced with complex and diverse real-world seismic data.

In contrast to traditional denoising approaches that rely on models, data-driven methods allow deep neural networks (DNNs) to learn intricate and useful image features directly from noisy data. Currently, most DNN-based denoising methods rely on a supervised learning framework, where models are optimized through end-to-end training and then validated on noisy seismic images. One example is the use of convolutional neural networks (CNNs) [19, 20], 3D-DnCNN (3D denoising convolutional neural network) [21, 22], generative adversarial networks (GANs) [23], and U-Net [24]. While these methods exhibit strong denoising performance, they heavily depend on high-quality, clean seismic data, making it difficult to obtain ground-truth seismic data for validation in real-world scenarios, thus limiting their generalizability and applicability. To address this challenge, recent research has explored unsupervised denoising methods based on autoencoders (AE) [25, 26], which can perform denoising using only noisy seismic images as input, without requiring clean data. Autoencoders leverage neural networks to learn latent features from the data, eliminating the need for labeled data, making them a powerful tool for handling tasks with unlabeled data. However, a major limitation of most DNN-based unsupervised denoising methods is their tendency to treat seismic data as 2D images, which overlooks the high-dimensional nature of seismic data and fails to capture its complex spatial structures when applied to 3D seismic data.

Moreover, various denoising techniques, including adaptive filtering [27, 28], morphological filtering [29], nonlinear diffusion filtering [30, 31], Bayesian inversion [32], total variation (TV) [33], and nonlocal means [34], have been widely applied to mitigate random noise in seismic signal processing. These methods extract effective signal structures from noisy data using various strategies, improving data quality. However, a common limitation across these techniques is the need for manual parameter selection and adjustment tailored to specific datasets. Given that noise characteristics in seismic data often fluctuate in both space and time, and its statistical properties are typically difficult to estimate accurately, determining the optimal parameters to adapt to these variations is challenging [22]. This dependence on manual tuning significantly restricts the denoising effectiveness and generalizability of these methods in complex seismic datasets.

We highlight that SURE-TCNN is a highly advanced method for 3D image denoising, particularly in handling noisy seismic data, which constitutes the central focus of this chapter. Compared to current deep neural network (DNN) -based denoising methods [19, 26], SURE-TCNN excels in ground-truth-free denoising of 3D seismic images, showcasing its superior capability in denoising tasks. This remarkable performance is attributed to a series of innovative approaches we have introduced, which will be thoroughly discussed in the subsequent sections as key contributions.

(1) Building upon the t-product framework based on transformations described in Sect. 6.5, We extend the conventional tensor neural network (t-NN) from the Fourier domain to a broader linear transform domain, including the time-frequency transform domain utilized in this research. Building on this extension, we introduce a SURE-TCNN framework, treating the 3D seismic noise reduction challenge as a nonlinear regression problem without relying on ground truth. As far as we are aware, this is the first research to employ SURE-based tensor deep learning techniques for denoising 3D seismic datasets without the need for ground truth.
(2) Leveraging the t-product framework based on transformations outlined in Algorithm 6.2, the SURE-TCNN approach decomposes the complex denoising problem into several independently solvable subproblems. Each subproblem is tackled by applying the SURE-CNN denoising network to the individual slices in the time-frequency space, such as the wavelet transform domain. Notably, there are two main distinctions between the SURE-CNN and the traditional U-Net architecture: first, it incorporates total variation (TV) regularization to better preserve image structures; Second, it employs Stein's unbiased risk estimator (SURE) to define the loss function., instead of the conventional mean squared error (MSE), thus eliminating the reliance on ground truth data.
(3) SURE-TCNN effectively connects deep learning (DL) with conventional wavelet thresholding techniques for noise reduction. Viewed through the lens of thresholding, SURE-TCNN can be considered a data-driven wavelet thresholding method that leverages deep learning and operates across multiple channels. This

unique positioning not only advances cutting-edge research in the field but also significantly enhances the application effectiveness of wavelet thresholding techniques in noise suppression.

Through extensive testing on synthetic data and two real seismic survey datasets, The SURE-TCNN model introduced in this work outperforms three current leading techniques in the field.

## 6.2 Seismic Random Noise Attenuation and Its Modelling

As mentioned earlier, conventional seismic data denoising methods have several limitations that constrain further performance improvements, which led to the introduction of the SURE-TCNN approach. To explain this innovation, We begin by outlining the denoising problem in Sect. 6.2.1, followed by a discussion of the solution in Sect. 6.2.2, we reformulate the problem as a supervised nonlinear regression task and address it using deep learning methods. Considering that traditional supervised learning models typically rely on large amounts of ground truth data for training, SURE-based regression models offer an effective solution for unsupervised denoising, which is discussed in detail in Sect. 6.2.3.

### *6.2.1 Problem Statement*

In the process of denoising 3D seismic images, the acquired seismic data $\mathcal{Y} \in \mathbb{R}^{N_1 \times N_2 \times N_3}$ is assumed to be the sum of the true 3-D seismic image $\mathcal{X} \in \mathbb{R}^{N_1 \times N_2 \times N_3}$ and noise $\mathcal{N} \in \mathbb{R}^{N_1 \times N_2 \times N_3}$, where the noise consists of both random noise and non-seismic components. Hence, the model for denoising 3D seismic data can be formulated as:

$$\mathcal{Y} = \mathcal{X} + \mathcal{N}. \tag{6.1}$$

The primary objective of denoising is to reconstruct the original seismic image (i.e., the denoised image) $\widehat{\mathcal{X}}$ from the noisy input, while minimizing the average discrepancy between the original and the predicted image, as shown below.

$$\mathcal{L}(\mathcal{X}_n, \widehat{\mathcal{X}}_n; \Theta) = \frac{1}{N} \sum_{n=1}^{N} \left\| \widehat{\mathcal{X}}_n - \mathcal{X}_n \right\|^2, \tag{6.2}$$

Restoring a clean seismic image $\mathcal{X}$ from Eq. (6.1) constitutes an ill-posed problem, which prevents the determination of a singular solution through direct minimization of Eq. (6.4). Two main strategies have been introduced for the denoising of seismic images: one operates in the spatial domain [3, 4], while the other works in the transform domain [1, 2]. Notably, the deep learning model proposed in this chapter is fundamentally a transform domain approach, which enables more effective noise reduction by leveraging the capabilities of domain transformations.

### 6.2.2 DL-Based Supervised Regression Model

Recent investigations have highlighted that deep learning-driven denoising methods significantly surpass traditional techniques in terms of effectiveness [19, 21, 22, 35]. In these models, the clean seismic image $\widehat{\mathcal{X}}$ is commonly represented through a nonlinear autoregressive framework.

$$\widehat{\mathcal{X}} = \boldsymbol{h}_{\Theta}(\mathcal{X}, \mathcal{Y}), \tag{6.3}$$

The denoising model $\boldsymbol{h}$ is represented by the parameters $\Theta$, and is approximated by training a deep neural network using clean seismic images $\mathcal{X}$.

Next, we substitute Eq. (6.4) into Eq. (6.3) to derive a loss function that measures the discrepancy between the denoising output and the true seismic image $\mathcal{X}$. This function quantifies the difference between the seismic image denoiser $\boldsymbol{h}(\mathcal{Y})$ and the desired image, as shown below.

$$\mathcal{L}(\mathcal{X}_n, \mathcal{Y}_n; \Theta) = \frac{1}{N} \sum_{n=1}^{N} \left\| \boldsymbol{h}_{\Theta}\left(\mathcal{Y}_n\right) - \mathcal{X}_n \right\|^2, \tag{6.4}$$

Based on the training dataset $\left\{\left(\boldsymbol{X}^{(1)}, \boldsymbol{y}^{(1)}\right), \ldots, \left(\boldsymbol{X}^{(N)}, \boldsymbol{y}^{(N)}\right)\right\}$, gradient descent methods are widely used to optimize the parameters $\Theta$ of deep learning (DL) networks $\boldsymbol{h}_{\Theta}(\cdot)$, with the goal of minimizing the loss function $\mathcal{L}(\cdot)$. Common gradient descent optimization methods include stochastic gradient descent (SGD) [36], nesterov accelerated gradient (NAG) [37], and the Adam optimization algorithm [38].

In seismic denoising tasks, obtaining clean seismic images often incurs high costs and may be impractical in certain scenarios. To tackle this issue, two approaches for obtaining noise-free seismic images are introduced. The first approach involves utilizing synthetic seismic data as "clean" references, where random noise is artificially added to create noisy training examples [20]. However, deep neural networks trained on such synthetic data may encounter performance limitations when applied to real seismic data. The second approach employs traditional noise reduction techniques to extract ground truth images from noisy seismic data [20], but this method may be constrained by the chosen noise suppression algorithm, potentially affecting the performance of the deep learning model.

### 6.2.3 DL-Based Ground Truth-Free Regression Model

To overcome the challenges mentioned earlier, we modify the deep learning regression framework presented in Eq. (6.4) into an unsupervised form by integrating Stein's Unbiased Risk Estimator (SURE) [39, 40]. This modification enables deep learning-based regression denoising to be performed effectively even without ground truth images. Unlike previous works [19, 26] that rely on true images, our approach operates without ground truth data and successfully conducts denoising tasks for 3-D seismic data within a high-dimensional framework. By combining the definitions of SURE presented in Eqs. (6.4) and (7.6), we can derive the following equation:

$$\mathcal{L}(\mathcal{Y}_n;\Theta) = \frac{1}{N}\sum_{n=1}^{N}\{\frac{1}{K}\|\mathcal{Y}_n - \boldsymbol{h}_\Theta(\mathcal{Y}_n)\|^2 - \sigma_n^2 + \frac{2\sigma^2}{K}\operatorname{tr}\left(\frac{\partial \boldsymbol{h}_\Theta(\mathcal{Y}_n)}{\partial \mathcal{Y}_n^\dagger}\right)\}. \tag{6.5}$$

By leveraging the properties of SURE, Eq. (6.5) is interpreted as an unbiased approximation of Eq. (6.4). When the final divergence term in Eq. (6.4) can be analytically calculated using Eq. (6.32), the resulting estimator for Eq. (6.4) is given by the following expression:

$$\begin{aligned}\mathcal{L}(\mathcal{Y}_n;\Theta) =& \frac{1}{N}\sum_{n=1}^{N}\left\{\frac{1}{K}\|\mathcal{Y}_n - \boldsymbol{h}_\Theta(\mathcal{Y}_n)\|^2 - \sigma_n^2 + \frac{2\sigma^2}{K\epsilon}\right. \\ &\left.\times(\tilde{\boldsymbol{n}}_n)^\top(\boldsymbol{h}_\Theta(\mathcal{Y}_n + \epsilon\tilde{\boldsymbol{n}}_n) - \boldsymbol{h}_\Theta(\mathcal{Y}_n))\right\},\end{aligned} \tag{6.6}$$

To optimize the objective function in Eq. (6.6), the primary goal is to construct a function $\boldsymbol{h}_\Theta(\cdot)$ that can accurately approximate $\mathcal{X}$, thereby achieving this objective through the training of the proposed SURE-TCNN model (see Eq. (6.3)). The main challenge in this process lies in handling high-dimensional seismic image data, which requires tensor operations rather than traditional matrix operations.

The optimization of the objective function aims to identify a function $\boldsymbol{h}_\Theta(\cdot)$ that can effectively reconstruct $\mathcal{X}$, with this process being carried out through the training of the SURE-TCNN model (see Eq. (6.3)). The key difficulty in solving this optimization problem lies in addressing the computational challenges posed by high-dimensional seismic images, necessitating the use of tensor operations instead of conventional matrix operations.

## 6.3 SURE Loss-Based Tensor Convolutional Neural Network (SURE-TCNNx)

To tackle the issues in the ground truth-independent regression model (6.6), we introduce a new tensor neural network (t-NN), which employs a transform-based tensor product (t-product) method to efficiently reduce random noise in 3D seismic data. The details of this approach are discussed in Sect. 6.3.1. Subsequently, the SURE-TCNN model is introduced, effectively connecting deep learning with noise reduction methods based on wavelet thresholding. This integration is elaborated in Sect. 6.3.2. By utilizing the characteristics of the transform-based tensor product, we break down the SURE-TCNN model into separate SURE-CNN solutions for each slice in the wavelet domain, ultimately deriving the complete solution for the SURE-TCNN. The mathematical derivations and details are provided in Sect. 6.3.3. Finally, the complete implementation process and detailed techniques of the SURE-TCNN model are thoroughly described in Sect. 6.3.4.

### 6.3.1 T-NN with the Transform-Based T-Product

With respect to noisy 3D seismic image processing, as outlined in Eq. (6.1), We introduce a denoising method that does not rely on ground truth, utilizing a TCNN, denoted as $\boldsymbol{h}\Theta(\cdot)$. This denoiser processes seismic images with noise as its input and produces their corresponding noise-free versions, described as $\widehat{\mathcal{X}} = \boldsymbol{h}\Theta(\mathcal{Y})$. The core idea is to generalize the traditional Stein's Unbiased Risk Estimator (SURE) approach into a tensor framework by optimizing the TCNN with 3D seismic images containing noise, employing the SURE loss function provided in Eq. (6.6). The proposed TCNN is inspired by the framework of t-NN, which apply tensor-product (t-product) algebra to extend traditional deep neural networks (DNNs) into higher-dimensional spaces [41]. The t-NN is defined as follows:

**Definition 6.1** (*t-NN* [41, 42]) Consider the tensors $\mathcal{X}_{\ell-1} \in \mathbb{R}^{n_{\ell-1} \times n_2 \times n_3}$, $\mathcal{W}_\ell \in \mathbb{R}^{n_\ell \times n_{\ell-1} \times n_3}$, and $\overrightarrow{\mathcal{B}}_\ell \in \mathbb{R}^{n_\ell \times 1 \times n_3}$. Accordingly, tensor forward propagation can be described as:

$$\mathcal{X}_\ell = \delta\left(\mathcal{W}_{\ell-1} * \mathcal{X}_{\ell-1} + \overrightarrow{\mathcal{B}}_\ell\right), \qquad \ell \in [K]. \tag{6.7}$$

Here, the tensor $\mathcal{W}_{\ell-1}$ represents the trainable weight parameters enabling the transformation between layer $\ell - 1$ and layer $\ell$. The function $\delta(\cdot)$ is a nonlinear activation function applied elementwise, playing a crucial role in modulating the training outcomes and enhancing the model's capacity. Additionally, the tensor $\overrightarrow{\mathcal{B}}_\ell$ denotes the bias term incorporated into each lateral slice of the product $\mathcal{W}_{\ell-1} * \mathcal{X}_{\ell-1}$.

The proposed t-NN framework builds upon the t-product and leverages the FFT for efficient implementation. Previous studies have demonstrated that time-frequency transforms, such as the wavelet transform, outperform traditional Fourier transforms in noise reduction [43]. Utilizing the recently introduced invertible linear transform-based t-product (refer to Sect. 6.5), we can naturally establish the generalized structure of t-NN within the time-frequency domain, as detailed below.

**Definition 6.2** (*t-NN with a transform-based t-product*) Consider the tensors $\mathcal{X}_{\ell-1} \in \mathbb{R}^{n_{\ell-1} \times n_2 \times n_3}$, $\mathcal{W}_\ell \in \mathbb{R}^{n_\ell \times n_{\ell-1} \times n_3}$, and $\overrightarrow{\mathcal{B}}_\ell \in \mathbb{R}^{n_\ell \times 1 \times n_3}$. Building on these concepts, we propose a tensor-tensor product (t-product) approach based on transforms, which results in a redefinition of Eq. (6.7) to enable a more efficient computational framework.

$$\mathcal{X}_\ell = \delta\left(\mathcal{W}_{\ell-1} *_L \mathcal{X}_{\ell-1} + \overrightarrow{\mathcal{B}}_\ell\right), \qquad \ell \in [K]. \tag{6.8}$$

In this context, invertible linear transformations can be expressed through various forms of linear transforms, including Fourier transforms, time-frequency transforms, and others. By incorporating a t-NN that utilizes a tensor-tensor product (t-product) approach based on transforms, we are able to effectively approximate $\boldsymbol{h}_\Theta(\mathcal{Y})$, as shown below:

$$\boldsymbol{h}_\Theta(\mathcal{Y}) = \delta\left(\mathcal{W}_\ell *_L \delta\left(\mathcal{W}_{\ell-1} *_L \dots\ \delta\left(\mathcal{W}^0 *_L \mathcal{Y} + \mathcal{B}^0\right)\dots \mathcal{B}_{\ell-1}\right) + \mathcal{B}_\ell\right) \tag{6.9}$$

where $\ell$ represents the last layer of the deep t-NN. By utilizing the characteristics of the $L$-transform-based tensor-tensor product (t-product) outlined in Algorithm 6.2, we can reformulate Eq. (6.10) as follows:

$$\boldsymbol{h}_\Theta(\widetilde{\mathcal{Y}}^{(k)}) = \delta\left(\widetilde{\mathcal{W}}_\ell^{(k)} \delta\left(\widetilde{\mathcal{W}}_{\ell-1}^{(k)} \dots\ \delta\left(\widetilde{\mathcal{W}}_0^{(k)} \widetilde{\mathcal{Y}}^{(k)} + \widetilde{\mathcal{B}}_0^{(k)}\right)\dots \widetilde{\mathcal{B}}_{\ell-1}^{(k)}\right) + \widetilde{\mathcal{B}}_\ell^{(k)}\right). \tag{6.10}$$

From the equation above, it is evident that the weight parameters of the t-NN are derived by applying conventional DNNs to every frontal slice $\widetilde{\mathcal{Y}}^{(k)}$ within the $L$-transform space. Building on our previous work [44], this approach can be refined into a streamlined and effective framework tailored for 3-D seismic noise reduction tasks.

$$\widehat{\mathcal{X}}^{(k)} = \boldsymbol{h}_\Theta(\widetilde{\mathcal{Y}}^{(k)}) = \mathrm{MDL}(\widetilde{\mathcal{Y}}^{(k)}, \Theta), \tag{6.11}$$

The main role of MDL($\cdot$) is to execute matrix deep learning (MDL) operations on each frontal slice. Although this approach could theoretically be applied to various deep learning (DL) frameworks, it falls outside the scope of this chapter, and thus, further discussion is omitted.

**Algorithm 6.1:** SURE-TCNN Algorithm

**Input**: $\tilde{\mathcal{Y}}_n^{(k)}, \sigma, \lambda_{\mathrm{TV}}, \alpha, N_{\mathrm{bs}}$.
**Output**: $\widehat{\mathcal{X}}_n$.
$\widetilde{\mathcal{Y}}_n = W(\mathcal{Y}_n) = \mathcal{Y}_n \times_3 W$.
**for** $k = 1, \cdots, \frac{N_3}{2}$ **do**
  $\tilde{\mathcal{X}}_n^{(k)} \leftarrow \text{SURE-CNN}\left(\tilde{\mathcal{Y}}_n^{(k)}, \sigma, \lambda_{\mathrm{TV}}, \alpha, N_{\mathrm{bs}}\right)$.
**end**
$\widehat{\mathcal{X}}_n = W^{-1}(\widetilde{\mathcal{X}}_n) = \widetilde{\mathcal{X}}_n \times_3 W^{-1}$.

### 6.3.2 *Bridging Tensor Deep Denoising and Wavelet Threshold Denoising*

To tackle the challenges associated with denoising 3-D seismic images, we have designed a dedicated solution framework; however, two critical issues persist. First, to implement the $L$ transform in our model, We employ a tensor-tensor product (t-product) approach based on transforms to incorporate the Mallat wavelet transform [45] into the t-NN. Second, it is essential to incorporate a U-Net based on SURE architecture with total variation (TV) regularization into the framework, which will be discussed in detail in the next section. After introducing the Mallat transform, we can reformulate Eq. (6.27).

$$\widetilde{\mathcal{Y}} = W(\mathcal{Y}) = \mathcal{Y} \times_3 W, \tag{6.12}$$

Here, the matrix $W$ is formed as a square matrix using different functions, such as the Haar wavelet, with its coefficients obtained from the one-dimensional discrete wavelet transform, which includes methods like the Mallat transform and wavelet packet decomposition. For simplicity, the Mallat transform is applied in Eq. (6.12) to calculate the wavelet coefficient volume $\widetilde{\mathcal{Y}}$, as illustrated in Fig. 6.2. This procedure can be realized using the MATLAB code below.

$$\widetilde{\mathcal{Y}}(i, j, :) = \text{wavedec}(\mathcal{Y}(i, j, :), m, \text{wname}), \tag{6.13}$$

The function wavedec$(\cdot)$ is used to perform multilevel discrete wavelet decomposition on the one-dimensional seismic signal $\mathcal{Y}(i, j, :)$ at level $m$, utilizing the specified wavelet wname. To provide a clearer understanding of our proposed model, we present several diagrams that highlight the connection between tensor-based deep denoising and conventional wavelet thresholding denoising for the coefficient volume $\widetilde{\mathcal{Y}}$, as illustrated below.

#### 6.3.2.1 Wavelet Threshold Denoising

For the wavelet coefficient tensor $\widetilde{\mathcal{Y}}$ affected by noise, wavelet thresholding methods are typically used to reduce random noise. This method involves using various thresholding strategies to select valid wavelet coefficients, with soft thresholding (often referred to as shrinkage) and hard thresholding being common approaches. After applying the thresholding operation, the inverse wavelet transform is performed on the remaining wavelet coefficients to reconstruct the denoised seismic image, denoted as $\widehat{\mathcal{X}} = \psi_T(\mathcal{Y})$, where $\psi_T(\cdot)$ represents the thresholding method used. In this case, the soft threshold is expressed as follows:

$$\psi_T(\widetilde{\mathcal{Y}}) = \text{sgn}(\widetilde{\mathcal{Y}}) \cdot \max(|\widetilde{\mathcal{Y}}| - T, 0), \tag{6.14}$$

In this expression, the function $\max(\cdot)$ is employed to select the largest value among the wavelet coefficients, while $\text{sgn}(\cdot)$ denotes a discontinuous sign function. This function modifies the input wavelet coefficient tensor $\widetilde{\mathcal{Y}}$ according to the threshold $T$, reducing its values towards zero. An alternative expression of the hard thresholding technique is given as follows:

$$\psi_T(\widetilde{\mathcal{Y}}) = \widetilde{\mathcal{Y}} \cdot 1\{|\widetilde{\mathcal{Y}}| > T\}, \tag{6.15}$$

where $1\{\cdot\}$ represents the indicator functions.

#### 6.3.2.2 Tensor Deep Denoising in the Wavelet Domain

From a thresholding perspective, our model can be considered a wavelet thresholding-based denoising method; however, the main difference is that the threshold is learned in a data-driven way through deep learning (DL). By combining Eqs. (6.6) and (6.11), we obtain the tensor-based deep denoising model in the wavelet domain, expressed as follows.

$$\begin{aligned} \mathcal{L}(\widetilde{\mathcal{Y}}_n^{(k)}; \Theta) = & \frac{1}{N} \sum_{nj=1}^{N} \left\{ \frac{1}{K} \left\| \widetilde{\mathcal{Y}}_n^{(k)} - \boldsymbol{h}_\Theta \left( \widetilde{\mathcal{Y}}_n^{(k)} \right) \right\|^2 - \sigma_n^2 + \frac{2\sigma^2}{K\epsilon} \right. \\ & \left. \times \left( \tilde{\boldsymbol{n}}_j \right)^{t} \left( \boldsymbol{h}_\Theta \left( \widetilde{\mathcal{Y}}_n^{(k)} + \epsilon \tilde{\boldsymbol{n}}_n \right) - \boldsymbol{h}_\Theta \left( \widetilde{\mathcal{Y}}_n^{(k)} \right) \right) \right\}, \end{aligned} \tag{6.16}$$

where $\sigma$ denotes the standard deviation of the noise, a key factor influencing the denoising effectiveness of the TCNN. In this study, $\sigma$ is adaptively estimated using the median absolute deviation method applied to the highest-frequency subband of the wavelet transform, as described below: [43].

$$\widehat{\sigma}_i = \frac{\text{median}\left( \left| \mathcal{W}_{(i)}^{\text{HH}} \right| \right)}{0.6745}, \quad i = 1, \ldots, B. \tag{6.17}$$

where $\text{median}(\cdot)$ is applied to determine the median of the coefficients in the wavelet transform. For each frontal slice, we first apply deep learning (DL)-based wavelet denoising using Eq. (6.16), followed by applying the inverse transform to reconstruct the denoised signal.

$$\mathcal{Y} = W^{-1}(\widetilde{\mathcal{Y}}) = \widetilde{\mathcal{Y}} \times_3 W^{-1}. \tag{6.18}$$

It is clear that deep learning (DL) outperforms traditional threshold-based wavelet denoising methods in terms of noise reduction. This is because DL leverages data-driven approaches to automatically learn valid features, enabling more accurate signal recovery compared to relying solely on thresholding techniques.

### 6.3.3 SURE-Based DnCNN with a TV Regularizer

Convolutional Neural Networks (CNNs) have demonstrated significant performance in image denoising tasks [19, 20], making the use of CNNs to approximate the function $\boldsymbol{h}_\Theta(\cdot)$ a natural choice in this study. However, compared to conventional CNNs methods, our model presents two key innovations. First, to perform noise reduction without ground truth data, we replace the traditional MSE loss function with the SURE loss function. Second, to mitigate the over-smoothing effect often encountered in standard CNNs, we introduce a total variation (TV) regularization term into the SURE framework, which helps preserve sharp edges and improves denoising performance. The definition of the TV regularization in this framework is as follows:

$$\mathcal{L}_{\text{TV}} = \left\|\nabla_x \delta(\widetilde{\mathcal{Y}}_j^{(k)})\right\|_2^2 + \left\|\nabla_y \delta(\widetilde{\mathcal{Y}}_j^{(k)})\right\|_2^2, \tag{6.19}$$

where $\nabla_x (\nabla_y)$ denotes the gradient operators in the inline and crossline directions, respectively. For each frontal slice $\widetilde{\mathcal{Y}}_j^{(k)}$ in the wavelet domain, the reconstruction loss can be formulated using Eq. (7.6) as shown below:

$$\mathcal{L}_{\text{sure}} = \eta_{\text{sure}}(\boldsymbol{h}_\Theta(\widetilde{\mathcal{Y}}_j^{(k)})). \tag{6.20}$$

In conclusion, the primary objective of our SURE-CNN is

$$\mathcal{L} = \mathcal{L}_{\text{sure}} + \lambda_{\text{TV}}\mathcal{L}_{\text{TV}}, \tag{6.21}$$

In this method, the parameter $\lambda_{\text{TV}}$ is adjusted to control the intensity of TV regularization, ensuring a balance between data fidelity and regularization effects. The model functions within an unsupervised learning framework, handling noisy 2-D slices in the wavelet domain directly for concurrent training and evaluation. Once the training converges, the denoised slices are output. To prevent overfitting, an early stopping mechanism is implemented, terminating the training process if the loss function shows no significant reduction within a predefined number of iterations. Finally, the denoised slices are converted from the wavelet domain to the time domain, yielding the final reconstruction of the 3-D seismic dataset.

### 6.3.4 Implementation Detail

After constructing the SURE-TCNN model, the focus shifts to its practical implementation. Algorithm 6.1 describes how the 3-D seismic denoising problem can be transformed into a 2-D image denoising task for each frontal slice in the wavelet space, a process performed by SURE-CNN. Unlike conventional convolutional neural networks (CNNs) that rely on ground truth for training, SURE-CNN leverages the SURE loss function in place of mean squared error (MSE). This design enables unsupervised denoising directly in the wavelet domain without requiring labeled data, simplifying the training process and enhancing the model's applicability in real-world scenarios.

$$\tilde{\mathcal{X}}_n^{(k)} \leftarrow \text{SURE-CNN}\left(\tilde{\mathcal{Y}}_n^{(k)}, \sigma, \lambda_{\text{TV}}, \alpha, N_{\text{bs}}\right), \tag{6.22}$$

In this approach, $\alpha$ represents the learning rate, which controls the step size for updating model parameters, while $N_{\text{bs}}$ denotes the batch size, specifying how many data points are processed by the SURE-CNN during each iteration. To process the noisy 3-D data cube $\tilde{\mathcal{Y}}_n^{(k)}$, a wavelet transform function $W(\cdot)$ is used to decompose the input data along the time axis, generating multi-scale frequency representations that are fed into the SURE-CNN for denoising. After the denoising process, the inverse wavelet transform $W^{-1}(\cdot)$ is applied to the denoised wavelet coefficients $\tilde{\mathcal{X}}_n^{(k)}$, converting them back to the time domain and yielding the final denoised 3-D seismic image.

In this approach, the transformation $W(\cdot)$ is implemented using one-dimensional Mallat transforms to decompose the seismic data tensor $\mathcal{Y}$ into the wavelet domain, as shown in Fig. 6.1. It is important to note that in the ground truth data, the wavelet coefficients corresponding to high-frequency components are predominantly sparse and almost zero (see Fig. 6.1b), and are generally considered as noise in the original signal. Based on this observation, the SURE operation is applied exclusively to the low-frequency components, thereby reducing unnecessary computations for high-frequency noise and significantly enhancing the denoising performance while lowering computational complexity.

For the function SURE $-$ CNN$(\cdot)$, we develop a convolutional U-Net-based architecture for denoising within the SURE-CNN framework, with $\tilde{\mathcal{Y}}_n^{(k)}$ as the input to generate the denoised output. The network follows a 19-layer U-Net architecture, as shown in Fig. 6.2. The design incorporates symmetric skip connections, stride convolutions, and max-pooling layers, which help efficiently capture multi-scale features and expand the receptive field. Additionally, each convolutional layer uses a $3 \times 3$ filter, after each convolutional layer, the ReLU (Rectified Linear Unit) activation function is employed, continuing through to the last layer.

## 6.4 Field Examples of SURE-TCNN

This section highlights the performance of the proposed SURE-TCNN approach by conducting detailed quantitative and qualitative assessments on both simulated and publicly available real-world datasets. A summary of the experimental configuration is presented, covering aspects such as the datasets, evaluation criteria, comparison algorithms, and network settings, with more detailed information found in Sects. 6.4.1.1, 6.4.1.2, 8.4.1.2, and 6.4.1.4. To comprehensively evaluate the denois-

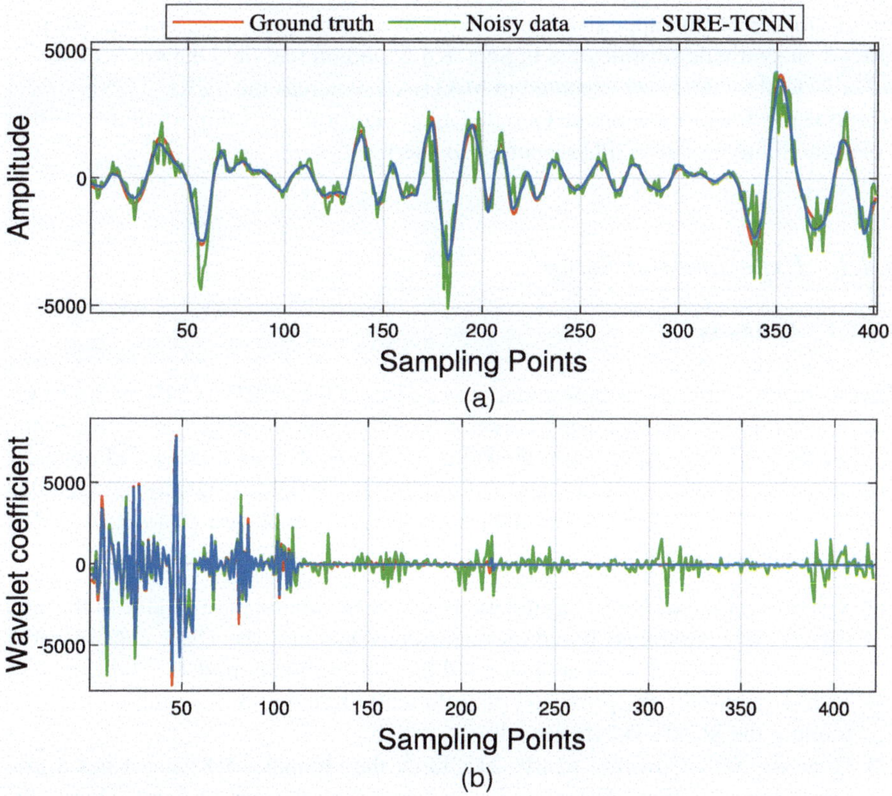


**Fig. 6.1** Single-trace analysis in the synthetic 3D physical model: **a** A comparison of the original data, noisy input, and the outcomes achieved using the SURE-TCNN method. **b** A comparison of the wavelet coefficients corresponding to (**a**)

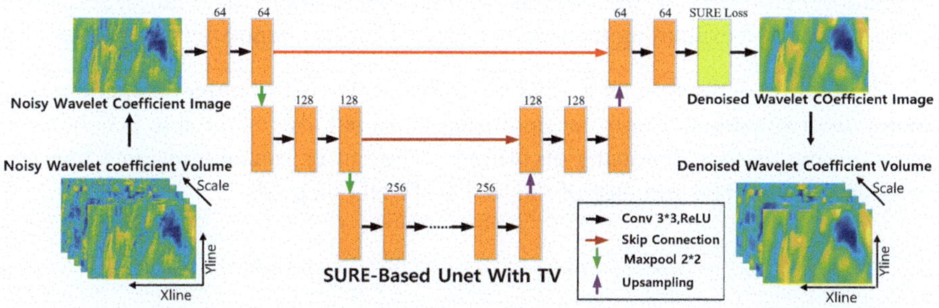


**Fig. 6.2** Schematic of the SURE-TCNN model for 3-D seismic noise attenuation in the absence of ground-truth data, where the weight parameters of the SURE-TCNN are derived by applying a matrix-based SURE-CNN to each frontal slice within the time-frequency domain, such as the wavelet domain

ing capability of this approach, both visual and quantitative assessments are performed on synthetic seismic data in Sect. 8.4.2, and on real field data in Sect. 8.4.3. All advanced methods are executed in MATLAB, whereas the SURE-TCNN model is built using TensorFlow on the Google Colab platform[1]. It is important to note that Colab randomly assigns GPU resources to users[2].

## 6.4.1 Experimental Setup

### 6.4.1.1 Datasets

This section explores three representative use cases of the SURE-TCNN method. One use case involves simulated datasets, while the other two are dedicated to processing real-world field data. Apart from the synthetic physical model dataset, all datasets utilized in this research are sourced from publicly available repositories. Detailed descriptions of these datasets and their corresponding applications are provided in the subsequent sections.

(1) *Synthetic physical model:* Synthetic physical model data are produced through simulation experiments based on various physical models. These data are more complex compared to those generated by convolutional models. The choice of such data ensures the provision of sufficiently rich and diverse information for training the SURE-TCNN model.
(2) *Opunake-3D*[3]: Covering nearly 215, $km^2$, the Opunake-3D dataset has a grid resolution of $6.25 \times 25.0$, m and a time sampling interval of 4, ms. Due to the presence of random seismic noise, it is frequently used to assess the reliability and performance of seismic attribute extraction methods.
(3) *F3 Netherlands*[4]: The F3 field data exhibit significantly higher levels of noise pollution compared to the Opunake-3D dataset, making them an ideal dataset for testing and evaluating the performance of various denoising algorithms.

Overall, these datasets span a range of complexities from simple to challenging, making them suitable not only for denoising tasks but also as reliable benchmarks for comparing various denoising algorithms. They offer valuable insights into the characteristics and performance of the SURE-TCNN method.

[1] Software available at https://www.tensorflow.org/.

[2] Colab available at https://colab.research.google.com/notebooks/gpu.ipynb.

[3] Data available at https://wiki.seg.org/wiki/Opunake-3D.

[4] Data available at https://terranubis.com/datainfo/F3-Demo-2020.

#### 6.4.1.2 Performance Metrics

Prior to performing the denoising tests, we begin by presenting two image quality metrics employed to assess the effectiveness of the denoising procedure. The first metric is the signal-to-noise ratio (SNR), measured in decibels (dB), which is commonly applied in synthetic experiments, especially when reference true images are provided. This metric quantifies the intensity ratio between the signal and noise and is defined as:

$$\mathrm{SNR} = -20\log_{10}\left(\frac{\|\mathcal{X} - \widehat{\mathcal{X}}\|}{\|\mathcal{X}\|}\right) \tag{6.23}$$

Where $\mathcal{X}$ refers to the original reference images, while $\widehat{\mathcal{X}}$ represents the denoised images obtained from the noisy input. In general, a higher signal-to-noise ratio (SNR) reflects better denoising performance.

In addition to the signal-to-noise ratio (SNR), local similarity is another metric used to evaluate the denoising performance. This metric evaluates the correlation between the denoised outcomes and the eliminated noise, as outlined in [46]. Local similarity helps determine whether useful signal information is preserved during noise removal [47, 48]. The local similarity ranges from 0 to 1, with lower values indicating better denoising performance. Further details can be found in Appendix 6.5.

#### 6.4.1.3 Baseline Algorithms

Building on the overview of 3-D seismic image denoising techniques in Sect. 8.1, we performed a comprehensive assessment and chose three representative baseline methods that have been extensively validated: multichannel singular spectrum analysis (MSSA), sequential generalized K-means (SGK), and tensor optimal shrinkage of singular value decomposition (t-OSSVD). These methods have demonstrated excellent performance in denoising and have been proven to be effective.

(1) *MSSA* [49][5]: Seismic datasets are usually structured into Hankel block matrices according to temporal and frequency characteristics. However, the presence of random noise often leads to an increase in the matrix rank, introducing unwanted interference. To address this issue, the MSSA method has been proposed as an effective rank reduction technique, aimed at removing noise from the raw data and recovering more accurate seismic images.
(2) *SGK* [13][6]: SGK is an innovative dictionary learning method specifically designed for denoising multidimensional seismic data. Unlike traditional K-SVD methods, SGK uses a sequential generalized K-means approach to update dictionary atoms by computing the arithmetic mean of the input signals, without

[5] Code available here: https://github.com/chenyk1990/reproducible_research/tree/master/usarray.

[6] Code available here: https://github.com/chenyk1990/reproducible_research/tree/master/sgk.

relying on the complex Singular Value Decomposition (SVD) process, thereby achieving more efficient noise suppression.

(3) *t-OSSVD* [17]: t-OSSVD This method extracts low-rank components from a tensor by applying singular value thresholding in the Fourier domain. By suppressing the corresponding singular vectors, it effectively mitigates noise and accurately recovers the low-rank structure of the data.

To ensure a fair comparison, each reference method underwent rigorous parameter optimization [13, 17, 49]. For instance, to achieve optimal denoising performance for MSSA and t-OSSVD (tensor optimal singular value thresholding), we incrementally adjusted the tensor rank starting from 2, increasing by steps of 2 up to 25, in order to determine the best rank for each method.

#### 6.4.1.4 Network Parameters

The SURE-TCNN significantly improves the denoising performance of 3D noisy data by incorporating a tensor network that does not require ground truth. The method optimizes two categories of parameters: model parameters and training parameters. For model parameters, the total variation (TV) tradeoff parameter in Eq. (8.8) is adjusted according to the data type: $\lambda_{TV}$ is set to $1 \times 10^5$ for the synthetic physical model, $1 \times 10^3$ for the Opunake-3D data, and $1 \times 10^4$ for F3 data. Additionally, the noise variance $\sigma$ in Eq. (6.17) is calculated as mean$(\widehat{\sigma}i)$, where mean$(\cdot)$ represents the mean of $\widehat{\sigma}_i$, as defined in Eq. (6.17). In terms of training, the SURE-U-Net model, incorporating a TV regularization term, is trained using the Adam optimizer [38]. The initialization of the SURE-TCNN model follows the method outlined in [50]. The best performance across all experiments was obtained using a minibatch size of 32 for 40 epochs, with an initial learning rate of $10^{-3}$ for the first 20 epochs, followed by a reduction to $10^{-4}$ for the remaining epochs.

### *6.4.2 Validation on Synthetic Seismic Data*

Initially, we applied the SURE-TCNN method to a simulated physical model dataset, which includes $785 \times 790$ seismic traces, each containing 400 time samples. To increase the complexity of the experiment, random noise was deliberately introduced into the data without providing ground truth, resulting in a noisy seismic image that serves as the input for SURE-TCNN, as shown in Fig. 6.5a, d. It is essential to emphasize that SURE-TCNN operates without direct access to the original seismic image. its primary function is to compute the signal-to-noise ratio (SNR). Nevertheless, we expect that SURE-TCNN will effectively recover the original noise-free image from the noisy 3D input. To thoroughly evaluate the effectiveness of SURE-TCNN in denoising, we devised three sub-experiments within a structured synthetic field data test.

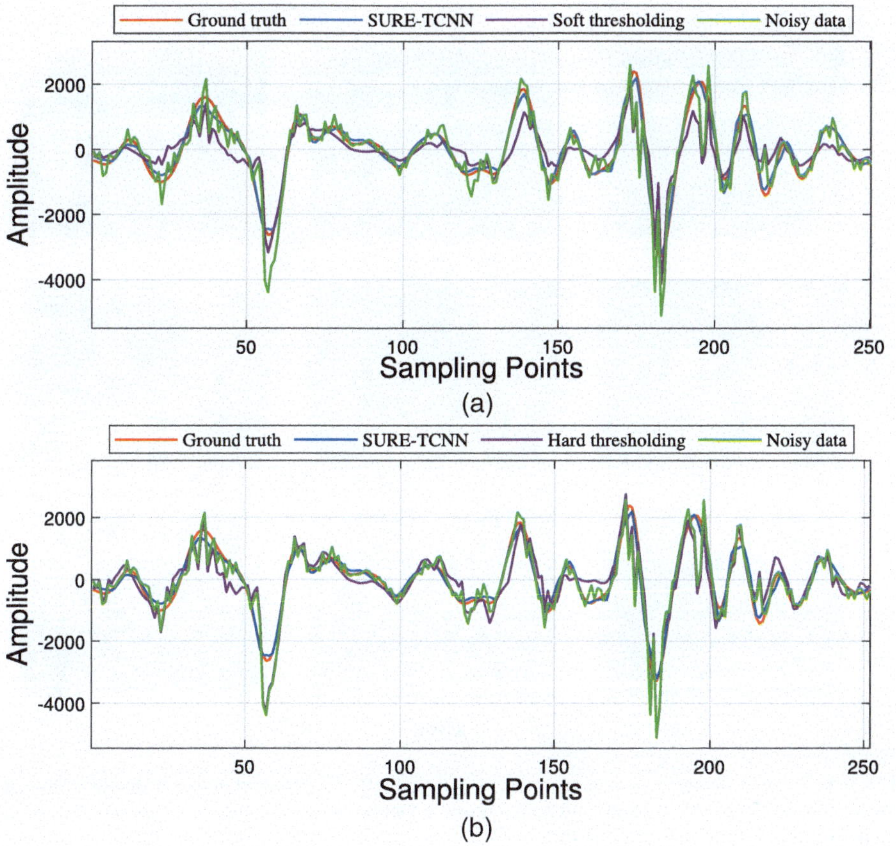


**Fig. 6.3** Single-trace comparison in the synthetic 3D physical model. **a** Comparison of the ground truth, noisy data, soft thresholding result, and SURE-TCNN denoising result. **b** Comparison of the ground truth, noisy data, hard thresholding result, and SURE-TCNN denoising result

The SURE-TCNN method presented is a deep learning-based, data-driven approach for wavelet thresholding, designed for multi-trace seismic signal denoising. To assess its effectiveness in denoising, we compared it with traditional wavelet thresholding methods. This comparison was implemented using the following MATLAB syntax: $\mathrm{XD} = \mathrm{wden}(W,' \mathrm{modwtsqtwolog}', \mathrm{SORH},' \mathrm{mln}', N, \mathrm{wname})$. Here, the function $\mathrm{wden}(\cdot)$[7] carries out an $N$-level wavelet transform of $W$, utilizing the chosen wavelet wname to calculate the wavelet features. The parameter SORH specifies the thresholding strategy, which can be implemented using either hard or soft thresholding methods. For this and all subsequent experiments, we set the parameters to $N = 3$ and wname = 'sym4'. Figure 6.3 presents a comparison of the denoising results obtained from the SURE-TCNN method and various wavelet thresholding

[7] https://ww2.mathworks.cn/help/wavelet/ref/wden.html.

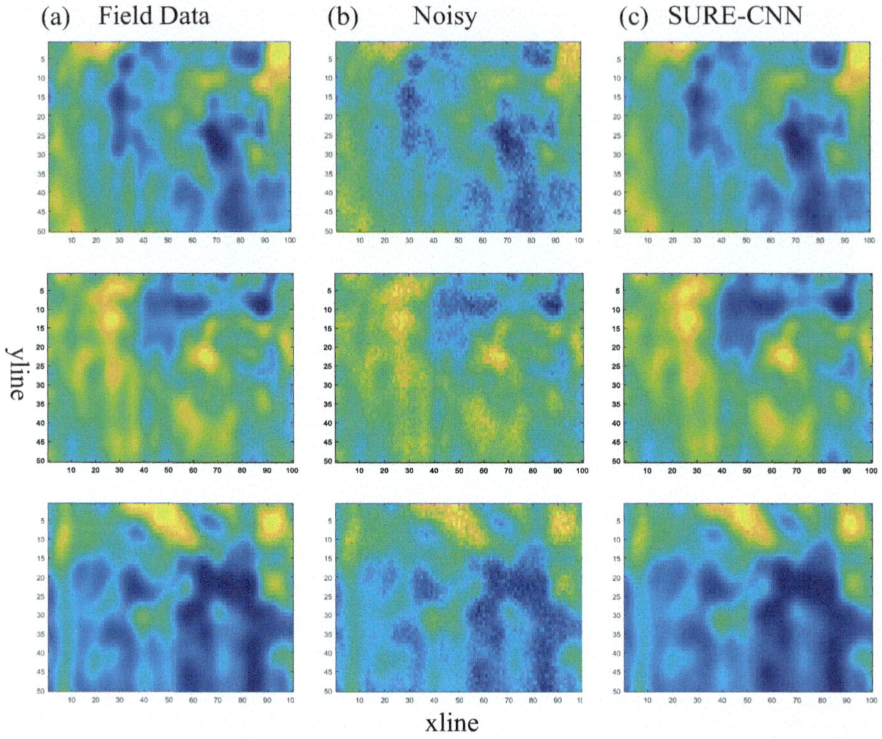


**Fig. 6.4** The denoising performance on slices of a synthetic 3D physical model is demonstrated. **a** Original noise-free data. **b** Data with added noise. **c** Data processed using the proposed denoising method combining SURE and TCNN

techniques for the same noisy signal. The comparison demonstrates that traditional wavelet thresholding methods produce denoised results that significantly deviate from the true waveform, while the SURE-TCNN method achieves results that closely align with the true waveform.

This section evaluates the performance of the SURE-CNN method in denoising spectral images. The denoising results of various spectral images processed by SURE-CNN are shown in Fig. 7.9, where each row corresponds to a different spectral image. Columns (a) and (b) display the noise-free true spectral images and their corresponding noisy versions, respectively, while column (c) presents the denoised spectral images. The results demonstrate that the SURE-CNN method successfully restores spectral details that are very close to the original true images, validating the effectiveness and superior performance of the proposed SURE-TCNN method in noise reduction (Fig. 6.4).

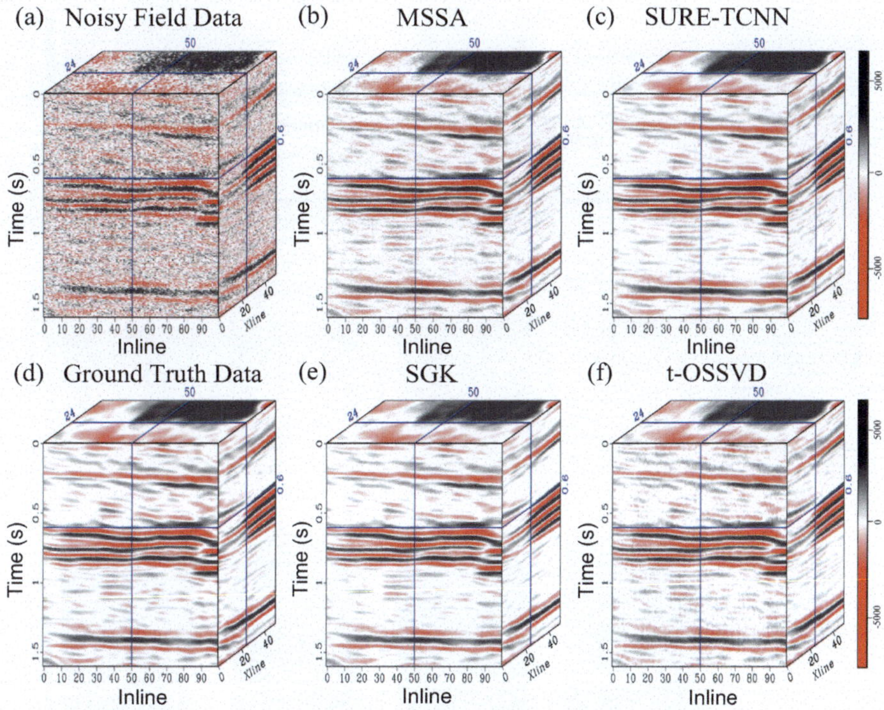


**Fig. 6.5** Evaluation of denoising effectiveness on synthetic physical model data: **a** The original noisy data, with an SNR of 1.75 dB, along with the denoised results achieved using the following methods: **b** MSSA, achieving an SNR of 13.96 dB; **c** the proposed SURE-TCNN method, resulting in an SNR of 17.32 dB; **e** SGK method with an SNR of 16.89 dB; **f** t-OSSVD method with an SNR of 16.32 dB. **d** The ground truth data without noise

Figures 6.5 and 6.6 present the denoising results of a 3D synthetic dataset processed at a noise level of SNR $= 6.02$dB using various methods. In Fig. 6.5, the noisy data volumes processed by SURE-TCNN and other advanced methods are visualized using the Madagascar software package[8]. In terms of signal-to-noise ratio (SNR), the SGK, t-OSSVD, and MSSA methods achieved SNR values of 16.89, 16.32, and 13.96dB, respectively, while the SURE-TCNN model yielded a higher SNR of 17.32dB, as shown in Fig. 6.5. Notably, the images generated by SURE-TCNN closely match the true spectral images, while the results from other advanced methods exhibit significant differences from the reference, as shown in Figs. 6.5 and 6.6. Furthermore, Fig. 6.6k–n display the local similarity maps comparing the denoised images with the extracted noise for both SURE-TCNN and the other methods. In contrast to leading methods, the local similarity map for SURE-TCNN reveals a smaller high-similarity region (blue area), clearly demonstrating that this approach is highly effective at preserving the key signal while efficiently eliminating noise.

[8] https://reproducibility.org/wiki/Main_Page.

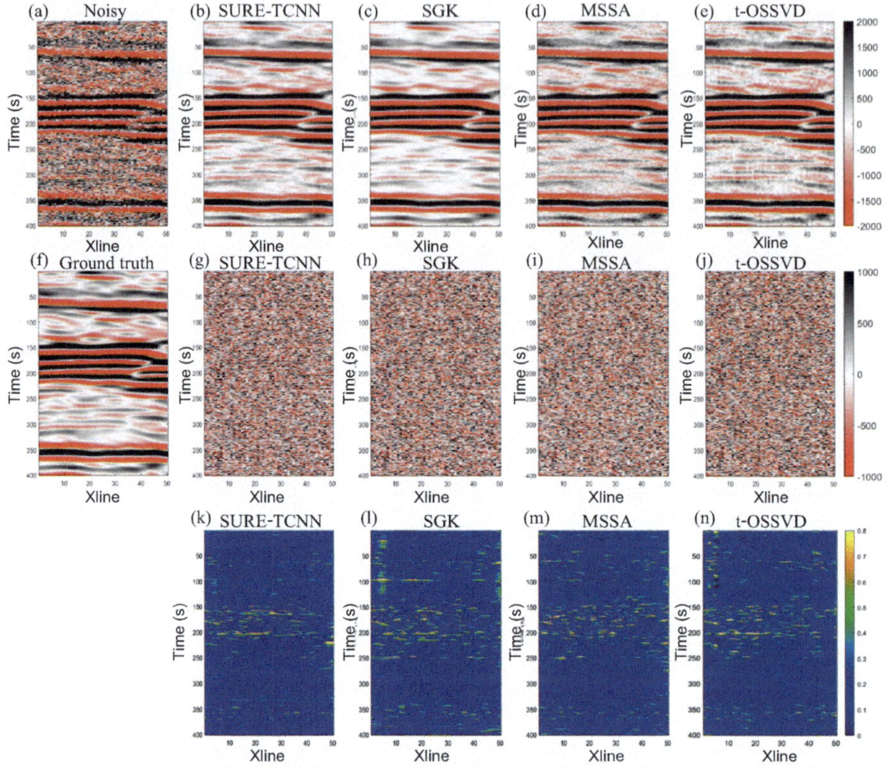


**Fig. 6.6** Denoising comparison for the initial inline section of the synthetic physical model data: **a** Original noisy data, with denoised outputs from **b** the proposed SURE-TCNN method, **c** SGK, **d** MSSA, and **e** t-OSSVD. **f** Ground truth data. **g–j** The noise removed by the methods in (**b**)–(**e**), respectively. **k–n** The local similarity maps calculated between the denoised results from (**b**)–(**e**) and the corresponding removed noise in (**g**)–(**j**)

## 6.4.3 Validation on Real Field Seismic Data

To evaluate the real-world performance of the method, we tested the SURE-TCNN model on two actual seismic datasets: the F3 Netherlands dataset and the Opunake-3D dataset. As described in Sect. 6.4.1.1, the Opunake-3D dataset (shown in Fig. 6.9a is mildly affected by noise, while the F3 dataset (shown in Fig. 6.9a is heavily contaminated with noise. This experimental setup effectively demonstrates the versatility of the SURE-TCNN model in handling varying noise levels and data characteristics. Since both the F3 and Opunake-3D datasets lack reliable ground truth data, the evaluation primarily relies on visual comparisons, focusing on the recovery results, residuals, and local similarity metrics. This contrasts with synthetic data experiments, where errors are quantified using signal-to-noise ratio (SNR) values.

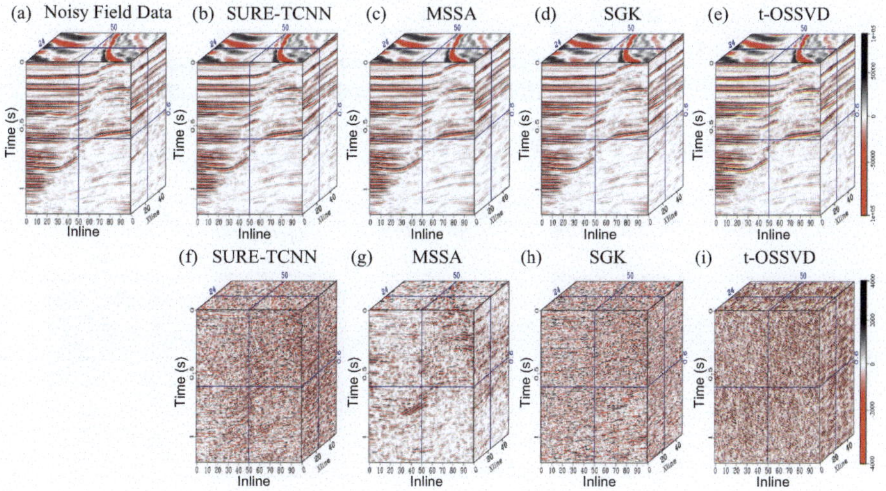


**Fig. 6.7** Denoising performance comparison for the Opunake-3D seismic dataset: **a** Noisy raw data, along with denoised results obtained using **b** MSSA, **c** the proposed SURE-TCNN, **e** SGK, and **f** t-OSSVD methods. **f–i** Display the noise components extracted by methods (**b**)–(**e**)

In the Opunake-3D test, we initially partitioned the noisy dataset into 3D cubes of size $50 \times 50 \times 450$. Next, we applied a 1D Mallat wavelet transform to each cube to create $50 \times 50$ spectral images, which served as the input data for the SURE-TCNN model.It is important to note that only spectral data points within the angular frequency range from 0 to $\pi/2$ in the wavelet domain were utilized. Figure 6.7a presents the original noisy seismic data, while Figs. 6.7b–i show the denoising results and corresponding noise removal outputs obtained with the SURE-TCNN approach and several advanced state-of-the-art (SOTA) methods.The results show that the MSSA method fails to retain important signal features in the denoising procedure, whereas the other methods typically maintain the shape and structure of the waveform. In contrast to synthetic data experiments, this study relies on visual comparison and qualitative analysis. Figure 6.8b–m present the recovery results, residuals, and local similarity comparisons for the various methods. From Fig. 6.8a–e, it can be seen that the residual noise left by the t-OSSVD method interferes with the signal measurements. Additionally, the local similarity analysis in Fig. 6.8j, k shows that, compared to the SGK method, the SURE-TCNN method exhibits fewer high-similarity regions, indicating that SURE-TCNN retains more meaningful signal information during the noise removal process. In summary, the results in Fig. 6.8j, k clearly demonstrate that the SURE-TCNN method effectively removes random noise while preserving key signal content, showing significant improvements over existing techniques.

In this study, we applied the SURE-TCNN method to the F3 Netherlands 3D field dataset, which is heavily contaminated with noise, following the same experimental setup as the Opunake-3D experiment. Figure 6.9 illustrates the outcomes achieved

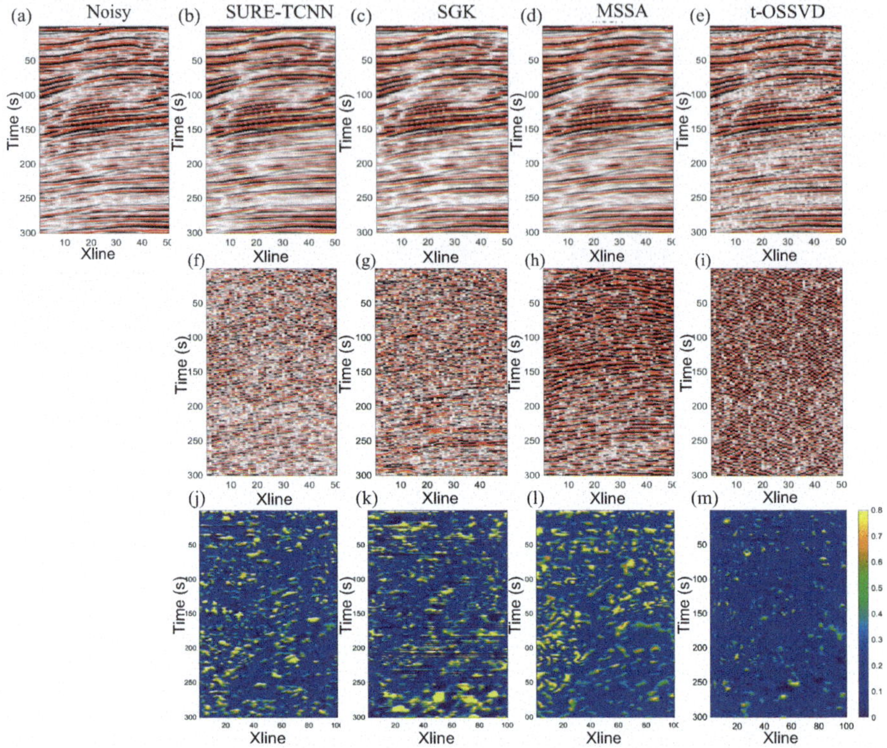


**Fig. 6.8** Denoising performance comparison for the initial inline section of the Opunake-3D seismic dataset: **a** Noisy data and denoised results obtained using **b** the proposed SURE-TCNN method, **c** SGK, **d** MSSA, and **e** t-OSSVD. **f–i** Show the noise removed corresponding to methods (**b**)–(**e**). **j–m** Display the local similarity maps obtained from the denoised data in (**a**)–(**e**) and the noise components shown in (**f**)–(**j**)

with the SURE-TCNN approach and other advanced state-of-the-art (SOTA) methods, including the cleaned images and their corresponding noise removal. The findings highlight that the SURE-TCNN method effectively reduces noise while maintaining essential signal features, demonstrating its superior noise suppression performance. To further compare the performance of SURE-TCNN with other methods, we zoomed in on a portion of the dataset and calculated the corresponding local similarity, as shown in Fig. 6.10. From this analysis, we observed that the MSSA method exhibited a significant deviation from the true signal, while the t-OSSVD method still retained considerable residual noise. The local similarity analysis in Fig. 6.10j, k reveals that the SGK method shows higher similarity in more regions compared to SURE-TCNN, indicating that SGK discarded more useful signal information during noise removal. Overall, as shown in Figs. 6.9 and 6.10, the SURE-TCNN method not only effectively removes noise but also retains more valuable signal details, outperforming existing denoising techniques.

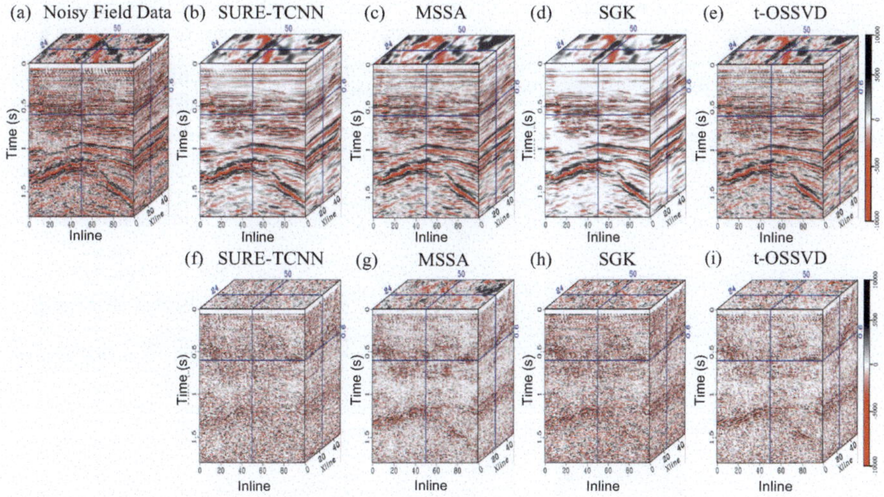


**Fig. 6.9** A comparison of denoising results using the F3 Netherlands seismic data. Panel **a** shows the noisy data, along with the denoised outputs achieved by **b** MSSA, **c** the proposed SURE-TCNN method, **e** SGK, and **f** t-OSSVD. Panels **f–i** display the noise removed corresponding to methods (**b**)–(**e**)

### *6.4.4 Discussion*

The SURE-TCNN method has demonstrated significant advantages in noise suppression across various datasets, including synthetic physical models, Opunake-3D, and F3 Netherlands, highlighting the superior performance of deep learning-based denoising techniques compared to traditional methods [1–4]. As a representative of unsupervised deep learning models, SURE-TCNN generally requires a longer training time when processing large-scale seismic data, particularly during the training of the CNN model. Based on the results in Table 6.1, while SURE-TCNN occasionally demonstrates faster computation times compared to SGK, it generally takes longer than MSSA and t-OSSVD, which is primarily related to the volume of training data. It should be noted that this comparison may not be completely accurate due to variations in hardware and platforms, but it offers an approximate evaluation of the computational efficiency of SURE-TCNN.

In real-world applications involving extensive seismic data processing, the SURE-TCNN method faces a trade-off between denoising performance and computational efficiency. For unsupervised learning models, although increasing the number of training samples may improve performance in some cases, once the model has reached convergence, adding more samples typically does not yield significant improvements. Instead, it often increases the computational load and training time. To address this issue, a more efficient approach is to first train the model on a small subset of noisy data, and then quickly apply the trained model to the remaining data

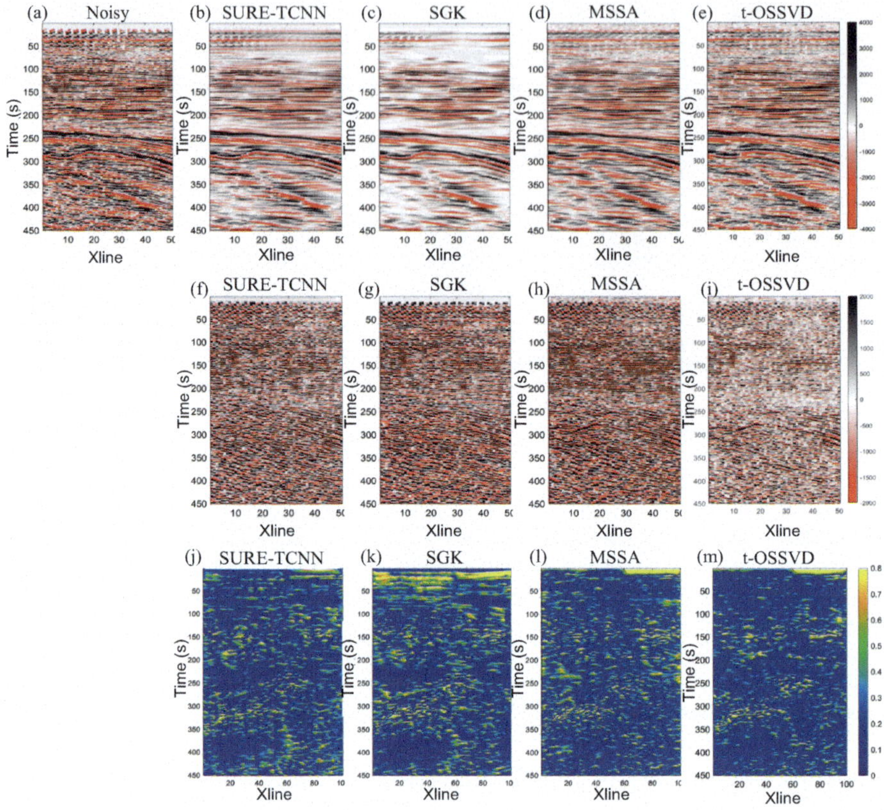


**Fig. 6.10** Comparison of denoising performance for the first inline section of the F3 Netherlands seismic dataset. **a** Noisy data and denoised results obtained using the following methods: **b** the proposed SURE-TCNN method, **c** SGK, **d** MSSA, **e** t-OSSVD. **f**–**i** Noise removal results corresponding to methods (**b**)–(**e**). **j**–**m** Local similarity maps obtained from the denoised data in (**a**)–(**e**) and the noise components removed in (**f**)–(**j**)

for denoising. This method not only enhances computational efficiency but also maintains high denoising performance, avoiding the negative effects of excessive training samples. Additionally, because the training and testing data are usually drawn from the same source (e.g., originating from the same seismic survey), this approach has a negligible effect on the model's overall noise reduction effectiveness.

The SURE-TCNN method introduces an innovative deep learning-driven global thresholding strategy that can be applied to all wavelet coefficients, effectively avoiding the excessive smoothing issues commonly encountered with traditional hard and soft thresholding techniques. By leveraging the structure of t-NN and its t-product operation in the transform domain, SURE-TCNN is not confined to wavelet transforms and can be easily adapted to other forms of time-frequency analysis. In theory, a more sparse time-frequency transform leads to more efficient noise reduction.

**Table 6.1** Comparing the computational time differences between the proposed method and traditional baseline approaches

| Dataset | MSSA (s) | SGK (s) | t-OSSVD (s) | SURE-TCNN (s) |
|---|---|---|---|---|
| Synthetic | 267.23 | 3927.24 | 22.53 | 425.95 |
| Opunake | 330.47 | 2010.36 | 70.92 | 2652.12 |
| F3 | 143.54 | 4817.27 | 32.58 | 1393.05 |

Nevertheless, due to the numerous time-frequency transform methods available in the image denoising domain, performing an exhaustive comparison of these techniques is not addressed in this work.

## 6.5 Summary

This chapter introduces a novel denoising method based on a TCNN deep learning framework with the SURE loss function, designed to denoise 3D seismic data without relying on ground truth. The method directly maps noisy seismic images to their denoised counterparts, avoiding the need for prior knowledge typically required in traditional approaches. We first reformulate the denoising task as an unsupervised regression problem and guide the model using the SURE loss function instead of the conventional mean squared error (MSE) loss. Next, within the t-NN framework, we propose an innovative SURE-driven TCNN architecture by incorporating a t-product based on transform to extend the standard t-NN structure. By leveraging the characteristics of the transform-based t-product, we break down the SURE-TCNN into matrix-level SURE-CNNs, each representing a slice along the frontal axis in the time-frequency space (e.g., wavelet domain). This decomposition significantly reduces the computational complexity, making the problem more tractable. Ultimately, a series of thorough experiments carried out on a synthetic physical model and two real seismic datasets show that the proposed method surpasses three advanced denoising techniques.

Prospective research could focus on enhancing the current SURE-TCNN model to address not only random noise but also more complex noise types, such as irregular and correlated noise. This could be achieved by further exploring the potential of the t-NN framework and leveraging its t-product based on transform mechanism. Additionally, future efforts may involve incorporating deeper convolutional neural network (CNN) architectures, such as Google's Inception Net or residual networks, to enhance the denoising capability of SURE-TCNN, thereby improving the model's performance and robustness in handling complex noise environments.

# Appendix

### Local Similarity

To provide a comprehensive evaluation of the denoising performance, this section reviews the local similarity metric introduced in Sect. 6.4.1.2, which is applied in Sect. 6.4 to evaluate the performance of SURE-TCNN in comparison with state-of-the-art (SOTA) methods. Local similarity is an effective technique for quantifying the similarity between two signals at each local region, capturing fine-grained differences between the signals. This metric was initially applied to reduce random noise in [46]. Given two signal vectors, $\boldsymbol{x}_1$ and $\boldsymbol{x}_2$, derived from reshaping a 3D tensor, the local similarity is characterized as described in [51].

$$s = \sqrt{s_1 \circ s_2}, \tag{6.24}$$

Here, $\boldsymbol{s}$ represents the calculated local similarity, where $\circ$ indicates the element-wise multiplication (Hadamard product), and $s_1$ and $s_2$ are the solutions derived from solving two linear least-squares inverse problems.

$$\begin{aligned} s_1 &= \arg\min_{\tilde{s}_1} \|\boldsymbol{x}_1 - X_2\tilde{s}_1\|_2^2, \\ s_2 &= \arg\min_{\tilde{s}_2} \|\boldsymbol{x}_2 - X_1\tilde{s}_2\|_2^2, \end{aligned} \tag{6.25}$$

The weighting matrices $X_1$ and $X_2$ are diagonal matrices, with the diagonal elements corresponding to the components of $\boldsymbol{x}_1$ and $\boldsymbol{x}_2$, respectively. Equation (6.25) is optimized by addressing a least-squares problem that includes regularization for shaping and a smoothness penalty, represented as $\mathcal{T}$.

$$\begin{aligned} s_1 &= \left[\lambda_1^2 I + \mathcal{T}\left(X_2^\top X_2 - \lambda_1^2 I\right)\right]^{-1} \mathcal{T} X_2^T \boldsymbol{x}_1, \\ s_2 &= \left[\lambda_2^2 I + \mathcal{T}\left(X_1^\top X_1 - \lambda_2^2 I\right)\right]^{-1} \mathcal{T} X_1^T \boldsymbol{x}_2, \end{aligned} \tag{6.26}$$

Here, $\lambda_1$ and $\lambda_2$ are two trade-off parameters that are dynamically adjusted during each iteration. They are computed as $\lambda_1 = \left\|X_2^\top X_2\right\|_2$ and $\lambda 2 = \left\|X_1^\top X_1\right\|_2$ [51].

### Tensor Notations and Transform-Based t-product

This section introduces some basic notations and definitions, which will provide the foundation for introducing new concepts and building the model. The notations follow the conventions used in [52, 53]. Scalars are represented by standard lowercase letters (e.g., $t$), vectors by bold lowercase letters (e.g., $\boldsymbol{t}$), matrices by bold uppercase letters (e.g., $\mathbf{T}$), and tensors by bold Euler script letters (e.g., $\mathcal{T}$). For third-order seismic data tensors with real values $\mathcal{T} \in \mathbb{R}^{n_1 \times n_2 \times n_3}$, the $(i, j, k)$-th entry is denoted as $\mathcal{T}_{ijk}$. MATLAB-style notations such as $\mathcal{T}(i, :, :)$, $\mathcal{T}(:, j, :)$, and $\mathcal{T}(:, :, k)$ are used to refer to the $i$-th horizontal slice, $j$-th lateral slice, and $k$-th frontal slice. For convenience, the frontal slice $\mathcal{T}(:, :, k)$ is often written as $\mathcal{T}^{(k)}$.

In tensor processing, the conventional t-product is often interpreted as a convolution-like procedure, which can be computed efficiently through the use of the Fast Fourier Transform (FFT). Nevertheless, the approaches in [54] and [52] enhance the t-product by introducing an invertible linear transformation $L$.This study aims to propose a denoising approach based on the t-product, incorporating an invertible linear transform, and does so without requiring ground truth. Specifically, given a linear transform $L : \mathbb{R}^{n_1 \times n_2 \times n_3} \rightarrow \mathbb{R}^{n_1 \times n_2 \times n_3}$, we obtain the transformed tensor $\widetilde{\mathcal{T}}$ by applying the linear transformation $L$ to each tube fiber $\mathcal{T}(i, j, :)$. Alternatively, we can obtain:

$$\widetilde{\mathcal{T}} = L(\mathcal{T}) = \mathcal{T} \times_3 L, \tag{6.27}$$

Here, $\times_3$ denotes The mode-3 (matrix) product involves multiplying a tensor $\mathcal{T}$ by an arbitrary invertible matrix $L \in \mathbb{R}^{n_3 \times n_3}$.The corresponding inverse transformation can be expressed as:

$$L^{-1}(\mathcal{T}) = \mathcal{T} \times_3 L^{-1}. \tag{6.28}$$

Drawing from the conventional formulation of the $t$-product, the $t$-product defined using the $L$ transform is given as follows.

**Definition 6.3** (*$L$ transform-based t-product*) [52] Consider an invertible linear transformation $L$, along with two third-order tensors $\mathcal{T}_1 \in \mathbb{R}^{n_1 \times n_2 \times n_3}$ and $\mathcal{T}_2 \in \mathbb{R}^{n_2 \times l \times n_3}$, the t-product based on transform $L$ is defined as $C = \mathcal{T}_1 *_L \mathcal{T}_2$, where $\mathcal{T}$ represents the tensor obtained by applying the transformation $L$ to $\mathcal{T}_1$ and $\mathcal{T}_2$. Specifically, $L(\mathcal{T}) = L(\mathcal{T}_1) \odot L(\mathcal{T}_2)$. Here, $\odot$ represents the face-wise product of two third-order tensors, which is defined as follows.

**Definition 6.4** (*Facewise product*) [54] Given two third-order tensors $\mathcal{T}_1$ and $\mathcal{T}_2$, their face-wise product $\mathcal{T}_1 \odot \mathcal{T}_2$ is defined by performing element-wise operations on the corresponding elements of the two tensors.

$$(\mathcal{T}_1 \odot \mathcal{T}_2)^{(k)} = \mathcal{T}_1^{(k)} \mathcal{T}_2^{(k)}, \quad k \in [n]. \tag{6.29}$$

Therefore, $L(\mathcal{T}) = L(\mathcal{T}_1) \odot L(\mathcal{T}_2)$ can be viewed as performing matrix multiplication on the slices of the tensors in the transformed domain, analogous to the form $\widetilde{\mathcal{T}}^{(k)} = \widetilde{\mathcal{T}}_1^{(k)} \widetilde{\mathcal{T}}_2^{(k)}$. This suggests that the $L$-transform-based t-product effectively translates the matrix multiplication between two frontal slices into the transformed domain. Building on this characteristic, the step-by-step procedure for computing the t-product is outlined in Algorithm 6.2.

**Definition 6.5** (*Tensor transpose*) [52] Consider an invertible linear transform $L$ and a third-order tensor $\mathcal{T} \in \mathbb{R}^{n_1 \times n_2 \times n_3}$, the tensor transpose is characterized as $\mathcal{T}^\dagger$, is defined such that for each $k \in [n_3]$, the following relation holds: $L\left(\mathcal{T}^\dagger\right)^{(k)} = \left(L(\mathcal{T})^{(k)}\right)^\dagger$.

**Algorithm 6.2:** $L$ transform-based t-product [52]

**Input**: $\mathcal{T}_1 \in \mathbb{R}^{n_1 \times n_2 \times n_3}$, $\mathcal{T}_2 \in \mathbb{R}^{n_2 \times l \times n_3}$, and $L : \mathbb{R}^{n_1 \times n_2 \times n_3} \rightarrow \mathbb{R}^{n_1 \times n_2 \times n_3}$.
**Output**: $\mathcal{T} = \mathcal{T}_1 *_L \mathcal{T}_2 \in \mathbb{R}^{n_1 \times l \times n_3}$.
$\widetilde{\mathcal{T}}_1 = L(\mathcal{T}_1), \widetilde{\mathcal{T}}_1 = L(\mathcal{T}_2)$.
**for** $k = 1, \cdots, n_3$ **do**
  $\widetilde{\mathcal{T}}^{(k)} = \widetilde{\mathcal{T}}_1^{(k)} \widetilde{\mathcal{T}}_2^{(k)}$.
**end**
$\mathcal{T} = L^{-1}(\widetilde{\mathcal{T}})$.

**SURE**

Inspired by [39], we replace the traditional MSE loss function with the SURE in a tensor-based deep learning framework to achieve 3D seismic data denoising without requiring ground truth data. To ensure the systematic and comprehensive nature of the method, we briefly review the basic concept of SURE [39]. Let $\mathbf{y}_i$ be a noisy vectorized image, and $\mathbf{x}_i$ its corresponding true noise-free version. We use the SURE-CNN [39] model to infer the denoised image $\widehat{\mathbf{x}}_i$ from $\mathbf{y}_i$, i.e., $\widehat{\mathbf{x}i} = \boldsymbol{h\theta}(\mathbf{y}i)$, where $\boldsymbol{h\theta}(\cdot)$ denotes the CNN-based denoising network. The SURE formulation for the denoising network is described in detail in [40].

$$\eta_{\text{sure}}(\boldsymbol{h}_\theta(\mathbf{y}_i)) = \frac{1}{N} \sum_{n=1}^{N} \{ \frac{1}{K} \|\mathbf{y}_i - \boldsymbol{h}_\theta(\mathbf{y}_i)\|_2^2 - \sigma_i^2 + \frac{2\sigma_i^2}{K} \operatorname{tr}\left( \frac{\partial \boldsymbol{h}_\theta(\mathbf{y}_i)}{\partial \mathbf{y}_i^\top} \right) \}, \quad (6.30)$$

In this context, $\top$ denotes the matrix transpose, and $\operatorname{tr}(\cdot)$ denotes the matrix trace. It is crucial to emphasize that SURE offers an unbiased approximation of the MSE, as demonstrated below:

$$\eta_{\text{mse}}(\boldsymbol{h}_\theta(\mathbf{y}_i)) = \frac{1}{KN} \sum_{i=1}^{N} \mathbb{E}\left[ \|\mathbf{x}_i - \boldsymbol{h}_\theta(\mathbf{y}_i)\|^2 \right], \quad (6.31)$$

Here, $\mathbb{E}[\cdot]$ represents the expected value considering the noise. To put it differently, the true MSE in Eq. (6.31) is equivalent to the expected value of the SURE term, eliminating the need for any ground truth data.

For any nonlinear estimator $\boldsymbol{h}_\theta(\cdot)$, the divergence component in Eq. (7.6) can be estimated through Monte Carlo simulations. Specifically, as discussed in [55], Stein's unbiased risk estimator (SURE) provides an effective way to estimate this term, circumventing the need for a direct computation of the complex divergence.

$$\operatorname{tr}\left( \frac{\partial \boldsymbol{h}_\theta(\mathbf{y}_i)}{\partial \mathbf{y}_i^T} \right) \approx \frac{1}{\epsilon} \tilde{\boldsymbol{n}}^\top (\boldsymbol{h}(y + \epsilon \tilde{\boldsymbol{n}}) - \boldsymbol{h}(y)), \quad (6.32)$$

In this context, $\tilde{\boldsymbol{n}}$ denotes a Gaussian random variable that is independent and identically distributed (i.i.d.) with a mean of zero and a variance of one, while $\epsilon$ is a small positive constant. Although SURE is typically based on the assumption of Gaussian noise, it is also capable of handling non-Gaussian noise [40]. In studies of seismic data denoising, it is commonly assumed that the noise follows a Gaussian distribution [56], a widely accepted assumption in many real-world applications.

## References


1. Jian M, Jinghuai G, Wenchao C (2006) On the denoising method of prestack seismic data in wavelet domain. In 2006 SEG annual meeting. OnePetro
2. Zhou Y, Yang J, Wang H, Huang G, Chen Y (2020) Statistics-guided dictionary learning for automatic coherent noise suppression. IEEE Trans Geosci Remote Sensing
3. Anvari R, Siahsar MAN, Gholtashi S, Kahoo AR, Mohammadi M (2017) Seismic random noise attenuation using synchrosqueezed wavelet transform and low-rank signal matrix approximation. IEEE Trans Geosci Remote Sens 55(11):6574–6581
4. Yangkang C, Yatong Z, Wei C, Zu S, Weilin H, Dong Z (2017) Empirical low-rank approximation for seismic noise attenuation. IEEE Trans Geosci Remote Sens 55(8):4696–4711
5. Hao S, Jianwei M, Huizhu Y (2009) Comparisons of wavelets, contourlets and curvelets in seismic denoising. J Appl Geophys 69(2):103–115
6. Li J-H, Zhang Y-J, Qi R, Liu QH (2017) Wavelet-based higher order correlative stacking for seismic data denoising in the curvelet domain. IEEE J Sel Top Appl Earth Obs Remote Sens 10(8):3810–3820
7. Andrzej G, Anna A, Michał M (2014) Application of curvelet denoising to 2d and 3d seismic data-practical considerations. J Appl Geophys 105:78–94
8. Dehui K, Zhenming P (2015) Seismic random noise attenuation using shearlet and total generalized variation. J Geophys Eng 12(6):1024–1035
9. Liu C, Wang D, Wang T, Feng F, Cheng H, Meng G (2014) Random seismic noise attenuation based on the shearlet transform. Acta Petrolei Sinica 35(4):692–699
10. Liu Y, Fomel S, Liu C (2015) Signal and noise separation in prestack seismic data using velocity-dependent seislet transform. Geophysics 80(6):WD117–WD128
11. Li J, Sacchi MD (2021) An LP-space matching pursuit algorithm and its application to robust seismic data denoising via time-domain radon transforms. Geophysics 86(3):V171–V183
12. Wei C, Jianyong X, Zu S, Shuwei G, Yangkang C (2016) Multiple-reflection noise attenuation using adaptive randomized-order empirical mode decomposition. IEEE Geosci Remote Sens Lett 14(1):18–22
13. Zhu L, Liu E, McClellan JH (2015) Seismic data denoising through multiscale and sparsity-promoting dictionary learning. Geophysics 80(6):WD45–WD57
14. Ouyang S, Lingling W, Hu X, Zhidan L (2021) Seismic denoising via truncated nuclear norm minimization. Geophysics 86(2):V153–V169
15. Chong W, Zhihui Z, Gu H, Wu X, Shuaiqi L (2018) Hankel low-rank approximation for seismic noise attenuation. IEEE Trans Geosci Remote Sens 57(1):561–573
16. Weilin H, Runqiu W, Yangkang C, Huijian L, Shuwei G (2016) Damped multichannel singular spectrum analysis for 3d random noise attenuation. Geophysics 81(4):V261–V270
17. Anvari R, Mohammadi M, Kahoo AR (2018) Enhancing 3-d seismic data using the t-SVD and optimal shrinkage of singular value. IEEE J Sel Top Appl Earth Obs Remote Sens 12(1):382–388
18. Feng J, Li X, Liu X, Chen C, Chen H (2021) Seismic data denoising based on tensor decomposition with total variation. IEEE Geosci Remote Sens Lett
19. Yu S, Jianwei M, Wenlong W (2019) Deep learning for denoising. Geophysics 84(6):V333–V350

20. Liuqing Y, Wei C, Hang W, Yangkang C (2021) Deep learning seismic random noise attenuation via improved residual convolutional neural network. IEEE Trans Geosci Remote Sens 59(9):7968–7981
21. Liu D, Wang W, Chen W, Wang X, Zhou Y, Shi Z (2018) Random noise suppression in seismic data: What can deep learning do? In: SEG technical program expanded abstracts 2018. Society of Exploration Geophysicists, pp 2016–2020
22. Dawei L, Wei W, Xiaokai W, Cheng W, Jiangyun P, Wenchao C (2019) Poststack seismic data denoising based on 3-d convolutional neural network. IEEE Trans Geosci Remote Sens 58(3):1598–1629
23. Yuan Y, Si X, Zheng Y (2020) Ground-roll attenuation using generative adversarial networksground-roll attenuation using gans. Geophysics 85(4):WA255–WA267
24. Sun J, Slang S, Elboth T, Greiner TL, McDonald S, Gelius L-J (2020) Attenuation of marine seismic interference noise employing a customized u-net. Geophys Prospect 68(3):845–871
25. Mi Z, Yang L, Yangkang C (2019) Unsupervised seismic random noise attenuation based on deep convolutional neural network. IEEE Access 7:179810–179822
26. Qiu C, Wu B, Liu N, Zhu X, Ren H (2021) Deep learning prior model for unsupervised seismic data random noise attenuation. IEEE Geosci Remote Sens Lett
27. Ristau JP, Moon WM (2001) Adaptive filtering of random noise in 2-d geophysical data. Geophysics 66(1):342–349
28. Yih J, Yi-Wei L, Chih-Sung C, Hsin-Yi C (2009) Adaptive filtering of random noise in near-surface seismic and ground-penetrating radar data. J Appl Geophys 68(1):36–46
29. Weilin H, Runqiu W (2018) Random noise attenuation by planar mathematical morphological filtering. Geophysics 83(1):V11–V25
30. Fehmers GC, Höcker CFW (2003) Fast structural interpretation with structure-oriented filtering. Geophysics 68(4):1286–1293
31. Kamel B, Jalal F, Tahar A, Noureddine D (2011) Seismic noise attenuation by means of an anisotropic non-linear diffusion filter. Comput Geosci 37(4):456–463
32. Sanyi Y, Shangxu W, Guofa L (2012) Random noise reduction using bayesian inversion. J Geophys Eng 9(1):60–68
33. Dehua W, Jinghuai G, Naihao L, Xiudi J (2020) Structure-oriented dtgv regularization for random noise attenuation in seismic data. IEEE Trans Geosci Remote Sens 59(2):1757–1771
34. David B, Mauricio S (2012) Denoising seismic data using the nonlocal means algorithm. Geophysics 77(1):A5–A8
35. Zhu W, Mousavi SM, Beroza GC (2019) Seismic signal denoising and decomposition using deep neural networks. IEEE Trans Geosci Remote Sens 57(11):9476–9488
36. Léon B et al (1998) Online learning and stochastic approximations. Online Learn Neural Netw 17(9):142
37. Nesterov Y (2008) A method of solving a convex programming problem with convergence rate o(1/k2). Sov Math Dokl 27
38. Kingma DP, Adam JB (2014) A method for stochastic optimization. arXiv preprint arXiv:1412.6980
39. Soltanayev S, Chun SY (2018) Training deep learning based denoisers without ground truth data. In Proceedings of the 32nd international conference on neural information processing systems, pp 3261–3271
40. Nguyen HV, Ulfarsson MO, Sveinsson JR (2020) Hyperspectral image denoising using sure-based unsupervised convolutional neural networks. IEEE Trans Geosci Remote Sens 59(4):3369–3382
41. Newman E, Horesh L, Avron H, Kilmer M (2018) Stable tensor neural networks for rapid deep learning. arXiv preprint arXiv:1811.06569
42. Horesh L, Newman E, Kilmer ME, Avron H (2020) Generating and managing deep tensor neural networks, 14 May 2020. US Patent App. 16/189, 744
43. Chang SG, Yu B, Vetterli M (2000) Adaptive wavelet thresholding for image denoising and compression. IEEE Trans Image Process 9(9):1532–1546

44. Feng Q, Zhangbo L, Yan W, Songjie L, Shengli P, Hu G (2021) Dtae: Deep tensor autoencoder for 3-d seismic data interpolation. IEEE Trans Geosci Remote Sens 60:1–19
45. Shensa MJ et al (1992) The discrete wavelet transform: wedding the a trous and mallat algorithms. IEEE Trans Signal Process 40(10):2464–2482
46. Chen Y, Fomel S (2015) Random noise attenuation using local signal-and-noise orthogonalization. Geophysics 80(6):WD1–WD9
47. Yangkang C, Dong Z, Zhaoyu J, Xiaohong C, Zu S, Weilin H, Shuwei G (2016) Simultaneous denoising and reconstruction of 5-d seismic data via damped rank-reduction method. Geophys J Int 206(3):1695–1717
48. Dong Z, Yatong Z, Hanming C, Wei C, Zu S, Yangkang C (2017) Hybrid rank-sparsity constraint model for simultaneous reconstruction and denoising of 3d seismic data. Geophysics 82(5):V351–V367
49. Yangkang C, Weilin H, Dong Z, Wei C (2016) An open-source matlab code package for improved rank-reduction 3d seismic data denoising and reconstruction. Comput Geosci 95:59–66
50. He K, Zhang X, Ren S, Sun J (2015) Delving deep into rectifiers: surpassing human-level performance on imagenet classification. In: Proceedings of the IEEE international conference on computer vision, pp 1026–1034
51. Sergey F (2007) Local seismic attributes. Geophysics 72(3):A29–A33
52. Lu C, Peng X, Wei Y (2019) Low-rank tensor completion with a new tensor nuclear norm induced by invertible linear transforms. In: Proceedings of the IEEE/CVF conference on computer vision and pattern recognition, pp 5996–6004
53. Lu C, Jiashi F, Yudong C, Wei L, Zhouchen L, Shuicheng Y (2019) Tensor robust principal component analysis with a new tensor nuclear norm. IEEE Trans Pattern Anal Mach Intell 42(4):925–938
54. Eric K, Misha K, Shuchin A (2015) Tensor-tensor products with invertible linear transforms. Linear Algebr Its Appl 485:545–570
55. Sathish R, Thierry B, Michael U (2008) Monte-carlo sure: a black-box optimization of regularization parameters for general denoising algorithms. IEEE Trans Image Process 17(9):1540–1554
56. Tie Z, Yue L, Wu N, Pengfei N, Baojun Y (2015) Statistical properties of the random noise in seismic data. J Appl Geophys 118:84–91

# Chapter 7
# Order-$p$ Tensor Deep Learning for Seismic Data Denoising

## 7.1 Introduction

From a machine learning standpoint, denoising seismic data is commonly framed as an ill-posed inverse problem, where the effectiveness of the solution largely hinges on the appropriate application of image priors or modeling techniques [1]. Seismic data denoising differs from traditional image denoising tasks in its added complexity, especially when handling multidimensional datasets like 3D poststack, 4D prestack, and 5D wide-azimuth prestack data, which involve a substantial number of variables. This brings up a key challenge in seismic data denoising: how to effectively capture and leverage information about the underlying multidimensional patterns from noisy observations. Accurately capturing and leveraging this prior information is crucial for effectively suppressing noise and reliably recovering true seismic reflections.

In the field of seismic data denoising, several approaches have been suggested to model and learn priors of seismic images to address the inherent challenges. These include methods based on sparsity, low-rank approximations, and advanced machine learning techniques. Seismic data denoising presents greater complexity than traditional image denoising due to the multidimensional characteristics of seismic datasets, such as 3D poststack, 4D prestack, and 5D wide-azimuth prestack data, which often encompass extensive parameter spaces. In this context, sparse representation has been widely adopted for seismic data denoising and has become an effective image prior. Sparse representation techniques are generally categorized into two types: analytical sparse transforms and data-driven sparse coding (SC). Analytical sparse transforms rely on the assumption that clean seismic images can be sparsely represented in a predefined transform domain, while noise does not exhibit this sparsity. Typical analytical methods encompass the Hilbert–Huang transform, among others [2], wavelet transform [3, 4], curvelet transform [5, 6], shearlet transform [7, 8], framelet transform [9, 10], seislet transform [11], and Radon transform [12]. However, using a single analytical transform for denoising can be limiting, as it may fail to capture the full range of image features effectively. In contrast, sparse

 
F. Qian et al., *Tensor Computation for Seismic Data Processing*,
Earth Systems Data and Models 6, https://doi.org/10.1007/978-3-031-78900-7_7

coding (SC) provides a more adaptable approach by learning a sparse encoding of the input seismic data, represented as a linear combination of fundamental components, known as a dictionary. This approach is better suited to capture the complex characteristics of seismic images. SC methods are highly adaptable and capable of capturing the diversity of seismic data. Notable examples of SC methods include fast sparse coding algorithms [13, 14], statistical approaches [15, 16], constrained optimization methods [17, 18], and multitask learning techniques [19, 20]. These methods provide new perspectives and tools for seismic data denoising, helping researchers extract valuable signals from complex multidimensional seismic data while effectively suppressing noise components.

In contrast, low-rank priors aim to capture the global structures present in multidimensional (M-D) seismic data, and as such, they are considered global image priors. These priors can be divided into two main types: matrix-based and tensor-based low-rank approximation (LRA). The matrix-based approach involves transforming M-D seismic data into a matrix, which is presumed to have a low rank. When noise contaminates the seismic data, the matrix rank typically increases. Therefore, rank reduction methods, such as general unstructured techniques [21, 22] and structured Hankel matrix-based LRA approaches [23, 24], can be employed to address the denoising challenge. On the other hand, tensor-based methods focus on representing the data as tensors rather than matrices. Well-known tensor-based low-rank approximation methods include tensor singular value decomposition (tSVD) [25, 26] and CANDECOMP/PARAFAC (CP) decomposition [27]. However, both sparse representation and low-rank models rely heavily on handcrafted image priors, which can pose limitations, especially when dealing with the complex geological features inherent in seismic data.

In contrast to conventional handcrafted image priors, deep prior learning directly learns a compact mapping function that converts noisy seismic data into latent seismic images, without imposing specific assumptions on the learning procedure. Currently, deep prior learning approaches can be broadly classified into two categories based on the dimensionality of the seismic data: 2-D and 3-D seismic image denoising.

The first approach, drawing inspiration from traditional image denoising methods, involves breaking down multidimensional (M-D) seismic data into 2-D slices or images. This allows for the application of various deep learning (DL) techniques for image denoising, such as convolutional neural networks (CNNs) [28, 29], generative adversarial networks (GANs) [30], autoencoders [31, 32], and U-Net architectures [33]. On the other hand, the second approach directly uses 3-D data cubes, which more effectively preserve the intrinsic spatial relationships within poststack seismic data. This category primarily includes two types of 3-D DL denoising methods: supervised 3-D CNNs [34, 35], and unsupervised patch-based 3-D autoencoders [36, 37]. A key limitation of patch-based deep learning denoising approaches is that the process of vectorizing or flattening the extracted M-D patches disrupts the original high-dimensional structure (e.g., 4-D prestack data), which reduces the efficacy of noise removal and limits their performance on more complex seismic datasets.

Compared to current deep learning (DL)-based noise reduction techniques, our FPTNN model has achieved three significant advancements in technical implementation, which provide a solid foundation for its practical application, as summarized below.

(1) *Key advancements:* The effectiveness of FPTNN stems from two major technological advancements. First, the conventional fast Fourier Transform (FFT) is substituted with the framelet transform, which offers enhanced precision and flexibility in signal representation. This change not only leads to improved signal reconstruction quality but also establishes a solid theoretical basis for the introduction of the framelet-based order-$p$ t-product. Building on this foundation, we enhance the tensor neural network (tNN) framework by integrating the framelet-based order-$p$ t-product, replacing the traditional t-product (tensor-tensor product) [38, 39], thus making the framework more efficient and robust for handling complex data. This novel approach is termed the order-$p$ tNN framework (order-$p$ t-NN).
(2) *Model:* Building on the order-$p$ tNN framework, we introduce an innovative tensor-based deep learning (DL) model that reframes the denoising process of multidimensional (M-D) seismic data as a nonlinear regression problem in a high-dimensional space. By utilizing the distinctive strengths of FPTNN, the model is capable of effectively recovering clean seismic images from noisy M-D tensors, exhibiting exceptional denoising performance. To the best of our knowledge, this is the first study to apply tensor deep learning (DL) methods for noise reduction in M-D seismic datasets, such as 4-D prestack data, making this approach a pioneering contribution.
(3) *Solution:* The framelet-based order-$p$ t-product can be computed efficiently through matrix multiplication in the framelet domain, as outlined in Definition 7.4. Specifically, when applying the order-$p$ t-product in the framelet domain, the optimal weighting parameters for the FPTNN are learned using deep learning (DL) techniques. These methods involve training a set of transformed matrix slices to further improve performance within the framelet representation. This approach has been validated in multiple studies, as demonstrated by [40–42].
(4) *Theory:* In the final contribution of this study, our approach notably departs from previous work by presenting FPTNN as an innovative deep learning (DL) model. Its distinctiveness lies in the fact that it is not simply an improvement of conventional DL models, but a fundamentally new architecture that leverages the high-dimensional nature of seismic data. This contribution not only opens new directions for future research but also offers crucial theoretical insights for advancing deep learning (DL) theory.

Through comprehensive experimental evaluation on both simulated and real-world seismic datasets, the proposed FPTNN effectively combines tensor-based deep learning (DL) with framelet transform techniques, fostering a synergistic interaction that boosts performance. This novel integration shows substantial improvements in denoising effectiveness, further validating its practical utility and overall performance.

## 7.2 Multidimensional Seismic Data Denoising and Its Modelling

### 7.2.1 Problem Statement

Seismic image denoising is a long-standing problem in the field of seismic data processing, which has attracted considerable focus in recent research. In the context of multi-dimensional (M-D) image denoising, a variety of approaches have been developed, including those based on dictionary learning [13, 43], as well as tensor decomposition techniques such as tensor singular value decomposition (t-SVD) [25, 26] and CANDECOMP/PARAFAC (CP) [27]. Despite their diversity, these methods are united by a common objective, which can be expressed as follows: formulated as

$$\mathcal{Y} = \mathcal{X} + \mathcal{N} \tag{7.1}$$

where $\mathcal{Y} \in \mathbb{R}^{n_1 \times n_2 \times \cdots \times n_p}$ represents the noisy M-D seismic image obtained by adding noise $\mathcal{N} \in \mathbb{R}^{n_1 \times n_2 \times \cdots \times n_p}$ to the original M-D image $\mathcal{X} \in \mathbb{R}^{n_1 \times n_2 \times \cdots \times n_p}$ (More related M-D tensor symbols are shown in Table 7.1).

Model-based noise reduction techniques focus on reconstructing $\mathcal{X}$ from $\mathcal{Y}$ as accurately as possible, as expressed in Eq. (8.1). In this framework, it is generally assumed that the noise $\mathcal{N}$ arises from a known physical process, or that $\mathcal{X}$ possesses a compact representation, often referred to as the image prior. In the context of seismic image denoising, $\mathcal{N}$ is frequently modeled as random noise. Consequently,

**Table 7.1** Notations related to multidimensional tensor

| Notation | Description |
|---|---|
| $x, \mathbf{x}, \mathbf{X}, \mathcal{X}$ | Scalar, vector, matrix, order-$p$ tensor. n>n |
| | The index set $\{1, 2, \cdots, n\}$. |
| $\mathcal{X}^{(\ell)}$ | The $\ell$th frontal tensor, i.e., $\mathcal{X}^{(\ell)} \equiv \mathcal{X}(:, :, \cdots, \ell)$. |
| $\mathcal{X}_{i_1 i_2 \cdots i_p}$ | The $i_1 i_2 \cdots i_p$-th entries of $\mathcal{X}$. |
| $\overrightarrow{\mathcal{X}}_j$ | The $j$ th lateral tensor. |
| $\overline{\mathcal{X}}$ | The frequency domain representation of $\mathcal{X}$: $\overline{\mathcal{X}} = \text{fft}(\widehat{\mathcal{X}}, [\ ], i)$, for $i = 3 : p$. |
| $\widetilde{\mathcal{X}}$ | The framelet domain representation of $\mathcal{X}$: $\widetilde{\mathcal{X}} = \text{fold}_p\left(\mathbf{W}\mathcal{X}_{(p)}\right)$, for $i = 3 : p$. |
| $\text{unfold}_n(\mathcal{X})$ | The mode-$n$ unfolding of $\mathcal{X}$. |
| $\text{fold}_n(\mathbf{X})$ | The folding operation of $\mathbf{X}$ along the mode-$n$. |
| $\Vert\mathcal{X}\Vert_F$ | The Frobenius norm of $\mathcal{X}$: $\mathcal{X}$: $\Vert\mathcal{X}\Vert_F = \sqrt{\sum_{ijk} x_{ijk}^2}$. |
| $\mathcal{X}^\dagger$ | The tensor transpose of $\mathcal{X}$: $\mathcal{X}^{\dagger(1)} = \mathcal{X}^{(1)\dagger}, \mathcal{X}^{\dagger(p+2-\ell)} = \mathcal{X}^{(\ell)\dagger}, 2 \leq \ell \leq p$. |

the optimal criterion for representing $\mathcal{X}$ with a given structure can be formulated as follows.

$$\mathcal{X} = \arg\min_{\mathcal{X}} \left\{ \frac{1}{2} \|\mathcal{X} - \mathcal{Y}\|_F^2 + \gamma \cdot \varphi(\mathcal{X}) \right\}, \tag{7.2}$$

where $\varphi(\mathcal{X})$ denotes the regularization function applied to the latent multidimensional (M-D) image $\mathcal{X}$, with the parameter $\gamma$ controlling the weight of this regularization term in the objective function. For example, in the case where Eq. (7.2) represents a low-rank model, we set $\varphi(\mathcal{X}) = |\mathcal{X}|_*$, where $|\cdot|_*$ refers to the tensor nuclear norm (TNN). As observed in Eq. (7.2), the denoising performance is significantly influenced by the regularization function $\varphi(\mathcal{X})$, which reflects the quality and adaptability of the selected image priors to some extent.

### 7.2.2 *Supervised Denoising Model Based on Deep Learning*

This study aims to investigate effective approaches to solving the multidimensional (M-D) seismic image denoising problem by utilizing recent developments in deep learning (DL). In DL-based denoising methods, there is usually no requirement to impose strict assumptions about the noise generation mechanism or the structure of the clean M-D images, which enhances the robustness of denoising performance in real-world scenarios. Within this framework, $\mathcal{X}$ is represented as a function parameterized by $\boldsymbol{h}(\mathcal{Y}; \Theta)$, where $\langle$ denotes the denoising network implemented via deep learning (DL), and the network is characterized by a large parameter set $\Theta$, as given below:

$$\widehat{\mathcal{X}} = \boldsymbol{h}_\Theta(\mathcal{X}, \mathcal{Y}), \tag{7.3}$$

The parameters $\Theta$ are learned through a heuristic approach by optimizing the loss function.

Next, by replacing Eq. (7.3) with Eq. (7.2) and ignoring the regularization term $\varphi(\mathcal{X})$, we derive a modified cost function that measures the discrepancy between the seismic M-D image denoiser $\boldsymbol{h}(\mathcal{Y})$ and the true seismic image $\mathcal{X}$, as shown below. Therefore,

$$\mathcal{L}(\mathcal{X}_n, \mathcal{Y}_n; \Theta) = \frac{1}{N} \sum_{n=1}^{N} \|\boldsymbol{h}_\Theta(\mathcal{Y}_n) - \mathcal{X}_n\|^2. \tag{7.4}$$

Given the training set $\{\mathcal{X}_n, \mathcal{Y}_n\}$, where $n \in [N]$, gradient descent is commonly used to optimize the deep learning (DL) model $h_\Theta(\cdot)$. Popular variants of gradient descent, such as stochastic gradient descent (SGD) [44, 45] and adaptive moment estimation (Adam) [46], are typically applied to minimize the objective function

$\mathcal{L}(\cdot)$, which is parameterized by the model parameters $\Theta$. With the regularization term $\varphi(\mathcal{X})$ excluded, the performance of the deep learning-based denoising approach is fully dependent on the success of this optimization strategy [47].

### 7.2.3 A Deep Learning-Based Denoising Model Without Ground Truth

However, the performance of supervised learning methods in seismic denoising is often constrained by the availability of ground truth, which, in practical applications, is frequently either inaccessible or unreliable. To address the challenge of ground truth scarcity, we reformulate Eq. (7.4) using an unsupervised objective function $\rho(\cdot)$, which denotes the expectation over the residual error. This alternative approach allows us to handle practical scenarios where ground truth data is inaccessible, thus shifting focus toward the unsupervised learning paradigm (Fig. 7.1). Specifically, we define:

$$\mathcal{L}(\mathcal{X}_n, \mathcal{Y}_n; \Theta) = \frac{1}{N} \sum_{n=1}^{N} \rho(\boldsymbol{h}_{\Theta}(\mathcal{Y}_n) - \mathcal{X}_n). \tag{7.5}$$

where $\rho(\cdot)$ denotes an unsupervised loss function, which may include methods like Stein's unbiased risk estimator (SURE) [48] or partially linear denoising techniques [49]. Building on previous advancements in unsupervised denoising [48, 50], we propose a SURE-based optimization criterion designed to more effectively assess solutions in the unsupervised learning context, as shown below.

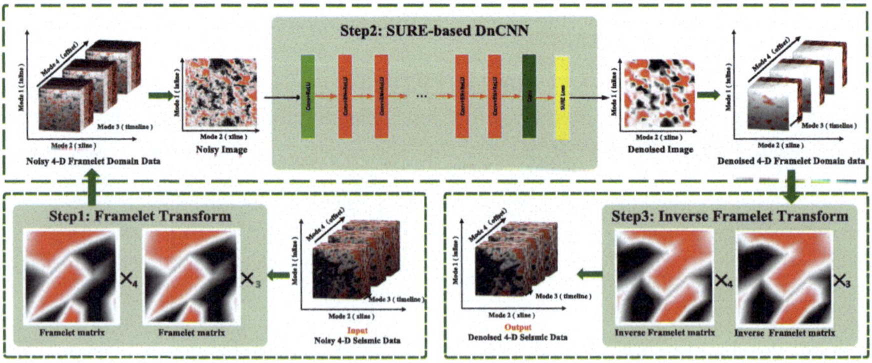


**Fig. 7.1** The diagram illustrates the evolution of the FPTNN model, where $\times_3$ and $\times_4$ denote the 3-mode and 4-mode operations, representing the interaction between a fourth-order tensor and a framelet matrix

$$\mathcal{L}(\mathcal{Y}_n;\Theta) = \frac{1}{N}\sum_{n=1}^{N}\left\{\frac{1}{K}\left\|\mathcal{Y}_n - \boldsymbol{h}_\Theta(\mathcal{Y}_n)\right\|^2 - \sigma_n^2 + \frac{2\sigma^2}{K}\operatorname{tr}\left(\frac{\partial \boldsymbol{h}_\Theta(\mathcal{Y}_n)}{\partial \mathcal{Y}_n^{\dagger}}\right)\right\}. \tag{7.6}$$

where $\operatorname{tr}(\cdot)$ denotes the trace of a matrix, $K$ refers to the total number of pixels in the input image, and $\sigma$ represents the noise variance. Given that Stein's unbiased risk estimator (SURE) offers an almost unbiased approximation of the mean square error (MSE), the unsupervised order-$p$ tNN model using the SURE-based loss function can achieve denoising results that are on par with those of the supervised tNN trained with MSE [48].

## 7.3 Framelet-Based Order-$p$ Tensor Neural Network (FPTNN)

Drawing inspiration from the benefits of the framelet transform over the FFT, we propose utilizing the framelet transform as a superior alternative to the FFT in the order-$p$ tensor product, as outlined in Sect. 7.3.1. By incorporating the framelet representation of the order-$p$ tensor product, we introduce the FPTNN model for seismic data denoising in Sect. 7.3.2. Further, in Sect. 7.3.3, we present the tensor backpropagation (TBP) algorithm to enhance the FPTNN model's performance. Finally, Section 8.3.3 offers a comprehensive guide to implementing the framelet-based FPTNN model. These sections are interrelated, as depicted in Fig. 7.2, which serves to help the reader navigate through the core concepts and organization of this chapter.

### 7.3.1 *Framelet-Based Order-p T-Product*

This subsection aims to derive the framelet-based order-$p$ t-product, a central element in constructing the FPTNN model, as illustrated in Fig. 7.2. Prior to formally introducing this innovative t-product, it is helpful to revisit some foundational concepts associated with the Fourier transform of M-D seismic data $\mathcal{X} \in \mathbb{R}^{n_1 \times n_2 \times \cdots \times n_p}$, which are detailed in Algorithm 7.2. For clarity, the Fourier transform is represented here in the form of matrix multiplication, as follows:

$$\overline{\mathcal{X}} = \operatorname{fold}_l\left(\mathbf{F}_{n_l}\mathcal{X}_{(l)}\right) \in \mathbb{R}^{n_1 \times n_2 \times n_l \times \cdots \times n_p}, \tag{7.7}$$

In this context, $\mathcal{X}_{(l)}$ denotes the unfolding of the tensor $\mathcal{X}$ along mode $l$, while $\mathbf{F}_{n_l}$ represents the Fourier transform matrix (FFT) of size $n_l$. This representation of the Fourier transform using matrix multiplication is designed to offer a more efficient framework for describing the framelet-based order-$p$ t-product.

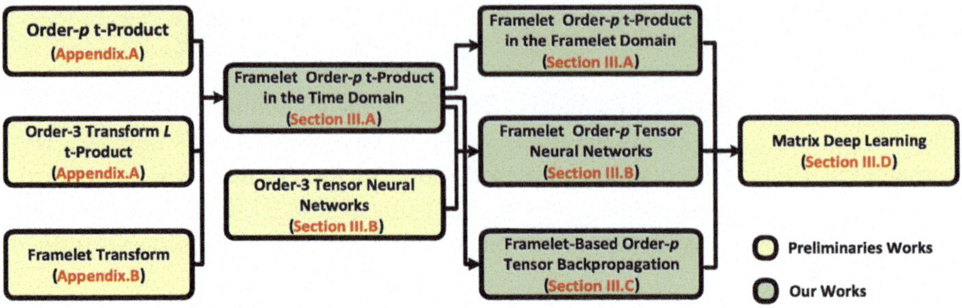


**Fig. 7.2** A block diagram is presented in this chapter to illustrate the interactions and collaborative mechanisms between the various components involved in the denoising process

To streamline the process, we substitute the Fourier transform (FFT) in Eq. (7.7) with the framelet transform. This substitution enables us to define the framelet representation of the order-$p$ t-product. As a result, the framelet transform of the M-D tensor along the $p$-mode is formulated as follows:

$$\widetilde{\mathcal{X}} = \text{fold}_l \left( \mathbf{W} \mathcal{X}_{(l)} \right) \in \mathbb{R}^{n_1 \times n_2 \times w_l n_l \times \cdots \times n_p}, \tag{7.8}$$

As detailed in Sect. 7.5, $\mathbf{W} \in \mathbb{R}^{w_l n_l \times n_l}$ represents the framelet transform matrix.

Building on Definitions 2 and 3, the order-$p$ t-product is represented as a matrix multiplication involving the frontal slices in the Fourier transform domain. By substituting the traditional Fourier transform with the framelet transform, we can similarly define the framelet-based order-$p$ t-product, as shown below:

**Definition 7.1** (*Framelet-based order-p t-product in the time domain*) [51, 52] Let $\mathcal{A}$ be a tensor of dimensions $n_1 \times r \times n_3 \times \cdots \times n_p$, and $\mathcal{B}$ be a tensor of dimensions $r \times n_2 \times n_3 \times \cdots \times n_p$. Their framelet-based $t$-product, denoted as $\mathcal{A} *_W \mathcal{B}$, produces a rank-$p$ tensor with dimensions $n_1 \times n_2 \times n_3 \times \cdots \times n_p$, and is expressed as follows:

$$\mathcal{A} *_W \mathcal{B} = \text{fold}(\text{circ}(\mathcal{A}) * \text{unfold}(\mathcal{B})). \tag{7.9}$$

Building upon Definition 2.1 from [53] and Definition 3 in Sect. 7.5, the framelet-based order-$p$ t-product can be reformulated as a matrix multiplication of the frontal slice in the framelet domain, effectively substituting the traditional operations performed in the Fourier domain. This relationship is outlined as follows:

**Definition 7.2** (*Order-p t-product in the framelet domain*) To streamline the implementation, the order-$p$ t-product described in Eq. (7.9) can be rewritten as a matrix product within the framelet space, as illustrated below:

$$\widetilde{\mathcal{C}}(:, :, \ell) = \widetilde{\mathcal{A}}(:, :, \ell) \cdot \widetilde{\mathcal{B}}(:, :, \ell), \; \ell \in \left[ n_3 n_4 \cdots n_p \right]. \tag{7.10}$$

Next, we present the FPTNN (Framelet-Based Order-$p$ Tensor Neural Network), a model tailored for handling M-D seismic tensors in the framelet representation.

**Algorithm 7.1:** Framelet-based order-$p$ tNN

**Input**: $\mathcal{Y}$, $\alpha$, $N_{\text{bs}}$, $\sigma$
**Output**: $\widehat{\mathcal{X}}$

// The framelet transform
**for** $l = 3$, *to*, $p$ **do**
  $\widetilde{\mathcal{Y}} = \text{fold}_l\left(\mathbf{W}\mathcal{Y}_{(l)}\right)$.
**end**
// Training and testing of the FPTNN model
**for** $i_3 = 1$, *to*, $n_3$ **do**
  **for** $i_4 = 1$, *to*, $n_4$ **do**
    ...
    **for** $i_p = 1$, *to*, $n_p$ **do**
      $\widetilde{\mathcal{X}}(:, :, i_3, i_4, \cdots, i_p) =$
        $\text{MDL}(\widetilde{\mathcal{Y}}(:, :, i_3, i_4, \cdots, i_p), \alpha, N_{\text{bs}}, \sigma)$.
    **end**
    ...
  **end**
**end**
// The inverse framelet transform
**for** $l = 3$ *to* $p$ **do**
  $\widehat{\mathcal{X}} = \text{fold}_l\left(\mathbf{W}^\top \widetilde{\mathcal{X}}\right)$.
**end**

The core computational process of this model revolves around the framelet-based order-$p$ t-product, as outlined in Definitions 7.1 and 7.2.

### 7.3.2 Framelet-Based Order-$p$ Tensor Neural Networks

In this section, we propose an innovative framework that incorporates the framelet-based order-$p$ t-product into tensor neural networks (TNNs), leading to the development of the FPTNN (Framelet-Based Order-$p$ Tensor Neural Network). A significant distinction of this model is that the order-$p$ t-product substitutes the traditional third-order t-product commonly used in classical TNNs [39], which are typically applied to third-order tensor data. This modification highlights a fundamental difference between the FPTNN and traditional TNNs. Moreover, the framelet transform may provide a more effective denoising approach in the transformed domain compared to the conventional Fast Fourier Transform (FFT) [41, 54]. Before delving into the detailed derivation of the FPTNN model, we first review key concepts from traditional TNNs, especially those utilizing the third-order t-product t-product.

**Definition 7.3** (*tNN with a third-order t-product*) [39] Given the tensors $\mathcal{X}_{\ell-1} \in \mathbb{R}^{n_{\ell-1} \times n_2 \times n_3}$, $\mathcal{W}_\ell \in \mathbb{R}^{n_\ell \times n_{\ell-1} \times n_3}$, and the bias term $\overrightarrow{\mathcal{B}}_\ell \in \mathbb{R}^{n_\ell \times 1 \times n_3}$, the forward propagation process of the tensor can be expressed as:

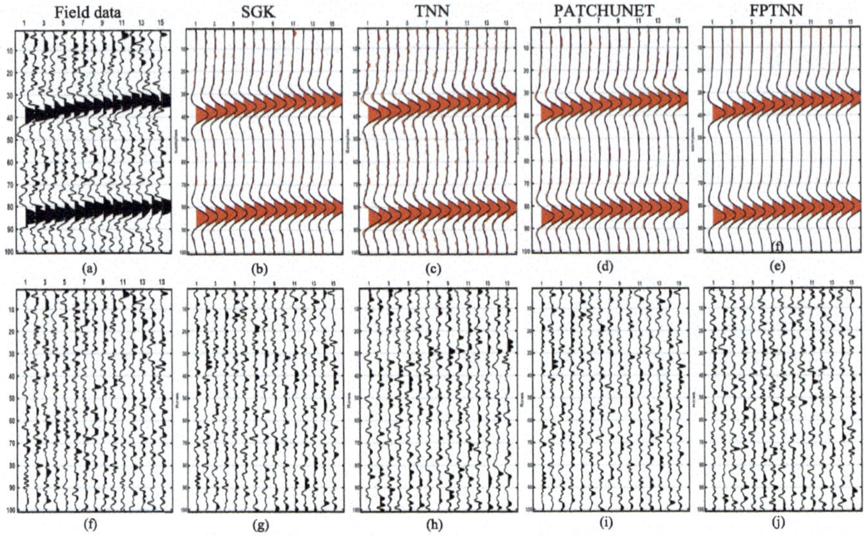


**Fig. 7.3** In the synthetic 5D data example, slice comparisons of denoising methods are presented: (a) shows the noisy data, while denoised results obtained by (b) SGK, (c) TNN, (d) PATCHUNET, and (e) the proposed FPTNN are displayed. Additionally, (f) illustrates the added noise, and the noise removal performance is shown and compared in (g) SGK, (h) TNN, (i) PATCHUNET, and (j) FPTNN

$$\mathcal{X}_\ell = \delta\left(\mathcal{W}_{\ell-1} * \mathcal{X}_{\ell-1} + \vec{\mathcal{B}}_\ell\right), \quad \ell \in [K]. \tag{7.11}$$

where $\ell$ denotes the final layer of the FPTNN model. It is clear that, based on Definition 7.3, the tNN (Tensor Neural Network) does not satisfy the solution conditions outlined in Eq. (8.1). As a result, in order to broaden the scope of the tNN's applicability, it becomes necessary to reformulate it by incorporating the order-$p$ tensor product (order-$p$ t-product), as defined below.

**Definition 7.4** (*tNN with a framelet-based order-p t-product*) Given the tensors $\mathcal{X}_{\ell-1} \in \mathbb{R}^{n_{\ell-1}\times n_2\times\cdots\times n_p}$, $\mathcal{W}_\ell \in \mathbb{R}^{n_\ell\times n_{\ell-1}\times\cdots\times n_p}$, and the bias term $\vec{\mathcal{B}}_\ell \in \mathbb{R}^{n_\ell\times 1\times\cdots\times n_p}$, the forward propagation process is given by the following formula:

$$\mathcal{X}_\ell = \delta\left(\mathcal{W}_{\ell-1} *_W \mathcal{X}_{\ell-1} + \vec{\mathcal{B}}_\ell\right), \quad \ell \in [K]. \tag{7.12}$$

It is essential to highlight that in a multi-layer tNN, multiple intermediate layers, commonly known as hidden layers, are positioned between the input and output layers. This configuration can be described as follows:

$$\boldsymbol{h}_\Theta(\mathcal{Y}) = (\mathcal{W}_\ell *_W \delta\,(\mathcal{W}_{\ell-1} *_W \ldots\ \delta\left(\mathcal{W}_0 *_W \mathcal{Y} + \vec{\mathcal{B}}_0\right)\ldots \vec{\mathcal{B}}_{\ell-1}\right) + \vec{\mathcal{B}}_\ell\Big), \tag{7.13}$$

The main challenge in optimizing the parameters $\Theta$ (where $\Theta = \{\mathcal{W}, \mathcal{B}\}$) in Eq. (7.13) lies in efficiently managing the order-$p$ t-product operations, which are essential for the implementation of the TBP algorithm [39, 42]. This challenge arises because existing TBP methods are primarily tailored for third-order tensor computations, and the handling of higher-order tensors remains an unresolved issue, as highlighted in [39].

### *7.3.3 Framelet-Based Order-p Tensor Backpropagation*

After developing the FPTNN model, the next crucial step involves the optimization process during training. In this stage, the loss function $\mathcal{L}(\mathcal{X}_n, \mathcal{Y}_n; \Theta)$ is a nonlinear function of the model parameters $\Theta$, and must be minimized using the order-$p$ TBP (Order-$p$ Backpropagation) algorithm. This algorithm is inspired by the traditional backpropagation approach, iteratively updating the parameters $\Theta$ of the tNN to minimize prediction errors. Finally, gradient descent is used to calculate the partial derivatives of the loss function with respect to each parameter, and the parameters $\mathcal{W}_{\ell-1}$ and $\mathcal{B}_\ell$ are adjusted based on these derivatives to improve the network's overall performance.

$$\begin{aligned} \mathcal{W}_{\ell-1} &\leftarrow \mathcal{W}_{\ell-1} - \alpha \frac{\partial \mathcal{L}(\mathcal{W}, \mathcal{B})}{\partial \mathcal{W}_{\ell-1}}, \\ \mathcal{B}_\ell &\leftarrow \mathcal{B}_\ell - \alpha \frac{\partial \mathcal{L}(\mathcal{W}, \mathcal{B})}{\partial \mathcal{B}_\ell}, \end{aligned} \tag{7.14}$$

where $\alpha$ is defined as the learning rate, which determines the step size for weight updates, effectively reducing the value of the loss function $\mathcal{L}(\mathcal{X}n, \mathcal{Y}n; \Theta)$. The following two expressions provide a deeper understanding of how the gradient descent method operates. First, the specific form of the partial derivative of the loss function with respect to the weights $\mathcal{W}_{\ell-1}$, denoted as $\partial\mathcal{L}/\partial\mathcal{W}_{\ell-1}$, is given by:

$$\frac{\partial \mathcal{L}}{\partial \mathcal{W}_{\ell-1}} = \frac{\partial \mathcal{L}}{\partial \widehat{\mathcal{X}}_{n,\ell}} \frac{\partial \widehat{\mathcal{X}}_{n,\ell}}{\partial \mathcal{W}_{\ell-1}}. \tag{7.15}$$

Next, we compute the partial derivative of Eq. (7.5) to obtain the gradient term with respect to $\widehat{\mathcal{X}}_{n,\ell}$, expressed as follows:

$$\frac{\partial \mathcal{L}}{\partial \mathcal{X}_{n,\ell}} = \frac{1}{N} \sum_{n=1}^{N} \rho'(\mathcal{X}_n - \widehat{\mathcal{X}}_{n,\ell}), \tag{7.16}$$

where $\rho'(\cdot)$ denotes the derivative of the SURE-based objective function $\rho(\cdot)$ with respect to the difference $\mathcal{X}_n - \widehat{\mathcal{X}}_{n,\ell}$. Likewise, the second formulation is concerned with computing $\partial\mathcal{L}/\partial\mathcal{B}_\ell$. In a similar manner to the calculation of $\delta\mathcal{W}_\ell$, the expression for $\partial\mathcal{L}/\partial\mathcal{B}_\ell$ can be derived as follows:

$$\frac{\partial\mathcal{L}}{\partial\mathcal{B}_\ell}=\frac{\partial\mathcal{L}}{\partial\widehat{\mathcal{X}}_{n,\ell}}\frac{\partial\widehat{\mathcal{X}}_{n,\ell}}{\partial\mathcal{B}_\ell}. \tag{7.17}$$

The remaining derivatives, $\partial\widehat{\mathcal{X}}_{n,\ell}/\partial\mathcal{W}_{\ell-1}$ and $\partial\widehat{\mathcal{X}}_{n,\ell}/\partial\mathcal{B}_\ell$, are obtained by taking the partial derivatives of Eq. (7.12) with respect to the weight matrix $\mathcal{W}_{\ell-1}$ and the bias term $\mathcal{B}_\ell$, respectively. Therefore, the expression for $\partial\widehat{\mathcal{X}}_{n,\ell}/\partial\mathcal{W}_{\ell-1}$ is given as follows:

$$\begin{aligned}\frac{\partial\mathcal{X}_{n,\ell}}{\partial\mathcal{W}_{\ell-1}}&=\frac{\partial\delta\left(\mathcal{W}_{\ell-1}*_W\mathcal{X}_{n,\ell-1}+\vec{\mathcal{B}}_\ell\right)}{\partial\mathcal{W}_{\ell-1}}\\&=\delta'\left(\mathcal{Z}_{n,\ell}\right)*_W\mathcal{X}_{n,\ell-1}^{\dagger},\end{aligned} \tag{7.18}$$

Here, $\mathcal{Z}_{n,\ell}$ is expressed as $\mathcal{W}_{\ell-1}*\mathcal{X}_{n,\ell-1}+\vec{\mathcal{B}}_\ell$, where $\delta'(\cdot)$ represents the derivative of the activation function. Then, the term $\frac{\partial\widehat{\mathcal{X}}_{n,\ell}}{\partial\mathcal{B}_\ell}$ can be written as:

$$\frac{\partial\mathcal{X}_{n,\ell}}{\partial\mathcal{B}_\ell}=\mathrm{sum}\left(\delta'\left(\mathcal{Z}_{n,\ell}\right),2\right). \tag{7.19}$$

Although the preceding derivations offer strong theoretical foundations for solving Eqs. (7.15) and (7.17), effectively managing the order-$p$ t-product operations in each training epoch of the order-$p$ TBP remains a challenging task. Ideally, the calculation of the order-$p$ t-product would be incorporated into the backpropagation (BP) process within TensorFlow to streamline the overall design of the order-$p$ tNN. However, implementing this approach in practice could involve several technical difficulties.

### 7.3.4 Implementation

To overcome the challenges in implementing the FPTNN, we explore the relationship between the order-$p$ tNN and matrix-based deep learning (DL), as outlined in Definition 7.2 of Sect. 7.3.1. The effective application of the TBP algorithm hinges on the accurate computation of derivatives, as described in Eq. (7.15). By utilizing the insights from Definition 6, we are able to express and optimize the formulation in (7.15) more succinctly and efficiently, as shown below:

$$\frac{\partial\widetilde{\mathcal{L}}^{(k)}}{\partial\widetilde{\mathcal{W}}_{\ell-1}^{(k)}}=\frac{1}{N}\sum_{n=1}^{N}\rho'(\widetilde{\mathcal{Y}}_n^{(k)}-\widetilde{\mathcal{X}}_{n,\ell}^{(k)})\left(\delta'\left(\widetilde{\mathcal{Z}}_{n,\ell}^{(k)}\right)\widetilde{\mathcal{X}}_{n,\ell-1}^{(k)\top}\right). \tag{7.20}$$

Equation (7.20) provides a significant approach for integrating the order-$p$ TBP with standard backpropagation algorithms in the framelet domain, enabling its effective implementation. Furthermore, by further extending Definition 7.2, the FPTNN can be restructured into a matrix-based deep learning (DL) model that incorporates the framelet transform, as described below.

$$\boldsymbol{h}_{\Theta}(\widetilde{\mathcal{Y}}^{(k)}) = \left(\widetilde{\mathcal{W}}_{\ell}^{(k)}\delta\left(\widetilde{\mathcal{W}}_{\ell-1}^{(k)} \cdots\ \delta\left(\widetilde{\mathcal{W}}_{0}^{(k)}\widetilde{\mathcal{Y}}^{(k)} + \widetilde{\mathcal{B}}_{0}^{(k)}\right)\cdots\widetilde{\mathcal{B}}_{\ell-1}^{(k)}\right) + \widetilde{\mathcal{B}}_{\ell}^{(k)}\right). \tag{7.21}$$

Thus, through the application of Definition 7.2, we have effectively established a strong linkage between the FPTNN and front-slice deep learning (DL) in the framelet domain. This connection is grounded in the two preceding analyses, which focus on the model's structural design and its solution approach. To accomplish this, we reformulate the FPTNN framework into a front-slice deep learning representation, as outlined below.

$$\widetilde{\mathcal{X}}^{(i)} = \mathrm{MDL}(\widetilde{\mathcal{Y}}^{(i)}, \alpha, N_{\mathrm{bs}}, \sigma). \tag{7.22}$$

The workflow of the FPTNN is outlined in Algorithm 7.1 and depicted in Fig. 7.1, comprising three primary computational phases. During the initial and final phases, the input data $\mathcal{Y}$ undergoes a forward framelet transform to produce $\widetilde{\mathcal{Y}}$, while in the final phase, we apply the inverse transform to $\widetilde{\mathcal{X}}$ to recover the denoised tensor $\mathcal{X}$ (see Fig. 7.3). In the middle phase, we optimize Eq. (7.5) using an unsupervised denoising convolutional neural network (DnCNN) [55], which operates on the frontal slices in the framelet domain, denoted as $\mathrm{MDL}(\cdot)$. Unlike conventional methods, the denoising process here utilizes Stein's unbiased risk estimator (SURE) instead of the typical mean squared error (MSE), a choice that has been proven to enhance denoising performance [48]. In fact, the unsupervised SURE-based DnCNN has been shown to deliver results on par with those of the supervised MSE-based DnCNN, as highlighted in [48]. For instance, Fig. 7.4 demonstrates the denoising outcomes for various x-line and y-line slices from a synthetic 4D vertical seismic profile (VSP) dataset in the framelet domain. In this figure, the noisy frontal slice (column (b)) in the framelet domain is fed into $\mathrm{MDL}(\cdot)$, and the corresponding denoised output (column (c)) is generated by the $\mathrm{MDL}(\cdot)$ process. It should be noted that the clean data (column (a)) is not input into $\mathrm{MDL}(\cdot)$; it is provided only for comparative purposes to illustrate the denoising results.

## 7.4 Field Examples of FPTNN

The main goal of this section is to assess the performance of the FPTNN method on synthetic datasets and compare its results with existing algorithms, highlighting its advantages. Furthermore, we will evaluate the robustness of the proposed method on real-world data. The experimental setup is outlined as follows: Sects. 7.4.1.1, 7.4.1.2, and 7.4.1.3 provide comprehensive descriptions of the baseline approaches, performance metrics, and experimental conditions, which form the foundation for

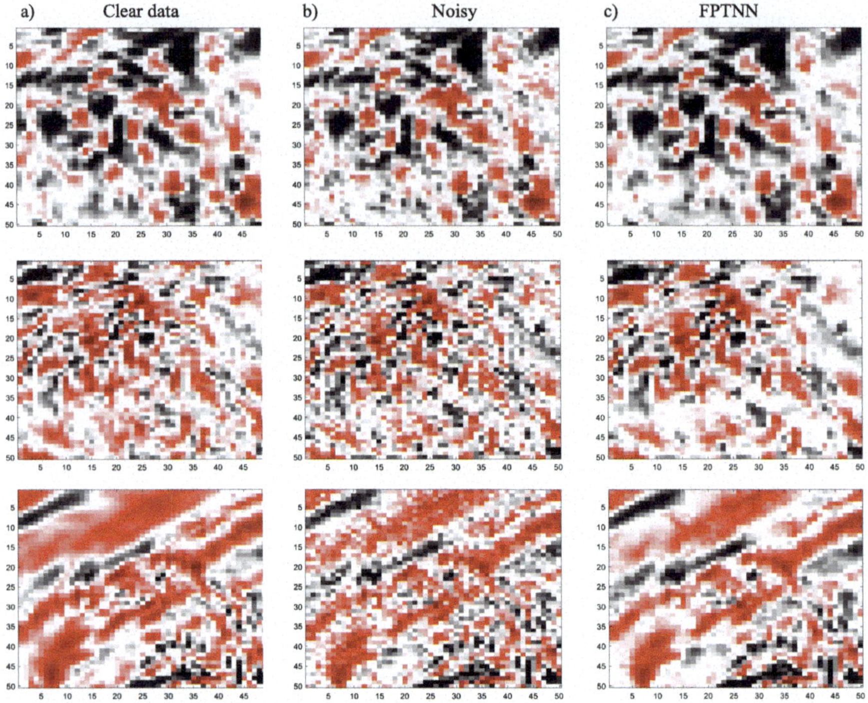


**Fig. 7.4** In the framelet domain, a comparison of denoising effects across different xline-yline slices is performed for a synthetic 4D VSP data example: (a) shows the original clean data, (b) presents the noisy data, and (c) displays the denoised results obtained using the proposed FPTNN method

the experiments. Sections 8.4.2 and 8.4.3 will present the results from synthetic and real datasets, respectively, and perform both qualitative and quantitative assessments to compare the proposed FPTNN with the baseline approaches. As for the computational environment, the algorithms are divided into two groups: deep learning-based approaches (including FPTNN and PATCHUNET) and traditional non-deep learning methods. All deep learning experiments were executed on the Google Colab platform utilizing the TensorFlow[1] framework[2]. It is important to highlight that GPU resources on Colab are dynamically allocated, which may lead to variations in computation time due to resource management. In contrast, the non-deep learning experiments were conducted on a 64-bit PC with an E7-4820 2.00 GHz CPU and 64 GB of RAM, using MATLAB R2016a.

[1] Official TensorFlow website: https://www.TensorFlow.org/.

[2] Google Colab access link: https://colab.research.google.com/notebooks/gpu.ipynb.

## 7.4.1 Experimental Setup

### 7.4.1.1 Benchmark Methods

The low-rank tensor techniques presented in Sect. 7.1 offer a robust approach for tackling the M-dimensional seismic data denoising challenge, especially when the clean seismic signals are represented as order-$p$ low-rank tensors, and noise causes an increase in tensor rank. In addition to this approach, dictionary learning (DcL) focuses on leveraging data-driven sparse representation learning to tackle this issue, demonstrating its potential in denoising. Therefore, through a comprehensive review of existing seismic data denoising techniques, this paper selects three benchmark approaches for comparison, which are outlined as follows:

1. *Tensor nuclear norm (TNN)* [40]: The TNN approach reformulates the denoising task for M-dimensional seismic data as a low-rank tensor recovery problem. In this framework, the clean signal is represented by a low-rank tensor, with its rank controlled by the tensor nuclear norm, while the noise leads to an increase in tensor rank. To address this, a transform-based alternating direction method of multipliers (ADMM) is employed, which efficiently solves the problem by isolating the low-rank component, thereby achieving effective denoising.
2. *PATCHUNET*[3] [56]: This method decomposes the noisy data into several smaller patches for use in both training and testing.Each patch is encoded to capture its essential features, and the decoder uses these features to reconstruct the original clean signal while removing unwanted noise.
3. *Sequential generalization of k-means (SGK)*[4] [13]: SGK has a different denoising mechanism compared to PATCHUNET and TNN, leveraging the sparse representation characteristics of seismic images to denoise M-dimensional seismic data. More specifically, clean seismic data can be sparsely represented using an appropriate set of overcomplete dictionaries, whereas noise remains dense in this dictionary domain.

When comparing the FPTNN with benchmark approaches, parameter optimization is a crucial factor for ensuring effective denoising performance. To ensure a fair evaluation, the parameters of each benchmark method must be carefully tuned to their optimal configurations, thereby ensuring the accuracy and fairness of the comparison results.

### 7.4.1.2 Performance Metrics

In experiments involving synthetic data, where the ground truth is known, the signal-to-noise ratio (SNR) serves as a metric to evaluate how well the denoised images

[3] https://github.com/omarmohamed15/PATCHUNET.

[4] https://github.com/chenyk1990/reproducible_research/tree/master/sgk5d.

**Fig. 7.5** The denoising performance of four methods on the second 4D synthetic dataset under various noise levels is depicted. The symbols represent each method as follows: " ◇ " for our FPTNN model, " × " for PATCHUNET, " ∇ " for TNN, and " ○ " for SGK

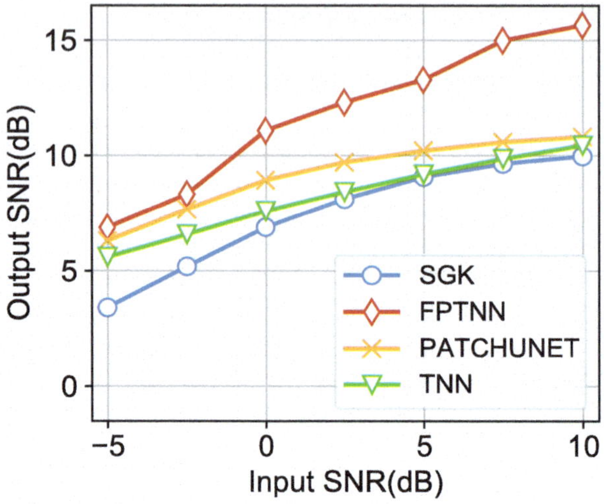


retain the original signal compared to the ground truth. A higher SNR value indicates more effective noise reduction (see Fig. 7.5). The SNR is defined as follows:

$$\mathrm{SNR[dB]} = -20 \log_{10} \left( \frac{\|\mathcal{X} - \hat{\mathcal{X}}\|}{\|\mathcal{X}\|} \right), \tag{7.23}$$

where $\widehat{\mathcal{X}}$ refers to the images obtained after denoising the noisy data, and $\mathcal{X}$ corresponds to the original, ground truth images

In experiments using real field data, where clean reference data is not available, the SNR cannot be directly applied to evaluate denoising performance. As a result, to assess the effectiveness of the FPTNN method and compare it with benchmark algorithms, we utilize the local similarity metric. This metric measures the degree of similarity between the denoised data and the noise that has been removed, without the need for clean reference data, as outlined in [57]. Local similarity provides insight into the amount of valid signal leakage during the denoising process, with values ranging from 0 to 1, where higher values indicate more significant leakage and greater similarity.

#### 7.4.1.3 Parameter Configuration

While the network architecture of the FPTNN model is outlined in Sect. 8.3.3, the choice of parameters for the objective function and the design of the training procedure are critical in determining its practical performance. These factors have a direct impact on the model's denoising capability. The upcoming section will discuss the process of selecting and fine-tuning the key parameters of the objective function and the training algorithm to enhance the denoising effectiveness of the FPTNN model.

*Determination of the objective function parameter:* According to Eq. (7.6) in the FPTNN model, the only parameter is the noise variance $\delta$, which will be adaptively adjusted using the following empirical formula:

$$\widehat{\sigma}_i = \frac{\text{median}\left(\left|\mathcal{F}_{(i)}^{\text{HH}}\right|\right)}{0.6745}, \quad i = 1, \ldots, B \tag{7.24}$$

The median operation, denoted as median$(\cdot)$, is applied to the framelet coefficients $\mathcal{F}_{(i)}$.

*Optimization of FPTNN training parameters:* In the FPTNN model, we substitute the standard MSE loss function with the SURE loss function and employ the Adam optimizer to adjust the network's training parameters. The primary parameter for the Adam optimizer is the learning rate $\alpha$ (as shown in Eq. (7.14)), which governs the update speed during optimization. For all seismic data experiments, we set the learning rate to $\alpha = 0.001$. Moreover, the FPTNN weights are updated using traditional backpropagation in batch mode, with a batch size of 32, applied to both synthetic datasets and the real WY field dataset.

### 7.4.2 Synthetic Data Tests

To assess the denoising effectiveness of the proposed FPTNN algorithm, we conduct two experiments using two typical synthetic datasets. The first dataset is a 5-dimensional synthetic set [4] generated using a Ricker wavelet with a sampling rate of 4ms, commonly employed for 5D reconstruction tasks [13]. The second experiment utilizes a more intricate 4-dimensional synthetic vertical seismic profile (VSP)[5] dataset. Both clean datasets are corrupted by significant band-limited Gaussian random noise, as shown in Fig. 7.3(a)(f), to create the noisy input data for the FPTNN model. It is important to note that clean data is not used during the FPTNN training process; rather, it is only employed for calculating the signal-to-noise ratio (SNR) to assess the performance of the denoising algorithms.

In the first synthetic 5D data experiment, Figs. 7.3(b)–(e) display a comparison of the denoising results from FPTNN and various benchmark algorithms applied to the noisy image shown in Fig. 7.3(a). The denoised waveforms (highlighted in red) are compared visually to the clean waveforms (in black), using the SeisLab toolbox[6], enabling direct visual comparison. The results clearly indicate that all benchmark algorithms, including TNN, PATCHUNET, and SGK, show noticeable discrepancies between their denoised outputs and the clean reference data. On the other hand, the denoised waveforms produced by FPTNN closely resemble the clean waveforms, with no visible residual noise or excessive smoothing. This is further corroborated by the quantitative results in Table 7.2, where the signal-to-noise ratios (SNR) for

[5] https://wiki.seg.org/wiki/3D_9C_synthetic_VSP_data.

[6] https://www.mathworks.com/MATLABcentral/fileexchange/15674-seislab-3-01.

**Table 7.2** A comparison of the denoising effectiveness between the proposed approach and existing methods is conducted using two synthetic datasets

| Method | Synthetic 5-D dataset (dB) | Synthetic VSP dataset (dB) |
|---|---|---|
| PATCHUNET | 17.49 | 11.10 |
| TNN | 15.41 | 9.23 |
| SGK | 18.99 | 9.10 |
| Proposed FPTNN | **26.67** | **13.31** |

**Table 7.3** Comparison of computation time between the proposed FPTNN and existing PATCHUNET methods

| Methods | Synthetic1 (s) | Synthetic2 (s) | Real WY (s) |
|---|---|---|---|
| PATCHUNET | 130.65 | 583.74 | 885.25 |
| FPTNN | 81.35 | 256.82 | 387.22 |

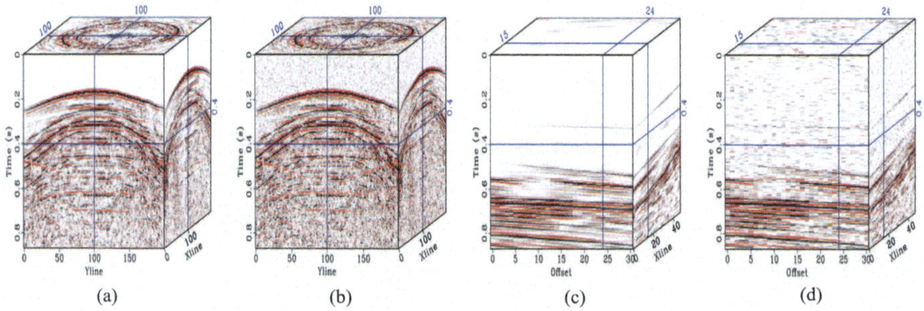

**Fig. 7.6** A second example using 4D synthetic VSP data. (a) Clean data volume for xline-yline-timeline. (b) Noisy data volume for xline-yline-timeline. (c) Clean data volume for xline-offset-timeline. (d) Noisy data volume for xline-offset-timeline

PATCHUNET, SGK, and TNN are 17.49, 18.99, and 15.41 dB, respectively, while FPTNN achieves an impressive SNR of 26.67 dB. From both the visual comparison in Fig. 7.3 and the quantitative data in Table 7.2, it is evident that FPTNN delivers superior denoising performance over the benchmark methods.

In the second 4D synthetic dataset experiment, a more intricate simulated VSP dataset with dimensions of $100 \times 10 \times 10 \times 10$ was utilized. To facilitate a clearer visualization of the denoising results, we extracted two 3D data cubes corresponding to the common midpoint gather (CMP) and common offset gather (COG), representing different perspectives of the 4D dataset, as shown in Fig. 7.6. These 3D visualizations were created using the open-source Madagascar software package[7]. Figures 7.7 and 7.8 show 3D visual comparisons of the denoising outcomes from FPTNN and

[7] https://reproducibility.org/wiki/Main.

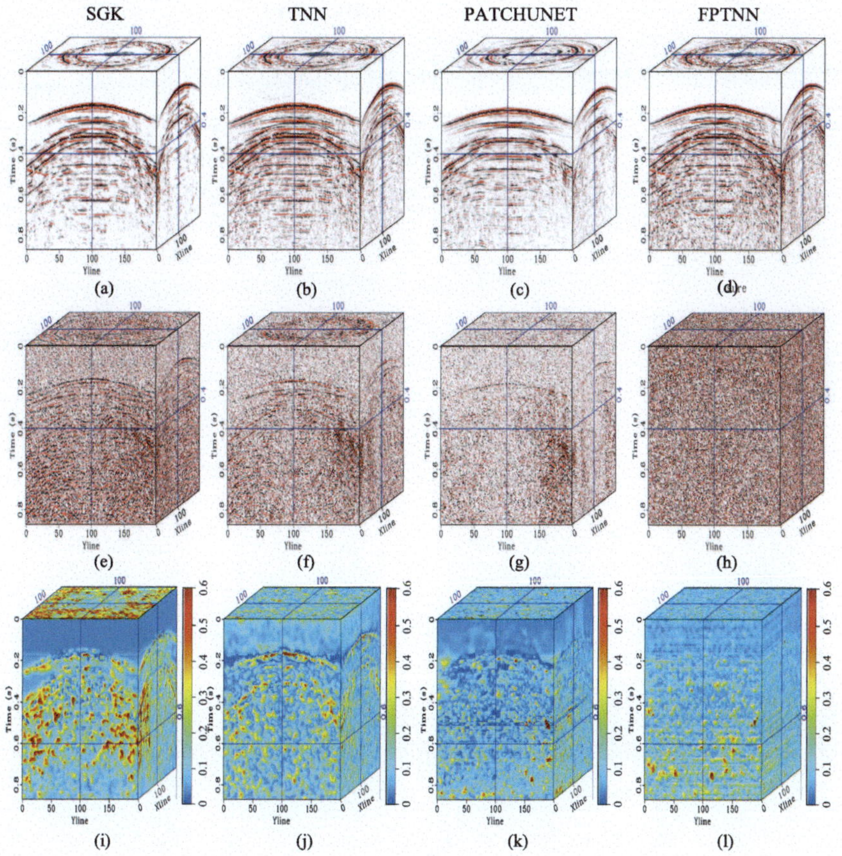


**Fig. 7.7** The denoising comparison for the xline-yline-timeline data volume (offset = 1) in the 4D synthetic VSP seismic data illustrates the results achieved by various methods: (a) SGK, (b) TNN, (c) PATCHUNET, and (d) the proposed FPTNN. Panels (e)–(h) display the noise removed by each method, while panels (i)–(l) present the local similarity maps calculated using the denoised data in (a)–(d) alongside the removed noise in (e)–(h)

the benchmark methods. While all methods, including FPTNN and the others, effectively remove random noise, a more detailed analysis of the noise-reduced images in Figs. 7.7(e)–(h) and 7.8(e)–(h), along with the corresponding local similarity volumes in Figs. 7.7(i)–(l) and 7.8(i)–(l), shows that FPTNN not only suppresses noise but also preserves critical signal components. This is further confirmed by the SNR values listed in Table 7.2, where the SNR values for TNN, PATCHUNET, and SGK are 9.23, 11.10, and 9.10 dB, respectively, while FPTNN achieves a higher SNR of 13.31 dB. Both the visual and quantitative results indicate that FPTNN significantly outperforms the other state-of-the-art methods in denoising.

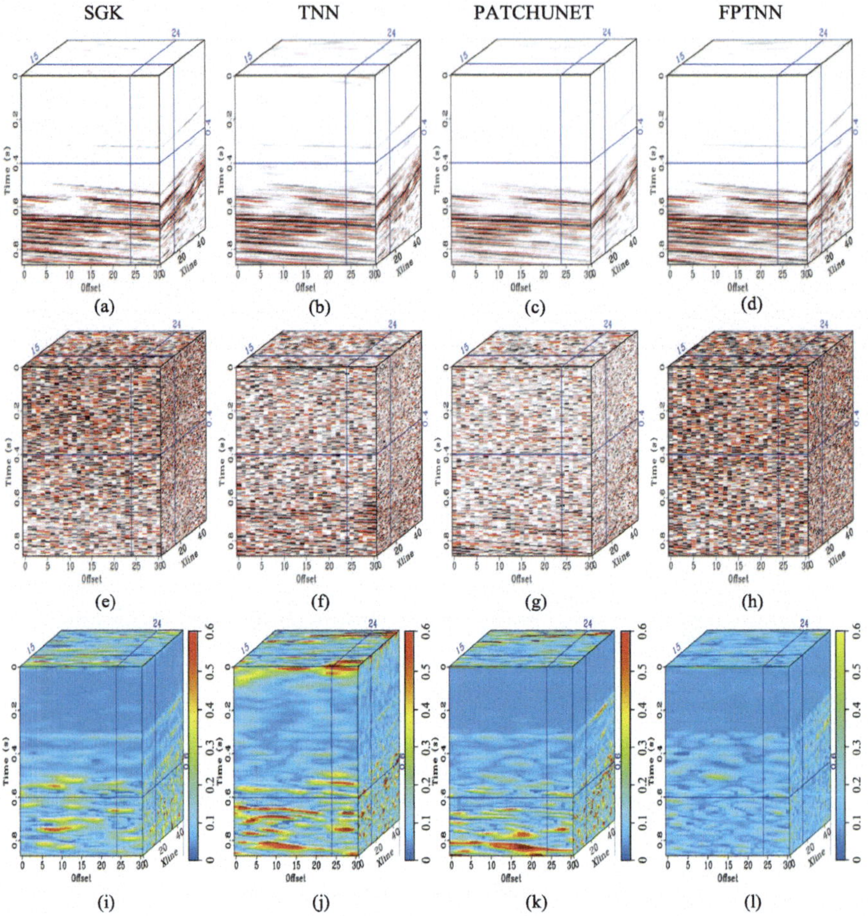


**Fig. 7.8** The denoising comparison for the xline-offset-timeline data volume (yline = 1) in the 4D synthetic VSP seismic data demonstrates the outcomes achieved by various methods. The denoised results for each approach are as follows: (a) SGK, (b) TNN, (c) PATCHUNET, and (d) the proposed FPTNN. The corresponding removed noise for the denoised data from methods (a) to (d) is shown in panels (e) to (h). Furthermore, panels (i) to (l) present the local similarity maps generated from the denoised results in (a) to (d) in conjunction with the removed noise in (c) to (h)

**Table 7.4** Comparison of accuracy and performance between two different loss functions for the FPTNN model on a 5D synthetic dataset

| Methods | −5 dB | −2.5 dB | 0 dB | 2.5 dB | 5 dB |
|---|---|---|---|---|---|
| FPTNN-SURE | 15.86 | 16.70 | 17.87 | 19.96 | 20.41 |
| FPTNN-MSE | 16.68 | 19.01 | 20.48 | 20.72 | 21.50 |

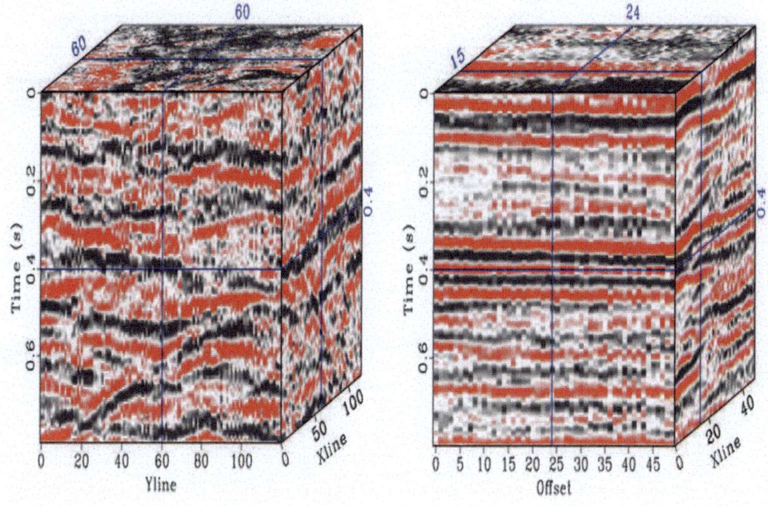


**Fig. 7.9** Third 4-D real WY example. (a) Noisy xline-yline-timeline data volume. (b) Noisy xline-offset-timeline data volume

In the previous two experiments, a noise level of 5 dB was added. To further validate the denoising performance of FPTNN and benchmark algorithms under varying noise intensities, we conducted a third experiment, setting the noise intensity to range from -5 dB to 10 dB in intervals of 2.5 dB. The results are shown in Fig. 7.5. It can be observed that, across all noise intensities, FPTNN consistently outperforms the existing benchmark algorithms, particularly in high SNR scenarios. In these cases, the key parameter $\sigma$ can be estimated more accurately using Eq. (7.24).

### *7.4.3 Field Data Tests*

Although the FPTNN algorithm has demonstrated significant denoising advantages over existing benchmark algorithms in synthetic data experiments, it is necessary to validate its robustness on real field dataset. The real dataset used in this study is a 4D seismic field dataset from the WY survey area, with dimensions of time, xline, yline, and offset. This dataset is heavily affected by random noise during the acquisition process, as shown in Fig. 7.9(a)(b), making it an ideal benchmark for testing denoising performance. Unlike synthetic data, real datasets do not have clean data for direct comparison of denoising results. In this case, qualitative analysis can only be performed through visual inspection of the recovery effects and residual noise. For quantitative analysis, we rely on local similarity as a metric, although it is less intuitive than SNR.

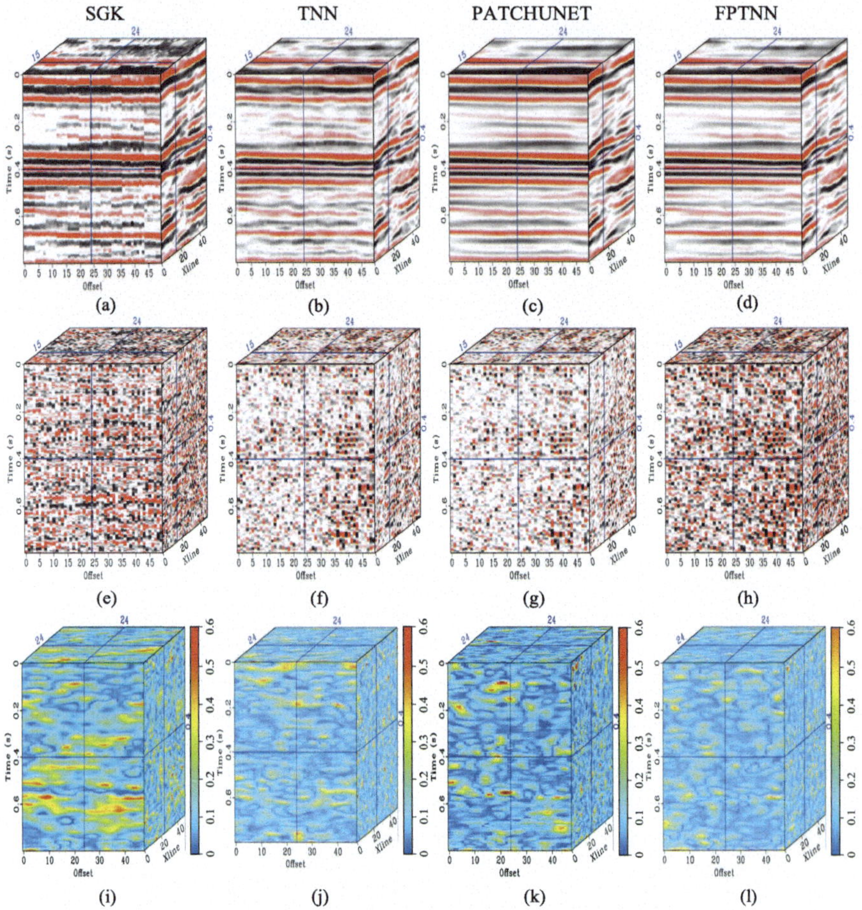

**Fig. 7.10** Denoising comparison for the xline-yline-timeline data volume (yline = 1) in the 4D real WY seismic dataset. Presented in (a) is the noisy data, followed by the denoised outputs from (a) SGK, (b) TNN, (c) PATCHUNET, and (d) the proposed FPTNN methods. Panels (f)–(i) display the corresponding removed noise from (e)–(h). Lastly, (i)–(l) show the local similarity maps calculated between the denoised data in (e)–(h) and the extracted noise in (f)–(j)

The denoising performance of FPTNN and benchmark algorithms on the 4D WY field dataset is shown in Figs. 7.11 and 7.10. Due to the complexity of the noise and data in real datasets, some benchmark algorithms are unable to effectively remove this noise, which is particularly evident in the prestack gather volume. In contrast, FPTNN effectively eliminates this real-world noise. Moreover, FPTNN not only removes noise but also preserves the valuable signal, as fully validated by the removed noise and corresponding local similarity in Figs. 7.11(e)–(l) and 7.10(e)–(l). From the local similarity maps, we observe that FPTNN has more blue

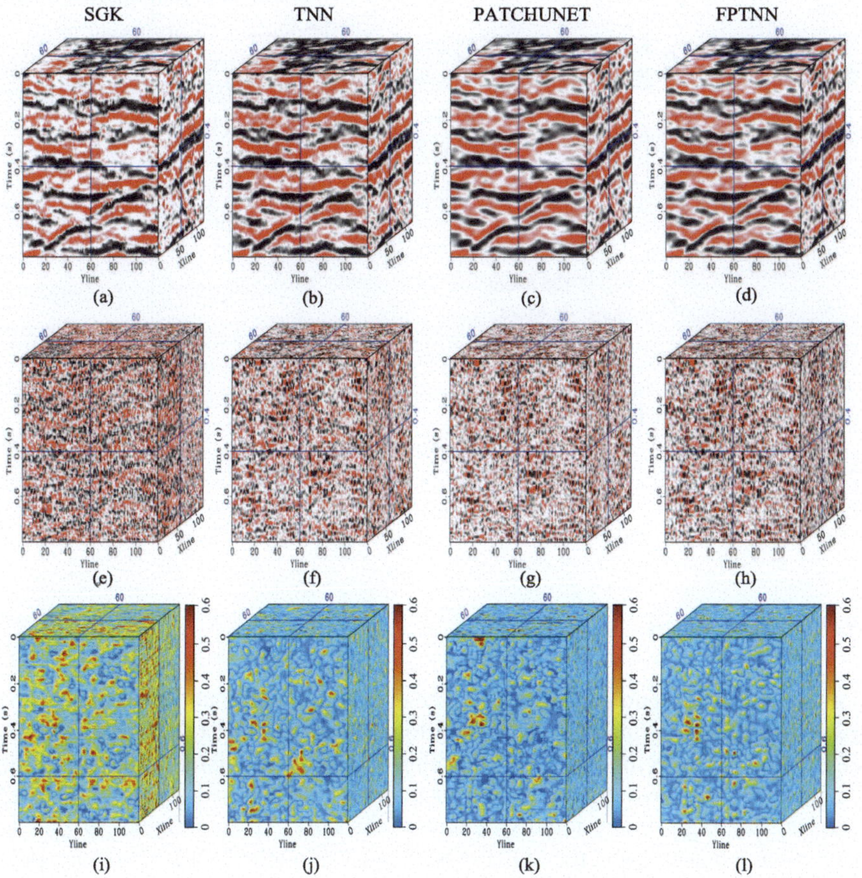


**Fig. 7.11** Comparison of denoising results for the xline-yline-timeline data volume (offset = 1) in the 4D real WY seismic dataset. The denoised outputs are generated by (a) SGK, (b) TNN, (c) PATCHUNET, and (d) the proposed FPTNN methods. The corresponding removed noise for each method in (a)–(d) is shown in (e)–(h). In addition, (i)–(l) display the local similarity maps derived from comparing the denoised data in (a)–(d) with the noise extracted in (e)–(h)

regions of low similarity and fewer red regions of high similarity compared to the benchmark algorithms, indicating less leakage of valuable signals into the removed noise. In summary, the denoising results, residuals, and local similarity presented in Figs. 7.11 and 7.10 lead to the conclusion that FPTNN demonstrates superior denoising performance on real field datasets compared to benchmark algorithms, aligning perfectly with the findings from the synthetic data experiments.

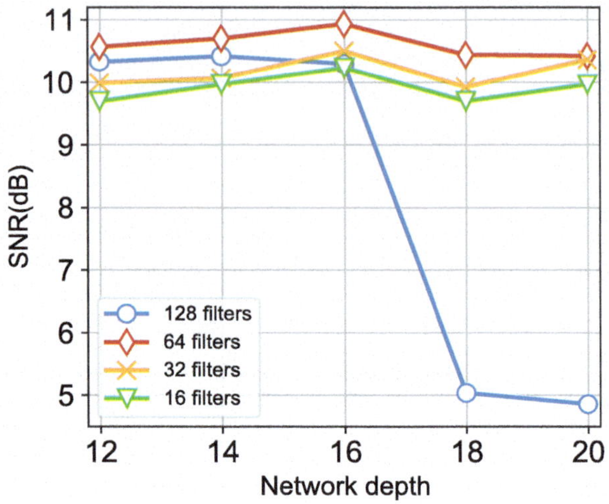


**Fig. 7.12** The impact of varying network depth and the number of filters per layer is illustrated. Symbols "○", "◇", "×", and "▽" represent filter counts of 128, 64, 32, and 16, respectively

## *7.4.4 Discussion*

As shown in Fig. 7.9 and Algorithm 7.1, the noise removal capability of the FPTNN model stems from its integration of traditional deep learning (DL) techniques for image denoising. This model transforms multi-dimensional seismic data into a series of framelet images, avoiding the need for flattening operations, thereby effectively preserving the high-dimensional structure of the data. These framelet images are then further denoised using the SURE-based DnCNN method [48]. While this paper does not delve into this aspect, it is clear that incorporating more advanced state-of-the-art (SOTA) deep learning models could potentially enhance the denoising performance of the FPTNN. Leveraging the powerful noise suppression capability of the SURE-DnCNN, the FPTNN demonstrates outstanding robustness across datasets with varying frequency and amplitude characteristics, as illustrated in Figs. 7.3, 7.7, 7.8, 7.10, and 7.11. The experimental results clearly indicate that the FPTNN excels in feature extraction compared to other existing advanced methods.

In the design of DnCNN, two key parameters influence the model's denoising performance: the number of layers in the network and the number of filters in each layer. Typically, optimizing these parameters involves time-consuming and computationally expensive experiments. To address this, we performed tuning experiments on the proposed FPTNN using a synthetic five-dimensional dataset. The results, presented in Fig. 7.12, demonstrate the denoising performance of the FPTNN under different configurations of network depth and filter counts. The best performance is observed when the network is configured with 16 layers, each containing 64 filters. It is worth noting that increasing the number of filters to 128 at this depth causes training to fail, resulting in experimental failure.

As a deep learning (DL) model, the FPTNN exhibits significant computational complexity, especially when compared to traditional methods like TNN. To offer a more relevant benchmark for assessing computational demands, we choose PATCHUNET as a reference model. Both models are based on deep learning and were run on the same Google Colab platform, as shown in Table 7.3. From the data presented, it is clear that FPTNN incurs lower computational costs than PATCHUNET. This discrepancy arises from the differing approaches used in sample generation. Specifically, PATCHUNET processes overlapping multidimensional patches from the image, whereas FPTNN operates on the entire framelet image, akin to conventional denoising methods. The use of overlapping patches in PATCHUNET leads to a greater number of samples, which significantly increases the computational load and overall complexity.

In the FPTNN model, we utilize the SURE criterion to estimate the MSE without relying on ground truth data. This criterion enables the DnCNN to function in an unsupervised learning mode, yielding denoising results that are comparable to those of MSE-DnCNN, which is trained in a supervised manner. However, additional experiments are needed to fully verify the robustness of this method. Table 7.4summarizes the denoising performance of FPTNN using both MSE and SURE criteria across different noise levels. While the supervised model (FPTNN-MSE) tends to deliver slightly better performance than the unsupervised model (FPTNN-SURE), the difference in SNR between the two approaches is minimal, even under noise conditions below 0 dB. This suggests that the SURE-based approach maintains stable denoising effectiveness, demonstrating its reliability under various noise conditions.

As previously mentioned, the noise reduction ability of our FPTNN model primarily stems from its reliance on deep learning techniques for image denoising. However, this reliance also implies that the FPTNN is somewhat limited when compared to DnCNN. As highlighted in the work by Zhang et al. [58], DnCNN faces two significant challenges: its inability to effectively handle real complex noise and its struggle to adapt to variations in noise levels.

## 7.5 Summary

This chapter introduces an innovative framework called FPTNN, which combines order-$p$ tensor operations with framelet transforms. This framework is specifically designed to tackle the challenge of extracting intricate structural details from high-dimensional seismic data. At the heart of FPTNN are two major components: the framelet-based order-$p$ tensor product and the order-$p$ tensor neural network architecture, with the latter being an extension of the former. By employing this order-$p$ tensor framework, the multi-dimensional denoising problem is transformed into a nonlinear regression task, making it suitable for deep learning (DL) techniques. The order-$p$ tensor operation is implemented via matrix multiplication in the framelet domain, allowing FPTNN to optimize its parameters through deep learning applied to the transformed matrix slices in this domain. Experimental evaluations show that

**Algorithm 7.2:** Order-$p$ tSVD [40, 59]

**Input**: $\mathcal{X}$.
**Output**: $\mathcal{U}, \mathcal{S}, \mathcal{V}$.
**for** $i = 3$ *to* p **do**
  $\overline{\mathcal{X}} \leftarrow \text{fft}(\mathcal{X}, i)$.
**end**
**for** $i_3 = 1$, *to*, $n_3$ **do**
  **for** $i_4 = 1$, *to*, $n_4$ **do**
    $\cdots$
    **for** $i_p = 1$, *to*, $n_p$ **do**
      $[\widehat{\mathbf{U}}, \widehat{\mathbf{S}}, \widehat{\mathbf{V}}] = \text{SVD}(\overline{\mathcal{X}}(:, :, i_3, i_4, \cdots, i_p))$.
      $\overline{\mathcal{U}}(:, :, i_3, i_4, \cdots, i_p) = \widehat{\mathbf{U}}$.
      $\overline{\mathcal{S}}(:, :, i_3, i_4, \cdots, i_p) = \widehat{\mathbf{S}}$.
      $\overline{\mathcal{V}}(:, :, i_3, i_4, \cdots, i_p) = \widehat{\mathbf{V}}$.
    **end**
    $\cdots$
  **end**
**end**
**for** $i = 3$ *to* p **do**
  $\mathcal{U} \leftarrow \text{ifft}(\overline{\mathcal{U}}, [\,], i)$;
  $\mathcal{S} \leftarrow \text{ifft}(\overline{\mathcal{S}}, [\,], i)$;
  $\mathcal{V} \leftarrow \text{ifft}(\overline{\mathcal{V}}, [\,], i)$.
**end**

FPTNN outperforms three advanced methods in terms of signal-to-noise ratio (SNR) and visual quality, on both synthetic and real-world seismic datasets.

There are several promising avenues for future research that should be explored. One key direction involves extending our model to address more diverse and complex noise types, including coherent and erratic noise. Furthermore, the adaptability of the FPTNN framework provides ample opportunity for enhancement. Future work could focus on integrating other sparse transforms, such as the shearlet and curvelet transforms, in addition to the current framelet transform. This would increase the model's versatility and may lead to improvements in its denoising performance.

## Appendix

### Multidimensional Tensor Notations

To enhance clarity, we begin by defining some fundamental tensor notations in Table 8.1. Next, we introduce three essential block-based operators [52] that are necessary for constructing our FPTNN schemes: the block circulant operator circ$(\cdot)$, the unfolding operator $(\cdot)$, and the folding operator $(\cdot)$. For a tensor $\mathcal{X} \in \mathbb{R}^{n_1 \times n_2 \cdots \times n_p}$, the notation $\mathcal{X}^{(\ell)}$ is used to denote the operation involving block circulants.

$$\text{circ}(\mathcal{X}) := \begin{bmatrix} \mathcal{X}^{(1)} & \mathcal{X}^{(n_p)} & \dots & \mathcal{X}^{(2)} \\ \mathcal{X}^{(2)} & \mathcal{X}^{(1)} & \dots & \mathcal{X}^{(3)} \\ \vdots & \vdots & \ddots & \vdots \\ \mathcal{X}^{(n_p)} & \mathcal{X}^{(n_p-1)} & \dots & \mathcal{X}^{(1)}. \end{bmatrix}. \tag{7.25}$$

**Framelet Transform**

A central concept explored in this chapter is the framelet transform, which provides better time-frequency localization than traditional global transforms, such as the Fourier transform [9, 10]. To support this discussion, we will offer a brief introduction to the framelet theory [60] for completeness. In the domain of multidimensional seismic data analysis, tight frames are well-suited for handling multivariate data. However, due to the unfolding operation applied to multidimensional tensors in Sect. 7.3.1, this chapter focuses on tight frames in a univariate context. This approach is chosen because the implementation involves tensor product frames. Specifically, we will examine a countable subset $\mathbf{X} \subset L_2(\mathbb{R})$, which is considered a tight frame if it meets the following condition for all $f \in L_2(\mathbb{R})$:

$$f = \sum_{g \in X} \langle f, g \rangle. \tag{7.26}$$

This is equivalent to

$$\|f\|^2_{L_2(\mathbb{R})} = \sum_{g \in X} |\langle f, g \rangle|^2. \tag{7.27}$$

where $|\cdot|$ represents the norm of the functions $f, g \in L_2(\mathbb{R})$, and $\langle \cdot, \cdot \rangle$ denotes their inner product. The notion of a tight frame relaxes the conventional requirement for $\mathbf{X}$ to be a strict basis, permitting some redundancy. This redundancy has been demonstrated to improve performance in a variety of applications related to seismic image processing (e.g., [9, 10]).

An affine (or wavelet) system $\mathbf{X}(\Psi)$ is a collection of function combinations obtained by applying different dilations and translations to each element in the set $\Psi = \{\psi_1, \dots, \psi_r\} \subset L_2(\mathbb{R})$.

$$\mathbf{X}(\Psi) = \left\{\psi_{j,k} = 2^{j/2}\psi\left(2^j x - k\right), \quad j \in \mathbb{Z}, k \in \mathbb{Z}, \psi \in \Psi\right\}. \tag{7.28}$$

When the system $\mathbf{X}(\Psi)$ constitutes a tight frame in the space $L^2(\mathbb{R})$, it is referred to as a tight framelet. Each individual function $\psi_l$ in this system is then termed a framelet element (or tight framelet element).

In image processing, the framelet transform of a vector $\mathbf{v} \in \mathbb{R}^n$ is typically represented by a matrix $\mathbf{W} \in \mathbb{R}^{wn \times n}$, which consists of $n$ filters and $l$ decomposition levels, where $w = (n-1)l + 1$ [60, 61]. Using the matrix $\mathbf{W}$, one can calculate the associated frame coefficient vector $\mathbf{u}$.

$$\mathbf{u} = \mathbf{W}\mathbf{v} \in \mathbb{R}^{wn}. \tag{7.29}$$

Once the transformation matrix $\mathbf{W}$ is determined, its inverse framelet transform can be derived by taking the transpose of $\mathbf{W}$, represented as $\mathbf{W}^\top$. This operation allows for the effective reconstruction of the original signal, effectively reversing the transformation process.

$$\mathbf{g} = \mathbf{W}^\top \mathbf{u}. \tag{7.30}$$

Typically, a shrinkage operator is utilized on the framelet coefficients for the purpose of reducing image noise [9, 10]. Since the shrinkage value needs to be predetermined based on the noise level, we propose an alternative method that employs deep learning to replace the conventional shrinkage operator, thereby enabling enhanced noise reduction capabilities.

## References

1. Zhang L, Zuo W (2017) Image restoration: from sparse and low-rank priors to deep priors [lecture notes]. IEEE Signal Process Mag 34(5):172–179
2. Chen W, Xie J, Zu S, Gan S, Chen Y (2016) Multiple-reflection noise attenuation using adaptive randomized-order empirical mode decomposition. IEEE Geosci Remote Sens Lett 14(1):18–22
3. Jian M, Jinghuai G, Wenchao C (2006) On the denoising method of prestack seismic data in wavelet domain. In: 2006 SEG annual meeting. OnePetro
4. Shan H, Ma J, Yang H (2009) Comparisons of wavelets, contourlets and curvelets in seismic denoising. J Appl Geophy 69(2):103–115
5. Li JH, Zhang YJ, Qi R, Liu QH (2017) Wavelet-based higher order correlative stacking for seismic data denoising in the curvelet domain. IEEE J Sel Top Appl Earth Obs Remote Sens 10(8):3810–3820
6. Górszczyk A, Adamczyk A, Malinowski M (2014) Application of curvelet denoising to 2d and 3d seismic data-practical considerations. J Appl Geophy 105:78–94
7. Kong Dehui, Peng Zhenming (2015) Seismic random noise attenuation using shearlet and total generalized variation. J Geophys Eng 12(6):1024–1035
8. Liu C, Wang D, Wang T, Feng F, Cheng H, Meng G (2014) Random seismic noise attenuation based on the shearlet transform. Acta Petrolei Sinica 35(4):692–699
9. Yang P, Gao J, Chen W (2012) Noise attenuation of the geophysical data using the framelet transform. In: 2012 IEEE international geoscience and remote sensing symposium. IEEE, pp 2300–2303
10. Liang J, Ma J, Zhang X (2014) Seismic data restoration via data-driven tight frame. Geophysics 79(3):V65–V74
11. Liu Y, Fomel S, Liu C (2015) Signal and noise separation in prestack seismic data using velocity-dependent seislet transform. Geophysics 80(6):WD117–WD128
12. Li J, Sacchi MD (2021) An $l_p$-space matching pursuit algorithm and its application to robust seismic data denoising via time-domain radon transforms. Geophysics 86(3):V171–V183
13. Wang H, Chen W, Zhang Q, Liu X, Zu S, Chen Y (2020) Fast dictionary learning for high-dimensional seismic reconstruction. IEEE Trans Geosci Remote Sens
14. Yangkang C (2017) Fast dictionary learning for noise attenuation of multidimensional seismic data. Geophy J Int 209(1):21–31
15. Zhou Y, Yang J, Wang H, Huang G, Chen Y (2020) Statistics-guided dictionary learning for automatic coherent noise suppression. IEEE Trans Geosci Remote Sens
16. Chen W, Saad OM, Oboué YA, Yang L, Chen Y (2021) Retrieving the leaked signals from noise using a fast dictionary learning method. Geophysics 87(1):1–86

17. Lina L, Jianwei M (2018) Structured graph dictionary learning and application on the seismic denoising. IEEE Trans Geosci Remote Sens 57(4):1883–1893
18. Xiaojing W, Jianwei M (2019) Adaptive dictionary learning for blind seismic data denoising. IEEE Geosci Remote Sens Lett 17(7):1273–1277
19. Siahsar MAN, Gholtashi S, Kahoo AR, Chen W, Chen Y (2017) Data-driven multitask sparse dictionary learning for noise attenuation of 3d seismic data. Geophysics 82(6):V385–V396
20. Zhu L, Liu E, McClellan JH (2015) Seismic data denoising through multiscale and sparsity-promoting dictionary learning. Geophysics 80(6):WD45–WD57
21. Anvari R, Siahsar MAN, Gholtashi S, Kahoo AR, Mohammadi M (2017) Seismic random noise attenuation using synchrosqueezed wavelet transform and low-rank signal matrix approximation. IEEE Trans Geosci Remote Sens 55(11):6574–6581
22. Ouyang S, Lingling W, Hu X, Zhidan L (2021) Seismic denoising via truncated nuclear norm minimization. Geophysics 86(2):V153–V169
23. Chong W, Zhihui Z, Gu H, Wu X, Shuaiqi L (2018) Hankel low-rank approximation for seismic noise attenuation. IEEE Trans Geosci Remote Sens 57(1):561–573
24. Yangkang C, Yatong Z, Wei C, Zu S, Weilin H, Dong Z (2017) Empirical low-rank approximation for seismic noise attenuation. IEEE Trans Geosci Remote Sens 55(8):4696–4711
25. Jun F, Xi L, Xiaoqin L, Xu W, Bin L (2021) Low-rank tensor minimization method for seismic denoising based on variational mode decomposition. IEEE Geosci Remote Sens Lett 19:1–5
26. Anvari R, Mohammadi M, Kahoo AR (2018) Enhancing 3-d seismic data using the t-SVD and optimal shrinkage of singular value. IEEE J Sel Top Appl Earth Obs Remote Sens 12(1):382–388
27. Feng J, Li X, Liu X, Chen C, Chen H (2021) Seismic data denoising based on tensor decomposition with total variation. IEEE Geoscie Remote Sens Lett
28. Yu S, Jianwei M, Wenlong W (2019) Deep learning for denoising. Geophysics 84(6):V333–V350
29. Liuqing Y, Wei C, Hang W, Yangkang C (2021) Deep learning seismic random noise attenuation via improved residual convolutional neural network. IEEE Trans Geosci Remote Sens 59(9):7968–7981
30. Yuan Y, Si X, Zheng Y (2020) Ground-roll attenuation using generative adversarial networksground-roll attenuation using gans. Geophysics 85(4):WA255–WA267
31. Saad OM, Chen Y (2020) Deep denoising autoencoder for seismic random noise attenuation. Geophysics 85(4):V367–V376
32. Feng Q, Wei G, Zhangbo L, Yu H, Gulan Z, Hu G (2022) Unsupervised erratic seismic noise attenuation with robust deep convolutional autoencoders. IEEE Trans Geosci Remote Sens 60:1–16
33. Sun J, Slang S, Elboth T, Greiner TL, McDonald S, Gelius L-J (2020) Attenuation of marine seismic interference noise employing a customized u-net. Geophy Prospect 68(3):845–871
34. Liu D, Wang W, Chen W, Wang X, Zhou Y, Shi Z (2018) Random noise suppression in seismic data: what can deep learning do? In: SEG technical program expanded abstracts 2018. Society of Exploration Geophysicists, pp 2016–2020
35. Dawei L, Wei W, Xiaokai W, Cheng W, Jiangyun P, Wenchao C (2019) Poststack seismic data denoising based on 3-d convolutional neural network. IEEE Trans Geosci Remote Sens 58(3):1598–1629
36. Yang L, Wang S, Chen X, Saad OM, Chen W, Oboué ASI, Chen Y (2021) Unsupervised 3-d random noise attenuation using deep skip autoencoder. IEEE Trans Geosci Remote Sens 60:1–16
37. Saad OM, Oboué YASI, Bai M, Samy L, Yang L, Chen Y (2021) Self-attention deep image prior network for unsupervised 3-d seismic data enhancement. IEEE Trans Geosci Remote Sens
38. Jen-Tzung C, Yi-Ting B (2017) Tensor-factorized neural networks. IEEE Trans Neural Netw Learn Syst 29(5):1998–2011
39. Newman E, Horesh L, Avron H, Kilmer M (2018) Stable tensor neural networks for rapid deep learning. arXiv preprint arXiv:1811.06569

40. Ely G, Aeron S, Hao N, Kilmer ME (2015) 5d seismic data completion and denoising using a novel class of tensor decompositions. Geophysics 80(4):V83–V95
41. Jiang TX, Ng MK, Zhao XL, Huang TZ (2020) Framelet representation of tensor nuclear norm for third-order tensor completion. IEEE Trans Image Process
42. Feng Q, Zhangbo L, Yan W, Songjie L, Shengli P, Hu G (2021) Dtae: deep tensor autoencoder for 3-d seismic data interpolation. IEEE Trans Geosci Remote Sens 60:1–19
43. Yangkang C, Jianwei M, Sergey F (2016) Double-sparsity dictionary for seismic noise attenuation. Geophysics 81(2):V103–V116
44. eon Bottou L et al (1998) Online learning and stochastic approximations. On-Line Learn Neural Netw 17(9):142
45. Zhou Y, Song X, Zhang Y, Liu F, Zhu C, Liu L (2021) Feature encoding with autoencoders for weakly-supervised anomaly detection. IEEE Trans Neural Netw Learn Syst
46. Kingma DP, Adam JB (2014) A method for stochastic optimization. arXiv preprint arXiv:1412.6980
47. Feng Q, Miao Y, Xiao-Yang L, Yao-Jun W, Lu C, Hu G-M (2018) Unsupervised seismic facies analysis via deep convolutional autoencoders. Geophysics 83(3):A39–A43
48. Soltanayev S, Chun SY (2018) Training deep learning based denoisers without ground truth data. In: Proceedings of the 32nd international conference on neural information processing systems, pp 3261–3271
49. Rihuan K, Carola-Bibiane S (2021) Unsupervised image restoration using partially linear denoisers. IEEE Trans Pattern Anal Mach Intell 44(9):5796–5812
50. Feng Q, Zhangbo L, Yan W, Yingjie Z, Hu G (2022) Ground truth-free 3-d seismic random noise attenuation via deep tensor convolutional neural networks in the time-frequency domain. IEEE Trans Geosci Remote Sens 60:1–17
51. Martin CD, Shafer R, LaRue B (2013) An order-p tensor factorization with applications in imaging. SIAM J Sci Comput 35(1):A474–A490
52. Jiang F, Liu X-Y, Lu H, Shen R (2018) Efficient multi-dimensional tensor sparse coding using t-linear combination. In: Proceedings of the AAAI conference on artificial intelligence, vol 32
53. Lu C, Peng X, Wei Y (2019) Low-rank tensor completion with a new tensor nuclear norm induced by invertible linear transforms. In: Proceedings of the IEEE/CVF conference on computer vision and pattern recognition, pp 5996–6004
54. Yan S, Xiaoyuan Y, Yuhua G (2013) Translation invariant directional framelet transform combined with gabor filters for image denoising. IEEE Trans Image Process 23(1):44–55
55. Kai Z, Wangmeng Z, Yunjin C, Deyu M, Lei Z (2017) Beyond a gaussian denoiser: Residual learning of deep CNN for image denoising. IEEE Trans Image Process 26(7):3142–3155
56. Saad, OM, Chen Y (2021) A fully unsupervised and highly generalized deep learning approach for random noise suppression. Geophys Prospect 69(4):709–726
57. Chen Y, Fomel S (2015) Random noise attenuation using local signal-and-noise orthogonalization. Geophysics 80(6):WD1–WD9
58. Kai Z, Wangmeng Z, Lei Z (2018) Ffdnet: toward a fast and flexible solution for CNN-based image denoising. IEEE Trans Image Process 27(9):4608–4622
59. Kilmer ME, Braman K, Hao N, Hoover RC (2013) Third-order tensors as operators on matrices: a theoretical and computational framework with applications in imaging. SIAM J Matrix Anal Appl 34(1):148–172
60. Cai J-F, Chan RH, Shen Z (2008) A framelet-based image inpainting algorithm. Appl Comput Harmon Anal 24(2):131–149
61. Tai-Xiang J, Ting-Zhu H, Xi-Le Z, Teng-Yu J, Liang-Jian D (2018) Matrix factorization for low-rank tensor completion using framelet prior. Inf Sci 436:403–417

# Chapter 8
# Robust Tensor Deep Learning for Seismic Erratic Noise Attenuation

## 8.1 Introduction

Incoherent or random noise poses a significant threat to seismic recordings, as it can severely degrade signal quality and lead to reflection records with diminished coherency. This includes both Gaussian and erratic noise. Erratic noise, in particular, is characterized by abnormally high amplitudes and a non-Gaussian distribution, making its elimination much more challenging than that of Gaussian noise [1]. Addressing this issue, several studies have proposed innovative approaches that have shown great promise. These approaches include dictionary learning (DcL) [2, 3], deep learning (DL) [4–6], and robust low-rank approximation (LRA) [7–9]. A key challenge highlighted across these studies is the need to develop effective representations or models that can accurately capture and identify erratic noise in contaminated records, thereby facilitating its removal.

In seismic literature, most denoising research has predominantly concentrated on addressing normal random noise. The corresponding methods can be broadly categorized into four main types: spatial domain filtering (SDF), sparse representation (SR), low-rank approximation (LRA), and deep learning (DL). Considering the correlation between a specific working position and its neighboring regions, SDF involves directly denoising seismic images by sliding a filter across each position. Representative SDF techniques include predictive filtering [10, 11], morphological filtering [12, 13], diffusion filtering [14–16], mean filtering [17], and median filtering [18], among others.

The SR method is based on the principle that clear seismic images can be encoded into a small number of coefficients within a known sparse transform domain, whereas noise cannot be encoded. Common transforms used in SR methods include wavelet transforms [19, 20], curvelet transforms [21, 22], seislet transforms [23, 24], shearlet transforms [25], and dictionary learning [2, 3, 26–28].

In LRA-based methods, the underlying assumption is that clear seismic images exhibit low-rank characteristics, and denoising is accomplished through various rank

 
F. Qian et al., *Tensor Computation for Seismic Data Processing*,
Earth Systems Data and Models 6, https://doi.org/10.1007/978-3-031-78900-7_8

reduction techniques. These include matrix LRA [8, 9, 29, 30], structured (Hankel) LRA [7, 31, 32], CANDECOMP/PARAFAC (CP) decomposition [33], tensor ring decomposition [34], and tensor singular-value decomposition (tSVD) [35–38].

DL offers a novel approach to designing highly effective denoisers for Gaussian noise by learning mappings between noisy and clean seismic images. DL networks used for this purpose include tensor DL networks [39, 40], reinforcement learning [41], Bayesian DL networks [42, 43], and generative adversarial networks (GANs) [44]. Various types of deep convolutional neural networks (CNNs) are also widely used, such as classic CNNs [4–6, 45], deep convolutional autoencoders (DCAEs) [46–49], U-Net [50], generator CNNs [51], and invertible neural networks [52].

Notably, specific methods must be designed to target the suppression of erratic noise, rather than simply applying standard Gaussian noise reduction techniques. This is because erratic noise is spatially non-Gaussian and manifests as abrupt pulses in 3-D seismic data. These unexpected pulses can arise from issues like malfunctioning geophones, electrical interference, or the use of machinery near geophones [1, 53]. Robust norm-based methods are commonly employed to address this type of noise, and filtering techniques that account for the non-Gaussian distribution of erratic noise have shown promising results [54, 55].

In denoising models, robust norms can be applied in two primary ways: either to directly target erratic noise or to impose an indirect constraint on data fidelity within the objective function. In the former approach, it is assumed that erratic noise is sparsely represented by the $\ell_l$ norm, and robust principal component analysis (RPCA) is used to decompose seismic images into a sparse component (representing erratic noise) and a low-rank component (the clean image) [56, 57]. In the latter approach, erratic noise is treated as statistical outliers, and the outlier-sensitive $\ell_2$-norm data fidelity is replaced with more robust norms, such as Tukey's biweight function [58], the Welsch function [59], or the Huber function [60, 61].

Simultaneously, useful signals are extracted from contaminated 3-D seismic data using various algorithmic approaches, including low-rank representations [1, 58, 60–63], sparse representations in transform domain [53, 64], deep learning [59], and dictionary learning [65, 66]. However, most existing methods approach the problem of suppressing erratic noise in a 2-D image-based manner. Modern seismic data, characterized by its high dimensionality, is not well suited to these approaches. Vectorization or matricization of the data fails to leverage the multilinear structure of seismic images, often resulting in suboptimal performance for erratic noise suppression.

In comparison to current approaches for reducing erratic noise, our RTDL model presents three key innovations that make it technically robust. These innovations form the core contributions of our work.

(1) *Erratic noise modeling*: The core of RTDL is the development of a third-order tensor sparse representation (SR) model, specifically a third-order tensor sparse $\ell_{1,1,2}$ norm, designed for erratic noise. This model leverages the spatial tubular sparse distribution characteristics of noise in 3-D space. As highlighted in the

literature review, our approach to modeling erratic noise distinguishes itself from all existing matrix-based noise removal methods.

(2) *RTDL model*: To fully leverage the multidimensional structure of seismic tensors, the RTDL model is introduced by decomposing the noisy data into tensor deep learning (DL) components and sparse components. This method is regarded as a generalization of robust tensor principal component analysis (RTPCA) and serves as a tensor adaptation of the robust deep autoencoder approach previously presented in [67–69]. This approach seeks to address a convex optimization problem, characterized by an objective function that combines two components: a tensor convolutional autoencoder and the sparse $\ell_{1,1,2}$-norm. To our knowledge, this chapter represents the first instance of applying tensor computations for unsupervised removal of erratic noise.

(3) *Solution*: To determine the new model parameters, an effective tensor optimization method can be derived based on alternating minimization (AltMin) for the RTDL model. This optimized method retains the low implementation cost advantages associated with AltMin.

Extensive experiments carried out on both synthetic and real field seismic data validate the effectiveness and quality of the proposed RTDL approach.

## 8.2 Seismic Erratic Noise Attenuation and Its Formulation

This section aims to address noise reduction using the RTDL model. First, we need to clarify the issue of 3-D erratic noise suppression in Sect. 8.2.1 to provide a foundational understanding of the model. Following this, our RTDL approach combines a tensor neural network (tNN) with a tensor sparse norm, as detailed in Sect. 8.2.2. Additionally, Table 8.1 provides the fundamental notations along with their corresponding descriptions.

**Table 8.1** Basic notations and their descriptions

| Notation | Description |
|---|---|
| $\mathcal{T}, T, t$ | tensor, matrix, vector. |
| $\mathcal{T}^{(k)}$ | $k$th frontal slice of a tensor $\mathcal{T}$: i.e., $\mathcal{T}^{(k)} \equiv \mathcal{X}(:, :, k)$. |
| $\widetilde{\mathcal{T}}$ | Fourier domain representation of a tensor $\mathcal{T}$. |
| $\mathcal{A} * \mathcal{B}$ | t-product of tensors $\mathcal{A}$ and $\mathcal{B}$. |
| $[N]$ | set $\{1, 2, \ldots, N\}$. |
| $\|T\|_F$ | Frobenius norm of a matrix $T$. |
| $\|\mathcal{T}\|_F$ | Frobenius norm of a tensor $\mathcal{T}$. |
| $T^\top$ | Transpose operation of a matrix $T$. |
| $\mathcal{T}^\dagger$ | Transpose operation of a tensor $\mathcal{T}$. |

### 8.2.1 Problem Statement

The problem of 3-D seismic erratic noise attenuation can be approximated as follows:

$$\mathcal{Y} = \mathcal{X} + \mathcal{E}, \tag{8.1}$$

where $\mathcal{Y} \in \mathbb{R}^{n_1 \times n_2 \times n_3}$ represents the observed 3-D seismic image, $\mathcal{X} \in \mathbb{R}^{n_1 \times n_2 \times n_3}$ denotes the unknown clean 3-D seismic image, and $\mathcal{E} \in \mathbb{R}^{n_1 \times n_2 \times n_3}$ signifies the erratic noisethat this work aims to eliminate through various methods, including low-rank representations [1, 60] and sparse representations (SRs) [53, 64, 70]. Equation (8.1) represents an ill-posed problem, meaning that we fail to produce a unique solution from the image noise model concerning erratic noise. Therefore, it becomes necessary for us to add constraints to obtain the actual solution. Consequently, Eq. (8.1) can be rewritten as follows:

$$\underset{\mathcal{X},\mathcal{E}}{\operatorname{argmin}} \|\mathcal{Y} - \mathcal{X} - \mathcal{E}\|_F + \gamma\varphi(\mathcal{X}) + \beta\psi(\mathcal{E}), \tag{8.2}$$

where $\varphi(\cdot)$ and $\psi(\cdot)$ represent the imposed constraints on $\mathcal{X}$ and $\mathcal{E}$, respectively. Thus, the key to solving Eqs. (8.1) and (8.2) lies in constructing the appropriate constraints for both $\mathcal{X}$ and $\mathcal{E}$.

### 8.2.2 Problem Modeling

In general, 3-D erratic noise exhibits two primary characteristics. First, due to the underlying causes of erratic noise [1], it appears in only a small portion of noisy prestack seismic records, naturally leading to column sparsity in 3-D space. Second, erratic noise that deviates significantly from the standard can be considered as column outliers and removed from noisy 3-D seismic data. To address this, we propose an RTDL model utilizing the $\ell_{1,1,2}$-norm, enhancing the algorithm's robustness to column outliers (as described in [71]), formally defined as follows:

$$\psi(\mathcal{E}) = \|\mathcal{E}\|_{1,1,2}, \tag{8.3}$$

where $\|\mathcal{E}\|_{1,1,2}$ is defined as $\sum_{i,j} \|\mathcal{E}(i, j, :)\|_F$ for 3-D seismic tensors $\mathcal{E}$.

In contrast to classical RTPCA, our approach does not strictly adhere to the low-tubal-rank structure, partly due to the absence of a column sparsity assumption for practical 3-D data, especially in cases involving complex geological structures. Given the potential of DL methods for seismic image denoising [46–48], DL can be leveraged to model clean 3-D data, thus eliminating the need for the low-tubal-rank assumption (i.e., removing the term $\varphi(\mathcal{X})$), which can be expressed as follows:

$$\widehat{\mathcal{X}} = \boldsymbol{h}_\Theta(\mathcal{Y} - \mathcal{E}), \tag{8.4}$$

**Algorithm 8.1:** Proposed RTDL method

**Input**: $\mathcal{Y}$, $\lambda$, $\beta$, $\eta$, $B$.
**Output**: $\mathcal{D}$ and $\mathcal{E}$.
Initialize: $\mathcal{E}_0$ and $\mathcal{D}_0$ are initialized to $\mathbf{0}$.
**while** *(converged=false)* **do**
  Update $\mathcal{D}_\ell = \mathcal{Y} - \mathcal{E}_\ell$.
  $\widetilde{\mathcal{D}}_\ell = \text{dct}(\mathcal{D}_\ell, [\,], 3)$
  **for** $k = 1, 2, \ldots, n_3$ **do**
    Update $\Theta^{(k)}$ using (8.10).
    Update $\widetilde{\mathcal{D}}^{(k)}_{\ell+1}$ using (8.11).
  **end**
  $\mathcal{D}_{\ell+1} = \text{idct}(\widetilde{\mathcal{D}}_{\ell+1}, [\,], 3)$.
  Update $\mathcal{E}_{\ell+1}$ using (8.13).
  Check for convergence:
  $\frac{\|\mathcal{Y}-\mathcal{D}_{\ell+1}-\mathcal{E}_{\ell+1}\|_F}{\|\mathcal{Y}\|_F} \leq \epsilon$.
**end**

where $h_\Theta(\cdot)$ represents the mapping function constructed using DL to model the relationship between the clean and observed tensor data. This process essentially functions as an unsupervised regression, generating an estimate of the clean signal. After developing models for both the erratic noise and useful 3-D seismic images, we can merge equations (8.3) and (9.6) into equation (8.8), leading to the following new expression:

$$\begin{aligned} &\underset{\Theta,\mathcal{E}}{\text{argmin}}\ \|\mathcal{Y} - \boldsymbol{h}_\Theta(\mathcal{Y} - \mathcal{E}) - \mathcal{E}\|_F + \beta\|\mathcal{E}\|_{1,1,2} \\ &\text{s.t. } \mathcal{Y} = \mathcal{X} + \mathcal{E}. \end{aligned} \tag{8.5}$$

Equation (8.5) represents the evolution of the original RTPCA into a comprehensive RTDL model, which combines the advantages of both data-driven and model-driven approaches. In this refined model, the optimization variables shift from $\mathcal{X}$ and $\mathcal{E}$ in RTPCA to $\Theta$ and $\mathcal{E}$, requiring the development of new solution algorithms instead of using the existing RTPCA methods. As a result, there is an immediate need to design the specific network architecture for $\boldsymbol{h}_\Theta(\cdot)$ and devise appropriate optimization algorithms for RTDL models.

## 8.3 Robust Tensor Deep Learning (RTDL)

To comprehensively address Problem (8.5), the RTDL model is proposed as an effective solution for suppressing 3-D erratic noise, as detailed in Sect. 8.3.1. Building on the principle of alternating minimization, we derive the solution algorithm step by step to identify the optimal parameters for this new model, as outlined in Sect. 8.3.2. Lastly, Sect. 8.3.3 provides the necessary details for implementing the RTDL model.

### 8.3.1 RTDL

As demonstrated in [46, 47, 59], and [48], the use of deep convolutional autoencoders (DCAEs) has proven effective for unsupervised seismic image denoising tasks. In this context, once erratic noise has been effectively separated using Eq. (9.6), the 3-D seismic image restoration process follows a similar approach as described in [72]. Thus, deep autoencoders are selected as the basis for $\boldsymbol{h}_{\Theta}(\cdot)$, with modifications expanding this framework to a tensor format, specifically tensor autoencoders, which were initially applied for seismic interpolation in [72]. With this introduction, Eq. (9.6) can be re-expressed as follows:

$$\widehat{\mathcal{X}} = \boldsymbol{h}_{\Theta}(\mathcal{Y} - \mathcal{E}) = \boldsymbol{D}_{\Theta}\left(\boldsymbol{E}_{\Theta}\left(\mathcal{Y} - \mathcal{E}\right)\right), \tag{8.6}$$

where $\boldsymbol{E}_{\Theta}(\cdot)$ represents the tensor encoder network, and $\boldsymbol{D}_{\Theta}(\cdot)$ denotes the tensor decoder network. By substituting Eq. (8.6) into Eq. (8.5), we establish an RTDL model designed specifically for unsupervised 3-D erratic noise denoising, the explicit formulation of which is as follows:

$$\begin{aligned} &\underset{\Theta,\mathcal{E}}{\operatorname{argmin}} \ \|\mathcal{D} - \boldsymbol{D}_{\Theta}\left(\boldsymbol{E}_{\Theta}\left(\mathcal{D}\right)\right)\|_F + \beta\|\mathcal{E}\|_{1,1,2}. \\ &\qquad \text{s.t. } \mathcal{Y} - \mathcal{D} - \mathcal{E} = 0. \end{aligned} \tag{8.7}$$

Based on the constraints of Eq. (8.7), the RTDL model divides the input noise tensor $\mathcal{Y}$ into two separate subsets, $\mathcal{D}$ and $\mathcal{E}$. Here, $\mathcal{D}$ serves as the input to the autoencoder $\boldsymbol{D}_{\Theta}\left(\boldsymbol{E}_{\Theta}(\mathcal{D})\right)$, which is trained by minimizing the reconstruction error $|\mathcal{D} - \boldsymbol{D}_{\Theta}\left(\boldsymbol{E}_{\Theta}\left(\mathcal{D}\right)\right)|F$ through backpropagation (BP). On the other hand, subset $\mathcal{E}$ contains erratic noise, represented using the $\|\cdot\|_{1,1,2}$ norm. The RTDL model minimizes the unconstrained objective function of the original problem (8.7) by reducing the average system error.

$$\begin{aligned} L(\Theta, \mathcal{E}) = \min_{\Theta,\mathcal{E}} \ &\|\mathcal{D} - \boldsymbol{D}_{\Theta}\left(\boldsymbol{E}_{\Theta}\left(\mathcal{D}\right)\right)\|_F + \beta\|\mathcal{E}\|_{1,1,2} \\ &+ \frac{\lambda}{2}\|\mathcal{Y} - \mathcal{D} - \mathcal{E}\|_F^2. \end{aligned} \tag{8.8}$$

Given the nonconvex nature of the full RTDL objective function in Eq. (8.8), achieving convergence to a global optimum poses significant challenges. However, as discussed in [39, 72], the tensor alternating minimization (Alt) algorithm can be effectively applied to obtain high-quality solutions that satisfy the 3-D erratic noise denoising requirements.

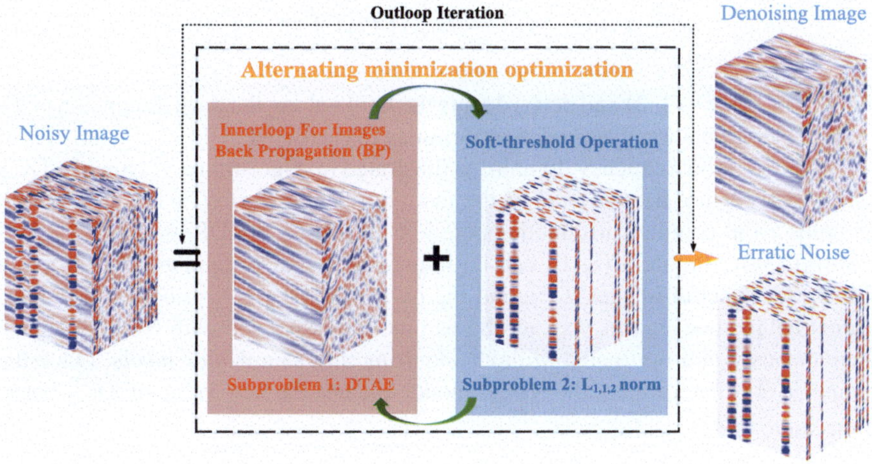


**Fig. 8.1** Schematic of our RTDL model

## 8.3.2 *Optimization of the RTDL Model*

The primary challenge in directly solving Eq. (8.8) lies in the presence of two variables, $\mathcal{E}$ and $\Theta$, within the proposed RTDL model. Drawing inspiration from [67, 68], a viable approach to tackle this RTDL issue is to implement an alternating (Alt) framework that decomposes the original problem into two subproblems: the $\mathcal{E}$-subproblem and the $\Theta$-subproblem, as illustrated in Fig. 8.1. Additionally, we will address several critical aspects of alternating optimization, including convergence criteria and convergence analysis.

### 8.3.2.1 Update $\mathcal{D}_{\ell+1}$

With a fixed $\mathcal{E}_\ell$, the parameter $\Theta_{\ell+1}$ can be updated by solving the following optimization problem:

$$\Theta = \min_{\Theta} \ \|\mathcal{D} - \boldsymbol{D}_{\Theta}\left(\boldsymbol{E}_{\Theta}\left(\mathcal{D}\right)\right)\|_F. \tag{8.9}$$

Similar to the problem addressed in our earlier works [39, 72], the issue outlined in (8.9) is essentially a classic tensor deep learning (TDL) or deep tensor convolutional autoencoder (DTCAE) problem. The theoretical training of these networks is conducted via tensor backpropagation (TBP). However, in practical applications, problem (8.9) can be tackled by resolving each frontal slice $\widetilde{\mathcal{X}}^{(k)}$ individually in the transformed domain, where $1 \leq k \leq n_3$.

$$\Theta^{(k)} = \min_{\Theta} \left\| \widetilde{\mathcal{D}}_{\ell}^{(k)} - \boldsymbol{D}_{\Theta^{(k)}} \left( \boldsymbol{E}_{\Theta^{(k)}} \left( \widetilde{\mathcal{D}}_{\ell}^{(k)} \right) \right) \right\|_F . \tag{8.10}$$

where $\widetilde{\mathcal{D}} = \text{dct}(\mathcal{D}, [\,], 3)$ and $\text{dct}(\cdot)$ denote the DCTs along the third dimension of $\mathcal{D}$. In fact, the computational approach of formula (15) is rooted in the characteristics of the $t$-product in the frequency domain, as illustrated in Algorithm 3.1. Notably, DCT is suggested for use here instead of the fast Fourier transform (FFT), an approach that helps avoid complex arithmetic operations in subsequent DL models.

Surprisingly, Problem (8.10) can be transformed into a standard DCAE-based problem for spectral images $\widetilde{\mathcal{D}}^{(k)}$, and the parameter $\Theta^{(k)}$ is learned using the traditional BP method. In this way, all the parameters associated with $\Theta$ can be tuned simultaneously and optimally, without requiring any samples or labels. Once the parameter $\Theta$ is determined, the RTDL model updates the new value of $\widetilde{\mathcal{D}}_{\ell+1}^{(k)}$ based on the value of $\widetilde{\mathcal{D}}_{\ell}^{(k)}$ as follows:

$$\widetilde{\mathcal{X}}^{(k)} = \widetilde{\mathcal{D}}_{\ell+1}^{(k)} = \boldsymbol{D}_{\Theta^{(k)}} \left( \boldsymbol{E}_{\Theta^{(k)}} \left( \widetilde{\mathcal{D}}_{\ell+1}^{(k)} \right) \right), \; k \in [n_3] . \tag{8.11}$$

Correspondingly, we use inverse DCT to compute $\mathcal{X}_{\ell+1}$; i.e., $\widehat{\mathcal{X}} = \text{idct}(\widetilde{\mathcal{X}}, [\,], 3)$.

#### 8.3.2.2 Update $\mathcal{E}_{\ell+1}$

When $\Theta_\ell$ is fixed, Eq. (8.8) reduces to a $\mathcal{E}_{\ell+1}$ minimization problem:

$$\mathcal{E}_{\ell+1} = \min_{\mathcal{E}} \; \beta \|\mathcal{E}\|_{1,1,2} + \frac{\lambda}{2} \|\mathcal{Y} - \mathcal{D} - \mathcal{E}\|_F^2 . \tag{8.12}$$

Equation (8.12) corresponds to the problem of sparse tensor optimization [71], the purpose of which is to find the sparsest solution to the underdetermined linear Eq. (8.1). Notably, sparse optimization constitutes a critical problem in a wide range of fields, such as pattern recognition, machine learning, and compressive sensing. As in [71], the analytical solution to Eq. (8.12) is given by

$$\mathcal{E}_{\ell+1}(i, j, :) = \mathcal{J}_{\frac{2\beta}{\lambda}} (\mathcal{Y}(i, j, :) - \mathcal{D}_{\ell+1}(i, j, :)), \; i \in [n_1] , j \in [n_2] , \tag{8.13}$$

where $\mathcal{J}_{\frac{2\beta}{\lambda}}(\cdot)$ denotes the soft-thresholding operator, which is characterized by a threshold parameter $\frac{2\beta}{\lambda} > 0$ and defined as follows:

$$\mathcal{J}_{\frac{2\beta}{\lambda}}(e) = \begin{cases} e - \frac{2\beta}{\lambda} \frac{e}{\|e\|_2}, & \|e\|_2 > \frac{2\beta}{\lambda} \\ 0, & \|e\|_2 <= \frac{2\beta}{\lambda} \end{cases} , \tag{8.14}$$

where $e$ is a column vector of tensor $\mathcal{E}$, represented using the following MATLAB notation: $e = \mathcal{E}(i, j, :)$.

The proposed RTDL model and all its steps are summarized in Algorithm 8.1. For completeness, here, we present the convergence condition, or the AltMin iteration, as follows:

$$\frac{\|\mathcal{Y} - \mathcal{D}_{\ell+1} - \mathcal{E}_{\ell+1}\|_F}{\|\mathcal{Y}\|_F} < \epsilon. \tag{8.15}$$

In this context, $\epsilon > 0$ represents a specified tolerance level. The convergence of Algorithm 8.1 is assured by the framework of alternate minimization, as outlined in [67, 68].

### 8.3.3 Implementation Details

Despite the completeness of the RTDL model, two implementation details remain unaddressed: the specific network architecture of the DCAE employed and the formulation of unsupervised training samples. Regarding the function $\boldsymbol{D}_{\Theta^{(k)}}(\boldsymbol{E}_{\Theta^{(k)}}(\cdot))$ in Eq. (8.11), a standard DCAE architecture can be utilized for the RTDL model, comprising two components: the encoder and decoder networks. In both the encoder and decoder networks, a convolutional layer utilizing a $3 \times 3$ kernel with ReLU activation is followed by a $2 \times 2$ max pooling layer. For reference, the parameters of the DCAE layers are detailed in Table 8.2.

**Table 8.2** Layer parameters of the DCAE network in the RTDL framework

| Order | Type | Kernel size |
|---|---|---|
| 1 | conv2D | $3 \times 3 \times 32$ |
| 2 | maxpooling2D | $2 \times 2$ |
| 3 | conv2D | $3 \times 3 \times 64$ |
| 4 | maxpooling2D | $2 \times 2$ |
| 5 | conv2D | $3 \times 3 \times 128$ |
| 6 | maxpooling2D | $2 \times 2$ |
| 7 | conv2D | $3 \times 3 \times 256$ |
| 8 | conv2D | $3 \times 3 \times 256$ |
| 9 | conv2D | $3 \times 3 \times 128$ |
| 10 | upsampling2D | $2 \times 2$ |
| 11 | conv2D | $3 \times 3 \times 64$ |
| 12 | upsampling2D | $2 \times 2$ |
| 13 | conv2D | $3 \times 3 \times 32$ |
| 14 | upsampling2D | $2 \times 2$ |
| 15 | conv2D | $3 \times 3 \times 1$ |

The second issue pertains to the generation of an adequate number of unsupervised training samples for the input 3-D seismic data, which are essential for deep learning techniques aimed at 3-D seismic data denoising. According to Eq. (8.11), the spectral image $\widetilde{\mathcal{D}}^{(k)}$ serves as input for eliminating erratic noise. In this context, the spectral image represents a single-frequency slice within the DCT domain and is derived via the discrete cosine transform applied to the time dimension of the noisy data $\mathcal{D}$. Subsequently, each spectral image $\widetilde{\mathcal{D}}^{(k)}$ is partitioned into either overlapping or non-overlapping patches, denoted as $\{\widetilde{\mathcal{D}}_n^{(k)}\}_{n=1}^{N}$, to ensure a sufficient quantity of training samples. Finally, after the well-trained RTDL model has effectively removed erratic noise from all samples, a reverse process is carried out to reconstruct the spectral images, and an inverse DCT transformation is applied to obtain the final output results.

## 8.4 Field Examples of RTDL

Once the RTDL model is developed, a series of comparative experiments are performed using both synthetic and real field data to assess the model's effectiveness compared to state-of-the-art (SOTA) methods in mitigating erratic noise. The initial section outlines the experimental setup, including performance metrics, baseline methods, and parameter configurations, which are detailed in Sects. 8.4.1.1–8.4.1.2. These experiments are conducted in two distinct phases. In the first phase, the validity of the improved method is confirmed, as detailed in Sect. 8.4.2. In the second phase, the robustness of the RTDL approach with real 3-D seismic data is demonstrated, as presented in Sect. 8.4.3. For a discussion regarding the two experiments, please refer to Sect. 8.4.4.

### *8.4.1 Experimental Setup*

#### 8.4.1.1 Performance Metrics

Two performance metrics can be employed to quantitatively evaluate the effectiveness of the RTDL approach in comparison to state-of-the-art (SOTA) methods, providing a basis for assessing the proposed model's robustness to both synthetic and real field data. For synthetic data experiments, where ground-truth data are available, it is appropriate to utilize both the signal-to-noise ratio (SNR) and local similarity to underscore the advantages of the RTDL algorithm in reducing erratic noise. The specific definitions of SNR and local similarity are as follows:

*SNR*: The SNR is a widely utilized quantitative metric for evaluating the strength of the desired seismic signal compared to the level of unwanted erratic noise. The specific definition is as follows:

$$\mathrm{SNR(dB)} = 20\log_{10}\left(\frac{\|\mathcal{T}\|_F}{\|\widehat{\mathcal{T}} - \mathcal{T}\|_F}\right), \tag{8.16}$$

where $\mathcal{T}$ and $\widehat{\mathcal{T}}$ represent the ideal 3-D seismic image and the result after removing erratic noise, respectively.

*Local Similarity*: Another metric utilized to assess the quality of the denoising process is local similarity, which emphasizes the correlation between the removed erratic noise and the denoised data. As detailed in [73], local similarity serves as an effective evaluation criterion for identifying potential signal leakage during the denoising process. Specifically, local similarity ranges from 0 to 1, with lower values indicating a minimal amount of useful signal energy present within the removed erratic noise.

#### 8.4.1.2 Baseline Algorithms

Logistically, selecting suitable baseline algorithms is an essential step for each experiment in this section to validate the SOTA performance of the proposed RTDL model. Based on the survey in Sect. 8.1, processing 3-D seismic data primarily relies on low-rank tensor decomposition, which yields promising results through RTPCA. Another approach is to employ a DTAE, which leverages the significance of the sparse $\ell_{1,1,2}$ norm in the TDL model for mitigating erratic noise. Finally, we include Pyseistr as the SOTA model for our evaluation, providing a robust baseline to demonstrate the effectiveness of our method in denoising tasks. The three selected methods are detailed below.

1. *DTAE* [72]: Utilizing a tNN framework, a DTAE can produce denoised tensors directly from corrupted 3-D seismic images in an unsupervised fashion. Similar to DCAEs, a DTAE features an hourglass-like architecture that includes both an encoder network and a decoder network.
2. *Pyseistr* [74][1]: Pyseistr is the inaugural Python package tailored for multichannel seismic data, serving applications in both exploration and earthquake seismology domains. It is crafted to leverage structural patterns within the data to enhance data processing.
3. *RTPCA* [75]: RTPCA represents an effective method for accurately recovering low-rank and sparse components from their combined representation. In this context, RTPCA utilizes an $\ell_1$-norm constraint to diminish erratic noise, while simultaneously applying a nuclear norm constraint to enhance the low rankness of the desired seismic data, thereby facilitating signal-noise separation under varying constraint conditions.

During the experimental phase, the implementations of Pyseistr and RTPCA were carried out using Python and MATLAB, respectively. In contrast, the RTDL model

[1] https://github.com/aaspip/pyseistr.

and DTAE were developed utilizing TensorFlow[2] on the Google Colab platform[3]. It is important to note that Colab assigns graphical processing units (GPUs) to users on a random basis, leading to variations in computational times.

#### 8.4.1.3 Network Parameters

The RTDL model encompasses two categories of parameters: those associated with the $\mathcal{E}$ subproblem and those related to the training algorithm parameters of the DCAE, denoted as $\Theta$. Unless stated otherwise, these parameters remain constant throughout the experiments.

*Parameters of the $\mathcal{E}$ Subproblem*: In this subproblem, the primary objective is to manage the threshold parameter in (8.12) to enhance confidence in identifying erratic noise. Consequently, the RTDL approach emphasizes the ratio of $\lambda$ to $\beta$ rather than their absolute values. For synthetic datasets, this ratio is established at $\lambda/\beta = 15$, while for real field data, it is set at $\lambda/\beta = 120$. The choice of $\lambda/\beta$ relies on existing knowledge regarding erratic noise, which is often unavailable in real dataset scenarios. Therefore, it is typical to adjust the $\mathcal{E}$ subproblem in a greedy manner, modifying only one of the two components at a time.

*Parameters of the DCAE Training Algorithm*: In this framework, there are two levels of iteration: the outer loop encompasses the entire RTDL process, while the inner loop focuses on training the DCAE. The outer loop is constrained to a maximum of 200 iterations, whereas the inner loop is limited to 30 iterations. During the inner loops, the DCAE is trained using the Adam optimizer with a learning rate set to 0.0001, and a batch size of 256 is used for both synthetic and real data experiments. It is worth noting that each inner iteration takes approximately 2.92 s, while each outer iteration requires about 88.01 s.

### *8.4.2 Synthetic Data Validation*

To verify the effectiveness of the proposed RTDL model, a synthetic data experiment is conducted using a simulated VSP dataset, which was initially produced by the Colorado School of Mines Reservoir Characterization Project during their Postle Field Study[4]. The simulated dataset comprises measurements taken at 31 underground locations spaced 16.667 m apart, with depths ranging from 1,350 to 1,850m. Following the approach in [53], noisy 3-D seismic images were generated by introducing erratic noise from non-Gaussian statistical distributions into the simulated

[2] https://www.tensorflow.org/.

[3] https://colab.research.google.com/.

[4] https://wiki.seg.org/wiki/3D_9C_synthetic_VSP_data.

VSP data (for further details, refer to the code[5] in [53]), as illustrated in Fig. 8.3(a) and (e). Notably, the RTDL method does not necessitate access to the original seismic images; its primary objective is to compute SNR values.

For qualitative assessments, Fig. 8.3 displays a comparison of the denoising capabilities of the RTDL model against state-of-the-art (SOTA) methods, using noisy 3-D synthetic data that has an SNR of 14.66dB. To delve deeper into the visual analysis of the denoising results, the data presented in Fig. 8.3 are sliced along the inline, x-line, and time axes, as demonstrated in Figs. 8.4 and 8.5. It is noteworthy that the results produced by the RTDL model closely match the ground-truth images, while the SOTA methods show noticeable deviations from the ground-truth data, as depicted in Figs. 8.5 and 8.4. Additionally, we reinforce this observation by presenting single-trace comparisons between the SOTA methods and the RTDL approach in Fig. 8.2.

In terms of quantitative analysis, it is crucial to evaluate the denoising effectiveness of the RTDL model against SOTA methods using the SNR metric, as detailed in Table 8.3. Our RTDL model attains an impressive 25.17dB, while the DTAE, Pyseistr, and RTPCA methods yield 2.30, 11.60, and 14.86dB, respectively. The findings from the local similarity metric corroborate the observations illustrated in Figs. 8.3d, g, j, m, 8.5d, g, j, m, and 8.4d, g, j, m. Regarding the SOTA methods, the local similarity map generated by the RTDL approach shows areas with lower similarity levels (indicated in blue). As expected, the data in Figs. 8.3, 8.5 and 8.4 and the results in Table 8.3 collectively suggest that both qualitative and quantitative assessments demonstrate the superior performance of the proposed RTDL method over the SOTA approaches.

### *8.4.3 Validation with Real Field Data*

To validate the practical applicability of the proposed method, we further implement our RTDL approach on a real 3-D seismic dataset, demonstrating its robustness against actual erratic noise. Figure 8.6a illustrates that this real seismic dataset is plagued by significant erratic noise, rendering it an appropriate benchmark for assessing the performance of both the RTDL and SOTA methods. Since acquiring ground-truth data for this experiment is impractical, we draw conclusions based on visual quality comparisons and local similarity values, rather than using SNRs as done in the synthetic data experiments.

Figure 8.6 illustrates the denoising performance comparisons between the proposed RTDL method and SOTA techniques, based on the entire real experimental dataset. To emphasize local details, we also present constant time slices, x-line sections, and inline views in Figs. 8.7 and 8.8, respectively. These figures clearly indicate that the proposed RTDL model effectively achieves optimal denoising while

[5] https://github.com/chenyk1990/reproducible_research/tree/f39feb1c94877b0d9704b5ee452eba958403a696/nonMada/erratic_mf.

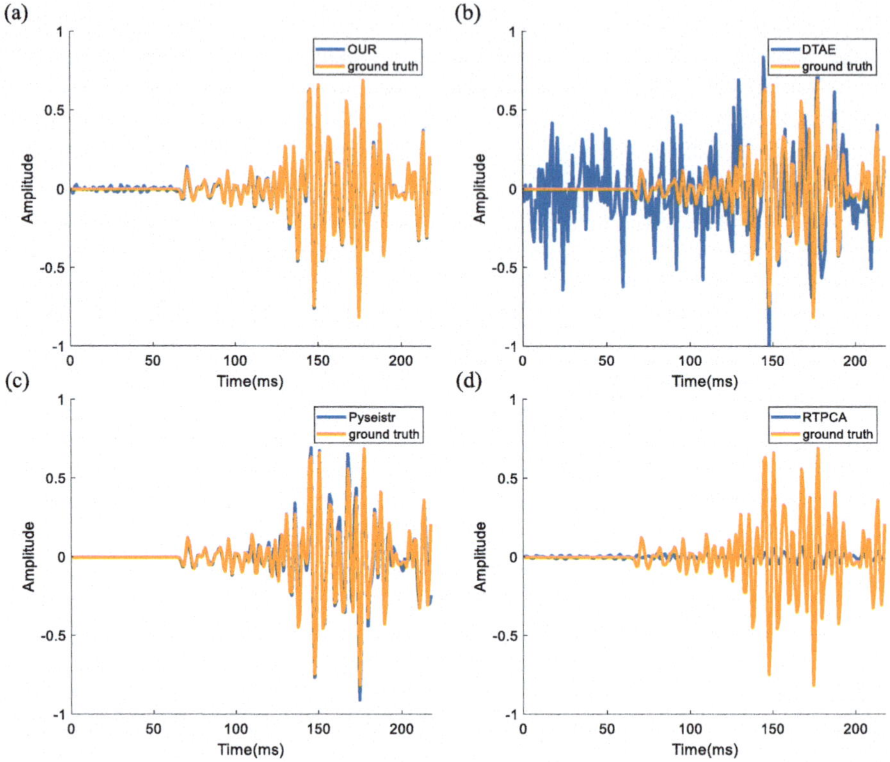


**Fig. 8.2** Single-trace comparison of synthetic 3-D data

preserving valuable signal information. To further substantiate this observation, the similarity maps in Figs. 8.6d, g, j, m,8.7d, g, j, m, and 8.8d, g, j, m provide an intuitive visual assessment of erratic noise removal performance, akin to the synthetic data experiments. The local similarity maps reveal that the RTDL results contain numerous blue areas and few yellow areas in the effective signal region, effectively illustrating the superiority of the proposed method over SOTA approaches.

### *8.4.4 Discussion*

The RTDL algorithm offers a mechanism for unsupervised attenuation of erratic noise, as evidenced by the experiments conducted above. As previously discussed, understanding this denoising mechanism hinges on the application of the sparse tensor norm, which serves a purpose similar to that of RTPCA. In contrast, a traditional DTAE does not effectively suppress erratic noise in the 3-D image data presented in Fig. 8.6, while the RTDL model performs admirably, highlighting the significance of

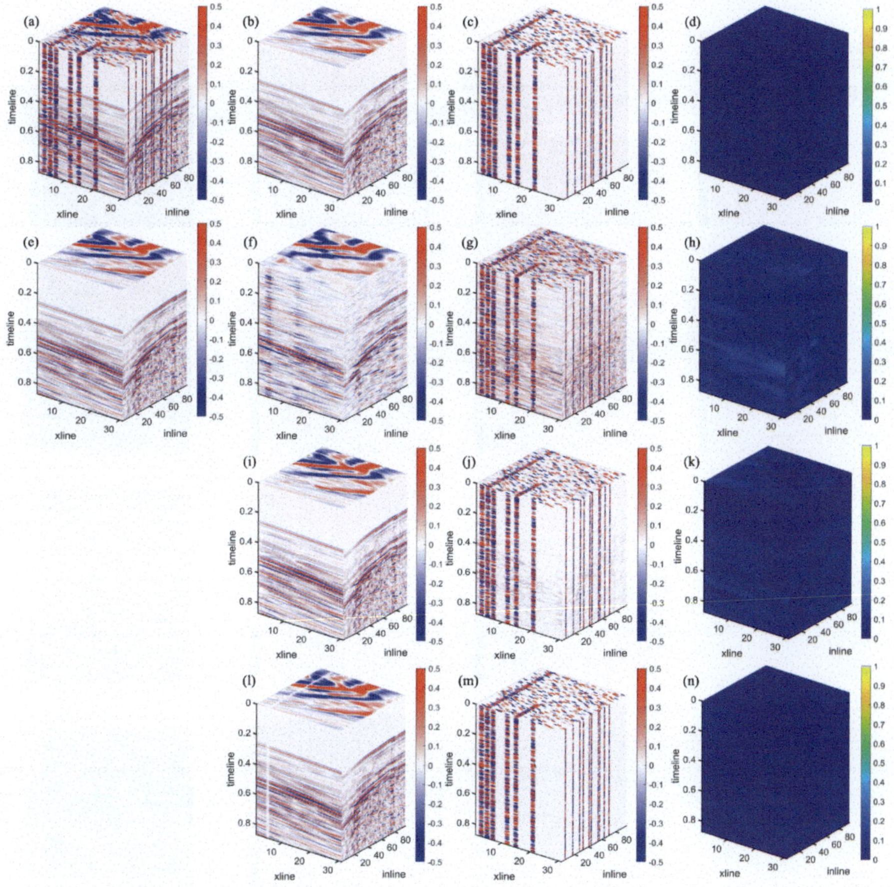


**Fig. 8.3** Synthetic data example. **a** Volume of noisy data. The denoised outputs include **c** results from the proposed RTDL method, **e** the ground-truth data volume, **g** outputs from the classic deep tensor autoencoder (DTAE) method, **j** results from the Pyseistr method, and **m** outputs from the robust tensor principal component analysis (RTPCA) method. Noise removal is illustrated with the outputs of **b** the proposed RTDL method, **f** DTAE method, **i** Pyseistr method, and **k** RTPCA method. Local similarity maps are constructed using **d** the proposed RTDL method, **h** DTAE method, **k** Pyseistr method, and **n** RTPCA method

the sparse tensor norm in achieving this objective. As noted in [46–48], it is indeed possible to reliably recover clear seismic images from noisy random observations through the use of the DCAE. Building on these two points, we lay a solid foundation for discussing the advantages of our RTDL approach in addressing erratic noise.

Additionally, a significant limitation of the RTDL model is that its computational efficiency is heavily influenced by the size of the network architecture employed, which is a common issue for DL models. While the RTPCA and Pyseistr methods are executed in MATLAB and Python, respectively, our RTDL model is implemented

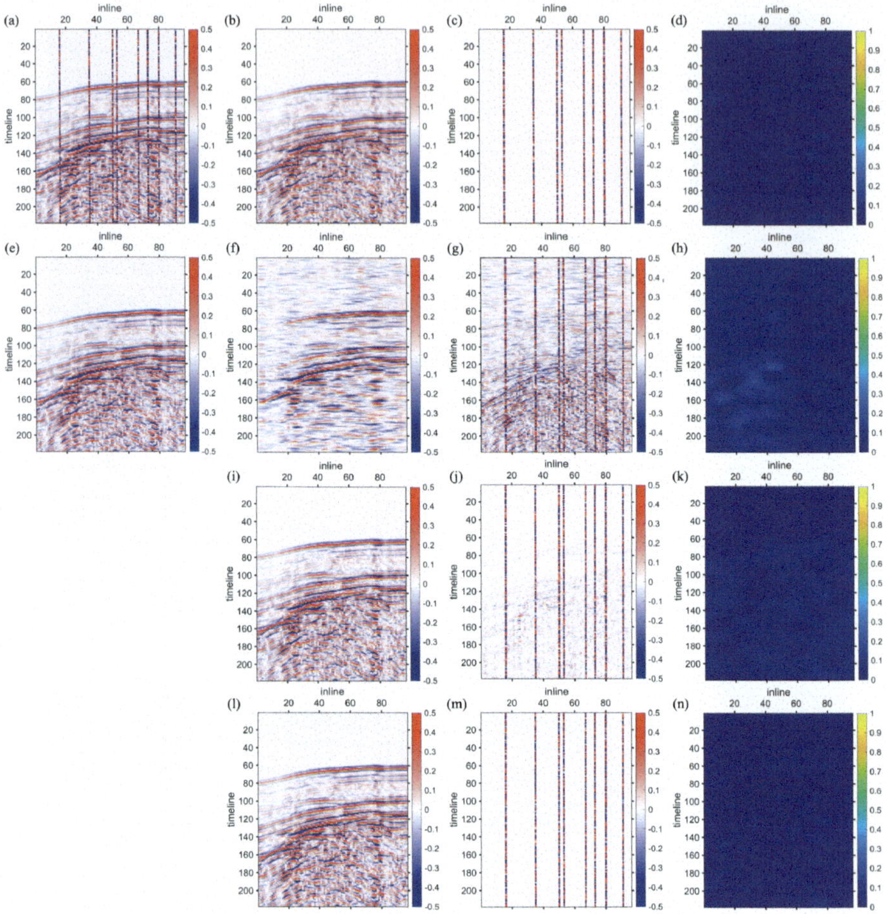

**Fig. 8.4** Synthetic data example. **a** Original 24th x-line section. The denoised results include **c** outputs from the proposed RTDL method, **e** the ground-truth data, **g** results from the DTAE method, **j** outputs from the Pyseistr method, and **m** results from the RTPCA method. Noise removal is demonstrated with **b** the proposed RTDL method, **f** DTAE method, **i** Pyseistr method, and **k** RTPCA method. Local similarity maps are generated using **d** the proposed RTDL method, **h** DTAE method, **k** Pyseistr method, and **n** RTPCA method

using TensorFlow. Consequently, the backpropagation algorithm in RTDL typically requires more time to perform noise suppression on the same image tensor. Furthermore, the RTDL model can essentially be viewed as an iterative DTAE that offers improved computational efficiency.

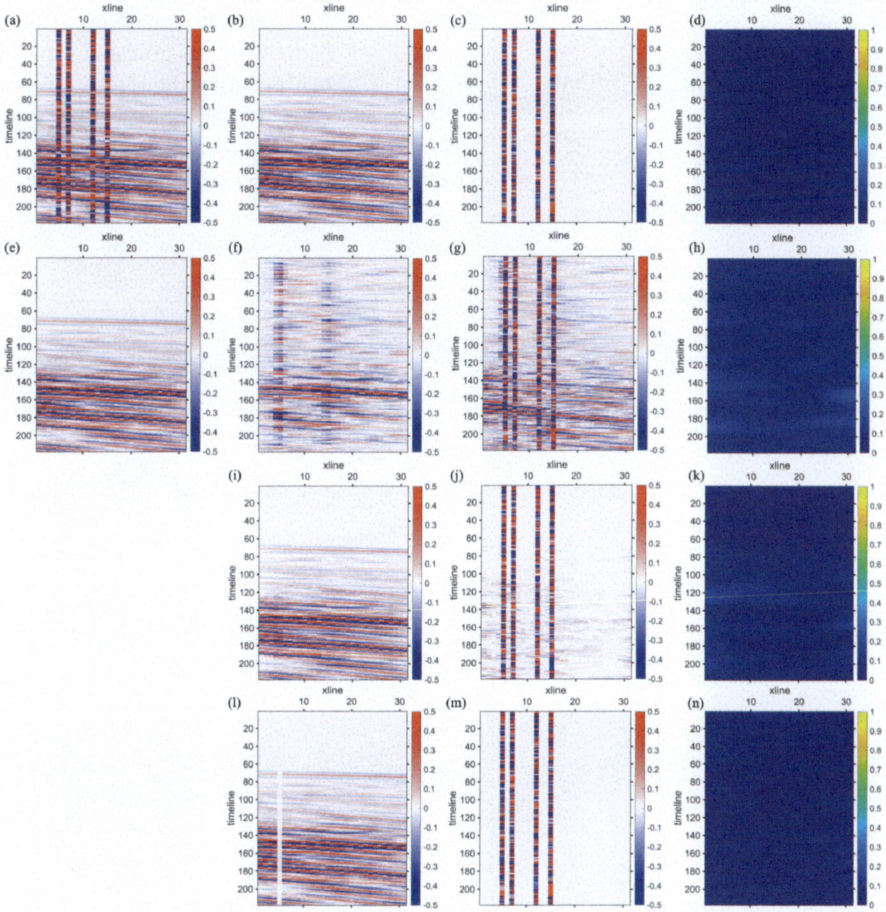


**Fig. 8.5** Synthetic data example. **a** Original 19th inline section. The denoised results include **c** outputs from the proposed RTDL method, **e** the ground-truth data, **g** results from the DTAE method, **j** outputs from the Pyseistr method, and **m** results from the RTPCA method. Noise removal is demonstrated with **b** the proposed RTDL method, **f** the DTAE method, **i** the Pyseistr method, and **k** the RTPCA method. Local similarity maps are generated using **d** the proposed RTDL method, **h** the DTAE method, **k** the Pyseistr method, and **n** the RTPCA method

**Table 8.3** Comparative assessment of noise reduction efficacy between SOTA methods and the proposed RTDL approach using synthetic datasets

| Method | Synthetic dataset |
|---|---|
| DTAE | 2.30 dB |
| Pyseistr | 11.60 dB |
| RTPCA | 14.86 dB |
| Proposed RTDL | **25.17** dB |

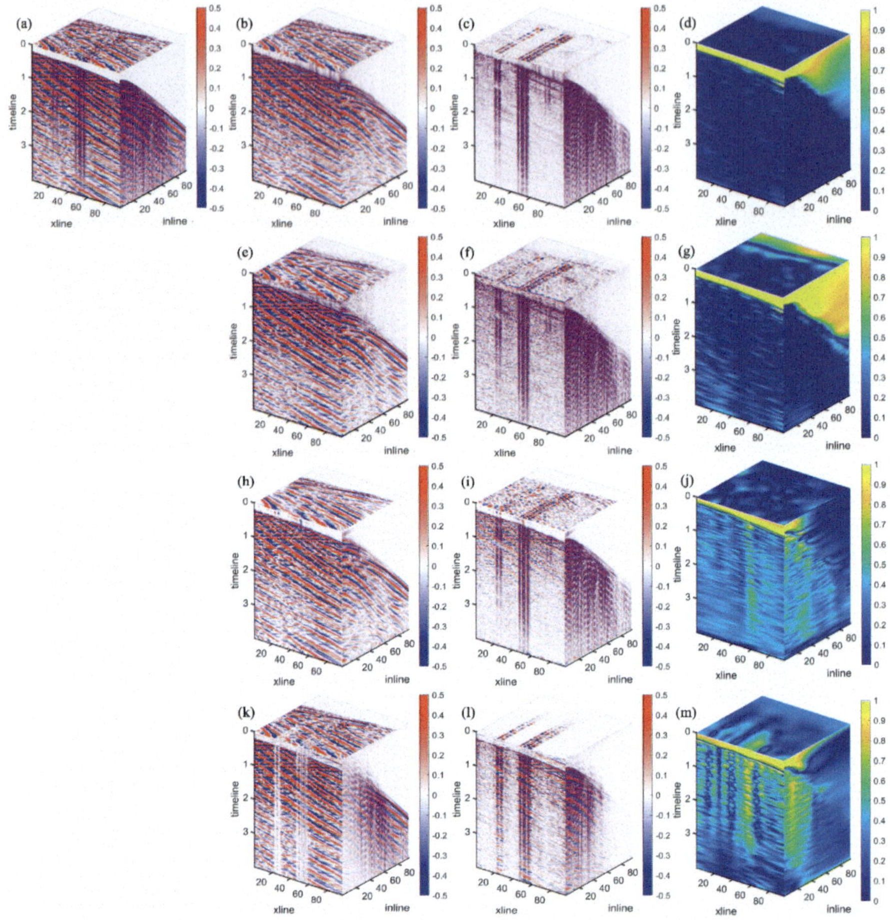

Fig. 8.6 Real 3D field example. **a** Noisy data volume. The data are denoised using **b** the proposed RTDL method, **e** DTAE method, **h** Pyseistr method, and **k** RTPCA method. Noise is removed using the **c** proposed RTDL method, **f** DTAE method, **i** Pyseistr method, and **l** RTPCA method. Local similarity maps are constructed using the **d** proposed RTDL method, **g** DTAE method, **j** Pyseistr method, and **m** RTPCA method

## 8.5 Summary

This chapter investigates a model-data-driven approach utilizing RTDL for enhancing the denoising of 3-D seismic images affected by erratic noise. The denoising challenge is framed within an RTDL model built on a tensor neural network (tNN) and tensor super-resolution (SR). To effectively model erratic noise, a robust tensor sparse norm is introduced, leveraging the spatial tubular sparse distribution characteristics inherent in 3-D space. By incorporating tensor sparse norm regularization into the objective function of a conventional tensor deep learning (TDL) network, the

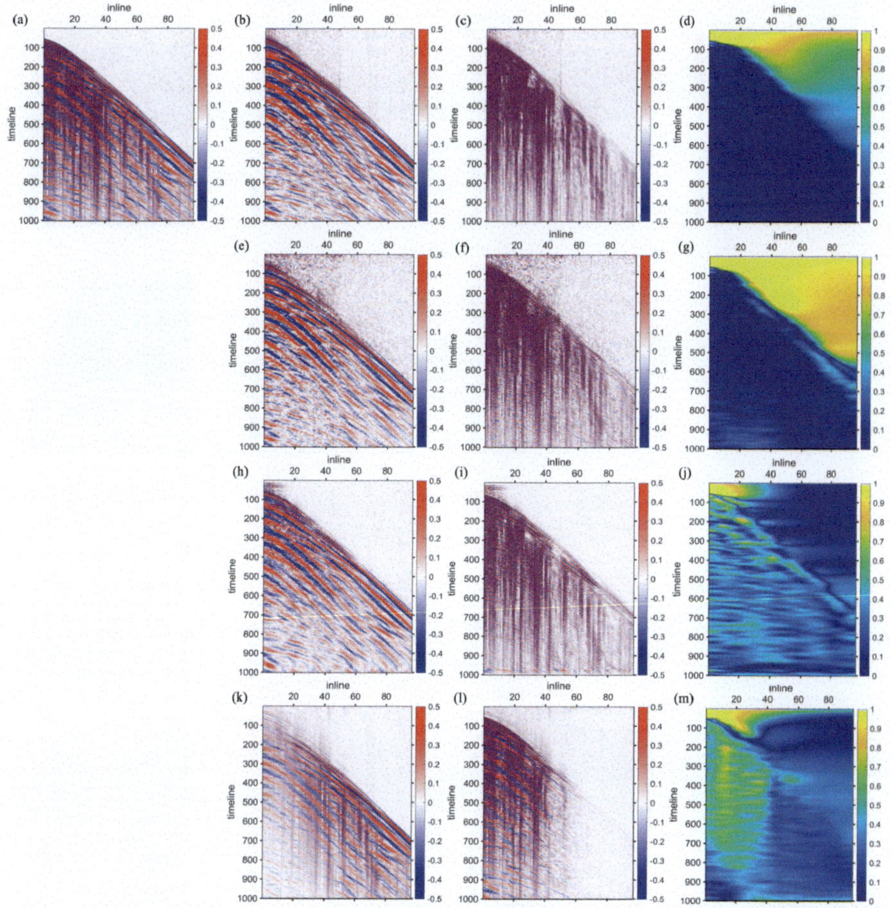


**Fig. 8.7** Real 3D field example. **a** Original 51st x-line section. The data are denoised using the **b** proposed RTDL method, **e** DTAE method, **h** Pyseistr method, and **k** RTPCA method. Noise is removed using the **c**proposed RTDL method, **f** DTAE method, **i** Pyseistr method, and **l** RTPCA method. Local similarity maps are constructed using the **d** proposed RTDL method, **g** DTAE method, **j** Pyseistr method, and **m** RTPCA method

proposed robust model is equipped to tackle the task of denoising erratic noise. To optimize the parameters of the RTDL model, an efficient tensor optimization method is derived based on alternating minimization (Alt). Through meticulously designed synthetic and real experiments, we empirically demonstrate that this method surpasses several state-of-the-art (SOTA) techniques.

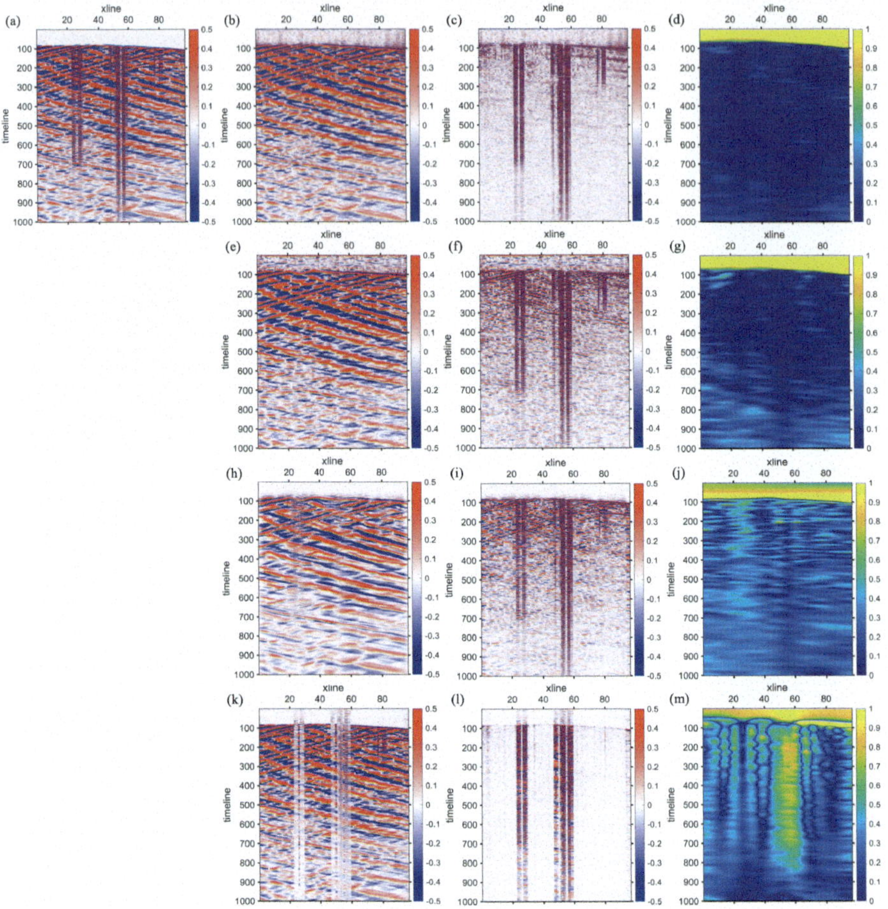

**Fig. 8.8** Real 3D field example. **a** Original 10th inline section. The data are denoised using the **b** proposed RTDL method, **e** DTAE method, **h** Pyseistr method, and **k** RTPCA method. Noise is removed using the **c**proposed RTDL method, **f** DTAE method, **i** Pyseistr method, and **l** RTPCA method. Local similarity maps are constructed using the **d** proposed RTDL method, **g** DTAE method, **j** Pyseistr method, and **m** RTPCA method

In our future work, we aim to leverage the extensible framework of the RTDL model to explore various new models tailored for different tasks. Specifically, by incorporating diverse regularization terms, such as the tensor nuclear norm, into tensor deep learning (TDL) models, we can create new practical models designed to address additional types of noise reduction or data recovery challenges.

# References

1. Chen K, Sacchi MD (2015) Robust reduced-rank filtering for erratic seismic noise attenuation. Geophysics 80(1):V1–V11
2. Xiaojing W, Jianwei M (2019) Adaptive dictionary learning for blind seismic data denoising. IEEE Geosci Remote Sens Lett 17(7):1273–1277
3. Sui Y, Wang X, Ma J (2023) Deep unfolding dictionary learning for seismic denoising. Geophysics 88(1):WA129–WA147
4. Yu S, Jianwei M, Wenlong W (2019) Deep learning for denoising. Geophysics 84(6):V333–V350
5. Liuqing Y, Wei C, Hang W, Yangkang C (2021) Deep learning seismic random noise attenuation via improved residual convolutional neural network. IEEE Trans Geosci Remote Sens 59(9):7968–7981
6. Liu D, Wang W, Chen W, Wang X, Zhou Y, Shi Z (2018) Random noise suppression in seismic data: What can deep learning do? In: SEG Technical Program Expanded Abstracts 2018. Society of Exploration Geophysicists, pp 2016–2020
7. Yangkang C, Yatong Z, Wei C, Zu S, Weilin H, Dong Z (2017) Empirical low-rank approximation for seismic noise attenuation. IEEE Trans Geosci Remote Sens 55(8):4696–4711
8. Anvari R, Siahsar MAN, Gholtashi S, Kahoo AR, Mohammadi M (2017) Seismic random noise attenuation using synchrosqueezed wavelet transform and low-rank signal matrix approximation. IEEE Trans Geosci Remote Sens 55(11):6574–6581
9. Xingye L, Xiaohong C, Jingye L, Yangkang C (2020) Nonlocal weighted robust principal component analysis for seismic noise attenuation. IEEE Trans Geosci Remote Sens 59(2):1745–1756
10. Fu C, Zhifei G, Lei C, Senlin Y, Lingli Z, Yangkang C (2022) 3-d structural complexity-guided predictive filtering: a comparison between different non-stationary strategies. IEEE Trans Geosci Remote Sens 60:1–15
11. Bin L, Fu C, Yuxiao R, Qingsong Z, Xu X, Yangkang C (2020) Structural complexity-guided predictive filtering. Geophys Prospect 68(5):1509–1522
12. Yanxin Z, Runqiu W, Weilin H (2020) Surface diffraction noise attenuation for marine seismic data processing with mathematical morphological filtering. IEEE Trans Geosci Remote Sens 59(9):8007–8021
13. Huijian L, Stewart G, Xu BL, Hao LQ, Yangkang C (2022) A generalized seismic attenuation compensation operator optimized by 2-d mathematical morphology filtering. IEEE Trans Geosci Remote Sens 60:1–15
14. Fehmers GC, Höcker CFW (2003) Fast structural interpretation with structure-oriented filtering. Geophysics 68(4):1286–1293
15. Qingbao Z, Jinghuai G, Zhiguo W, Kexue L (2015) Adaptive variable time fractional anisotropic diffusion filtering for seismic data noise attenuation. IEEE Trans Geosci Remote Sens 54(4):1905–1917
16. Yushu Z, Hongbo L, Yue L, Haitao M (2019) Seismic signal enhancement and noise suppression using structure-adaptive nonlinear complex diffusion. IEEE Trans Geosci Remote Sens 58(3):2198–2211
17. Chuangjian L, Suping P, Xiaoqin C, Du W, Peng L (2023) Seismic data enhancement based on common-reflection-surface-based local slope and trimmed mean filter. IEEE Trans Geosci Remote Sens 61:1–9
18. Shuwei G, Shoudong W, Yangkang C, Xiaohong C, Kui X (2016) Separation of simultaneous sources using a structural-oriented median filter in the flattened dimension. Comput Geosci 86:46–54
19. Jian M, Jinghuai G, Wenchao C (2006) On the denoising method of prestack seismic data in wavelet domain. In: 2006 SEG annual meeting. OnePetro
20. Hao S, Jianwei M, Huizhu Y (2009) Comparisons of wavelets, contourlets and curvelets in seismic denoising. J Appl Geophys 69(2):103–115

21. Li J-H, Zhang Y-J, Qi R, Liu QH (2017) Wavelet-based higher order correlative stacking for seismic data denoising in the curvelet domain. IEEE J Sel Top Appl Earth Obs Remote Sens 10(8):3810–3820
22. Song J, Li Z, Wang G, Lei G, Yang J (2023) Weak seismic signal enhancement using curvelet transform and compressive sampling. IEEE Trans Geosci Remote Sens
23. Dehui K, Zhenming P (2015) Seismic random noise attenuation using shearlet and total generalized variation. J Geophys Eng 12(6):1024–1035
24. Chengming L, Deli W, Tong W, Fei F, Hao C, Gege M (2014) Random seismic noise attenuation based on the shearlet transform. Acta Petrolei Sinica 35(4):692–699
25. Liu Y, Fomel S, Liu C (2015) Signal and noise separation in pre-stack seismic data using velocity-dependent seislet transform. Geophysics 80(6):WD117–WD128
26. Zu S, Hui Z, Wu R, Maocai J, Yangkang C (2019) Dictionary learning based on dip patch selection training for random noise attenuation. Geophysics 84(3):V169–V183
27. Yang Y, Jinghuai G, Zhiguo W, Guowei Z, Xiangxiang Z (2019) 2-d seismic random noise attenuation via self-paced nonnegative dictionary learning. IEEE J Sel Top Appl Earth Obs Remote Sens 12(12):5391–5401
28. Kuruguntla L, Dodda VC, Elumalai K (2021) Study of parameters in dictionary learning method for seismic denoising. IEEE Trans Geosci Remote Sens 60:1–13
29. Ouyang S, Lingling W, Hu X, Zhidan L (2021) Seismic denoising via truncated nuclear norm minimization. Geophysics 86(2):V153–V169
30. Lin P, Peng S, Xiang Y, Li C, Cui X, Zhang W (2023) Structure-oriented cur low-rank approximation for random noise attenuation of seismic data. IEEE Trans Geosci Remote Sens
31. Chong W, Zhihui Z, Gu H, Wu X, Shuaiqi L (2018) Hankel low-rank approximation for seismic noise attenuation. IEEE Trans Geosci Remote Sens 57(1):561–573
32. Yijie Z, Haoran Z, Yang Y, Naihao L, Jinghuai G (2021) Seismic random noise separation and attenuation based on MVMD and MSSA. IEEE Trans Geosci Remote Sens 60:1–16
33. Feng J, Li X, Liu X, Chen C, Chen H (2021) Seismic data denoising based on tensor decomposition with total variation. IEEE Geosci Remote Sens Lett
34. Liu D, Sacchi MD, Chen W (2022) Efficient tensor completion methods for 5-d seismic data reconstruction: low-rank tensor train and tensor ring. IEEE Trans Geosci Remote Sens 60:1–17
35. Jun F, Xi L, Xiaoqin L, Xu W, Bin L (2021) Low-rank tensor minimization method for seismic denoising based on variational mode decomposition. IEEE Geosci Remote Sens Lett 19:1–5
36. Ely G, Aeron S, Hao N, Kilmer ME (2015) 5d seismic data completion and denoising using a novel class of tensor decompositions. Geophysics 80(4):V83–V95
37. Anvari R, Mohammadi M, Kahoo AR (2018) Enhancing 3-d seismic data using the t-SVD and optimal shrinkage of singular value. IEEE J Sel Top Appl Earth Obs Remote Sens 12(1):382–388
38. Feng Q, Cangcang Z, Lingtian F, Lu C, Gulan Z, Hu G (2020) Tubal-sampling: Bridging tensor and matrix completion in 3-d seismic data reconstruction. IEEE Trans Geosci Remote Sens 59(1):854–870
39. Feng Q, Zhangbo L, Yan W, Yingjie Z, Hu G (2022) Ground truth-free 3-d seismic random noise attenuation via deep tensor convolutional neural networks in the time-frequency domain. IEEE Trans Geosci Remote Sens 60:1–17
40. Feng Q, Yan W, Bingwei Z, Zhangbo L, Yingjie Z, Hu G (2022) Multidimensional seismic data denoising using framelet-based order-p tensor deep learning. IEEE Trans Geosci Remote Sens 60:1–18
41. Liang C, Lin H, Ma H (2023) Reinforcement learning based denoising model for seismic random noise attenuation. IEEE Trans Geosci Remote Sens
42. Qiao Z, Wang D, Zhang L, Liu N (2023) Random noise attenuation of seismic data via self-supervised bayesian deep learning. IEEE Trans Geosci Remote Sens
43. Wang D, Qiao Z, Zhang L, Liu N (2023) Total variation regularized self-supervised Bayesian deep learning for seismic random noise attenuation. IEEE Trans Geosci Remote Sens
44. Liu D, Niu W, Wang X, Sacchi MD, Chen W, Wang C (2023) Improving vertical resolution of vintage seismic data by a weakly supervised method based on cycle generative adversarial network. Geophysics 88(6):V445–V458

45. Dawei L, Wei W, Xiaokai W, Cheng W, Jiangyun P, Wenchao C (2019) Poststack seismic data denoising based on 3-d convolutional neural network. IEEE Trans Geosci Remote Sens 58(3):1598–1629
46. Saad OM, Chen Y (2020) Deep denoising autoencoder for seismic random noise attenuation. Geophysics 85(4):V367–V376
47. Mi Z, Yang L, Yangkang C (2019) Unsupervised seismic random noise attenuation based on deep convolutional neural network. IEEE Access 7:179810–179822
48. Yang L, Wang S, Chen X, Saad OM, Chen W, Oboué YASI, Chen Y (2021) Unsupervised 3-d random noise attenuation using deep skip autoencoder. IEEE Trans Geosci Remote Sens 60:1–16
49. Saad OM, Oboué YASI, Bai M, Samy L, Yang L, Chen Y (2021) Self-attention deep image prior network for unsupervised 3-d seismic data enhancement. IEEE Trans Geosci Remote Sens
50. Dawei L, Sacchi MD, Wang X, Chen W (2023) Unsupervised deep learning for ground roll and scattered noise attenuation. IEEE Trans Geosci Remote Sens
51. Dawei L, Zheyuan D, Cheng W, Xiaokai W, Wenchao C (2020) An unsupervised deep learning method for denoising Prestack random noise. IEEE Geosci Remote Sens Lett 19:1–5
52. Meng C, Gao J, Tian Y, Li Z (2023) Learning to decouple and generate seismic random noise via invertible neural network. IEEE Trans Geosci Remote Sens
53. Qiang Z, Du Q, Xufei G, Yangkang C (2018) Signal-preserving erratic noise attenuation via iterative robust sparsity-promoting filter. IEEE Trans Geosci Remote Sens 56(6):3547–3560
54. Guangtan H, Min B, Qiang Z, Wei C, Yangkang C (2021) Erratic noise suppression using iterative structure-oriented space-varying median filtering with sparsity constraint. Geophys Prospect 69(1):101–121
55. Oboué YASI, Chen Y, Bai M, Chen W, Chen Y (2022) Erratic and random noise attenuation using adaptive local orthogonalization. Geophysics 87(4):V381–V396
56. Cheng J, Chen K, Sacchi MD (2015) Application of robust principal component analysis (RPCA) to suppress erratic noise in seismic records. In: SEG technical program expanded abstracts 2015. Society of Exploration Geophysicists, pp 4646–4651
57. Wu B, Yu J, Haoran R, Yihuai L, Naihao L (2019) Seismic traffic noise attenuation using $l\_\{p\}$ -norm robust PCA. IEEE Geosci Remote Sens Lett 17(11):1998–2001
58. Yaoguang S, Siyuan C, Siyuan C, Liuqing Y, Su Y (2023) Simultaneous random plus outlier attenuation and deblending based on l1-norm misfit function. Geophys Prospect 71(6):983–997
59. Qian F, Yue Y, He Y, Yu H, Zhou Y, Tang J, Hu G (2023) Unsupervised seismic footprint removal with physical prior augmented deep autoencoder. IEEE Trans Geosci Remote Sens
60. Weilin H (2023) A two-step singular spectrum analysis method for robust low-rank approximation of seismic data. Geophysics 88(1):V9–V19
61. Trickett S, Burroughs L, Milton A (2012) Robust rank-reduction filtering for erratic noise. In: SEG Int Expo Annu Meet. SEG, pp SEG–2012
62. Brox DS, Sacchi MD (2022) Robust vector MSSA for SNR enhancement of seismic records. IEEE Trans Geosci Remote Sens 60:1–6
63. Mi Z, Yang L, Haoran Z, Yangkang C (2020) Incoherent noise suppression of seismic data based on robust low-rank approximation. IEEE Trans Geosci Remote Sens 58(12):8874–8887
64. Xu W, Yanhui Z, Xiaokai W, Wenchao C (2020) MOG-based robust sparse representation for seismic erratic noise suppression. IEEE Geosci Remote Sens Lett 19:1–5
65. Kuruguntla L, Dodda VC, Mandpura AK, Elumalai K (2022) Erratic noise attenuation using double sparsity dictionary learning method. IEEE Trans Geosci Remote Sens 60:1–13
66. Chen W, Oboué YASI, Chen Y (2023) Retrieving useful signals from highly corrupted erratic noise using robust residual dictionary learning. Geophysics 88(1):WA55–WA64
67. Zhou C, Paffenroth RC (2017) Anomaly detection with robust deep autoencoders. In: Proceedings of the 23rd ACM SIGKDD international conference on knowledge discovery and data mining, pp 665–674
68. Chalapathy R, Menon AK, Chawla S (2017) Robust, deep and inductive anomaly detection. In: Machine learning and knowledge discovery in databases: European conference, ECML PKDD 2017, Skopje, Macedonia, 18–22 Sept 2017, Proceedings, Part I 10. Springer, pp 36–51

69. Qian F, Hua H, Wen Y, Pan S, Zhang G, Hu G (2024) Unsupervised 3-d seismic erratic noise attenuation with robust tensor deep learning. IEEE Trans Geosci Rem Sens
70. Yu FQ, Yue HY, Yingjie Z, Wu B, Hu G (2023) Improved low-rank tensor approximation for seismic random plus footprint noise suppression. IEEE Trans Geosci Remote Sens 61:1–19
71. Zhang Z, Ely G, Aeron S, Hao N, Kilmer M (2014) Novel methods for multilinear data completion and de-noising based on tensor-SVD. In: Proceedings of the IEEE conference on computer vision and pattern recognition, pp 3842–3849
72. Feng Q, Zhangbo L, Yan W, Songjie L, Shengli P, Hu G (2021) DTAE: deep tensor autoencoder for 3-d seismic data interpolation. IEEE Trans Geosci Remote Sens 60:1–19
73. Chen Y, Fomel S (2015) Random noise attenuation using local signal-and-noise orthogonalization. Geophysics 80(6):WD1–WD9
74. Chen Y, Savvaidis A, Fomel S, Chen Y, Saad OM, Oboué YASI, Zhang Q, Chen W (2023) Pyseistr: a python package for structural denoising and interpolation of multichannel seismic data. Seism Soc Am 94(3):1703–1714
75. Lu C, Jiashi F, Yudong C, Wei L, Zhouchen L, Shuicheng Y (2019) Tensor robust principal component analysis with a new tensor nuclear norm. IEEE Trans Pattern Anal Mach Intell 42(4):925–938

# Chapter 9
# Tensor Dictionary Learning for Seismic Data Super-Resolution

## 9.1 Introduction

Due to the constraints associated with seismic acquisition and processing, field exploration data frequently exhibit low resolution and noise interference, which create challenges for subsequent seismic interpretation. These limitations hinder our capacity to extract crucial information from seismic data, such as reservoir distribution, stratigraphic features, and fault details. Consequently, enhancing the quality of seismic data is vital for effective data interpretation in oil and gas exploration, as well as in geological studies. Optimal acquisition quality is essential for processing techniques—such as static correction and frequency extension—to yield their best outcomes. Two widely adopted approaches to tackle these issues are resolution enhancement and denoising. While denoising methods have been thoroughly covered in Chaps. 4, 6, 7, and 8, this chapter primarily emphasizes improving the resolution of seismic data.

The resolution of seismic data pertains to the accuracy of spatial measurements concerning subsurface structures and can be divided into vertical and horizontal resolution. Various techniques have been developed by researchers to improve the resolution of seismic images, which can be generally categorized into two main types: high-density acquisition [1, 2] and broadband seismic [3–5]. The first approach typically enhances horizontal resolution by employing a greater number of sources and receivers to obtain densely sampled datasets. In contrast, the second approach improves vertical resolution by capturing a wider frequency spectrum, encompassing both low and high frequencies. However, both methods come with substantial costs in terms of data acquisition and processing. Increasing horizontal resolution demands more sources and receivers to enhance sampling density, while enhancing vertical resolution requires sophisticated equipment capable of recording a broader frequency range during acquisition.

To minimize the costs associated with seismic data acquisition, researchers have devised various algorithms to enhance the resolution of seismic data without alter-

 
F. Qian et al., *Tensor Computation for Seismic Data Processing*,
Earth Systems Data and Models 6, https://doi.org/10.1007/978-3-031-78900-7_9

ing the existing acquisition equipment. Improving vertical resolution is a traditional processing technique focused on expanding the frequency bandwidth and elevating the dominant frequency, fundamentally aimed at recovering the true amplitude of weak signals. Common methods can be classified into three categories: deconvolution, absorption compensation, and frequency extension based on the time-frequency spectrum. Deconvolution techniques enhance resolution by compressing the seismic wavelet to extract or restore the original signal from the recorded output, thereby broadening the frequency spectrum [6–9]. Absorption compensation is mainly accomplished through inverse Q filtering, where the extended stability factor method of inverse Q filtering [10] incorporates parameters and gain constraints to achieve the desired high resolution. Frequency extension techniques primarily rely on time-frequency decomposition as a foundational approach, processing both high- and low-frequency components in the time-frequency domain to compress wavelets and facilitate broadband signals [11].

Compared to the research on vertical resolution, studies on horizontal resolution are relatively limited. This chapter presents an innovative three-dimensional seismic trace interpolation method based on tensor sparse coding [12], which effectively captures and utilizes the spatial correlations inherent in seismic data. The core objective of this method is to perform data reconstruction by training a pair of dictionaries for high-granularity and low-granularity seismic data. First, tensor sparse coding is applied to both low-granularity and high-granularity seismic data to construct suitable sparse representations at different granularity levels. Next, we assume that the structural features in both low- and high-granularity seismic data are similar, allowing them to share the same sparse representation. This assumption enables the use of sparse coefficients extracted from low-granularity data in conjunction with the high-granularity dictionary to accurately reconstruct high-granularity seismic data. Finally, experiments conducted on actual field seismic data demonstrate that the proposed method significantly improves the resolution of seismic data imaging, showcasing its practical applicability and value.

## 9.2 Preliminaries in Seismic Data Super-Resolution

### 9.2.1 *Problem Definitions*

The objective of seismic data super-resolution is to reconstruct the corresponding high-granularity seismic data from low-granularity seismic data, as illustrated in Fig. 9.1. In general, the low-granularity seismic data $\mathcal{Y}_l \in \mathbb{R}^{N_1 \times N_2 \times N_3}$ can be described as the result of the following degradation process:

$$\mathcal{X}_h = \mathbb{D}\left(\mathcal{Y}_l; \theta\right) \tag{9.1}$$

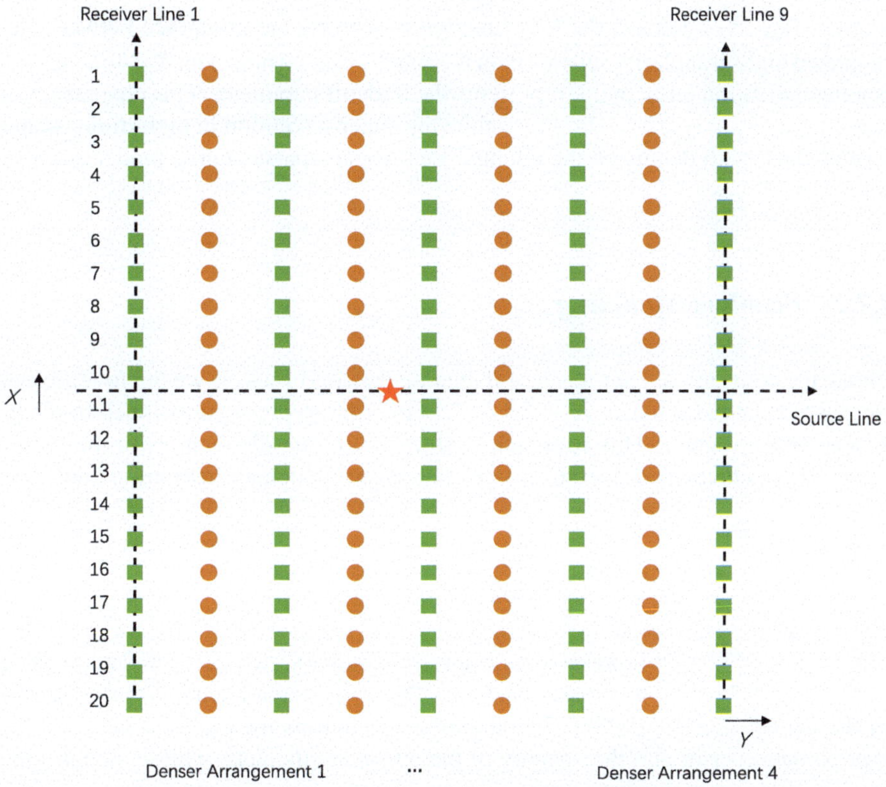


**Fig. 9.1** Creating training pairs for low- and high-granularity seismic data

where $\mathbb{D}(\cdot)$ represents the degradation mapping function, $\mathcal{X}_h$ denotes the corresponding high-granularity seismic image, and $\theta$ signifies the parameters of the degradation process. Typically, both the degradation process (i.e., $\mathbb{D}(\cdot)$ and $\theta$) are unknown, and only low-granularity seismic data is available. This scenario is referred to as blind seismic data super-resolution, where the goal is to recover an approximation $\widehat{\mathcal{X}}_h$ of the high-granularity seismic data $\mathcal{X}_h$ from the low-granularity seismic data $\mathcal{Y}_l$. The process follows these steps:

$$\widehat{\mathcal{X}}_h = \mathbb{F}\left(\mathcal{Y}_l; \theta\right) \tag{9.2}$$

Although the degradation process is not well understood and can be affected by various factors, including sensor noise and environmental interference, researchers seek to model the degradation mapping using techniques like tensor sparse coding. Consequently, the objective function for seismic data super-resolution is defined as follows:

$$\hat{\theta} = \arg\min_{\theta} \mathbb{L}\left(\widehat{\mathcal{X}}_h, \mathcal{X}_h\right) + \lambda\Phi(\theta), \tag{9.3}$$

where $\mathbb{L}\left(\widehat{\mathcal{X}}_h, \mathcal{X}_h\right)$ denotes the loss function measuring the difference between the generated high-granularity seismic data $\widehat{\mathcal{X}}_h$ and the true seismic data $\mathcal{X}_h$, $\Phi(\theta)$ serves as a regularization term, and $\lambda$ represents the tradeoff parameter. This chapter seeks to leverage tensor sparse coding for achieving super-resolution in high-dimensional seismic data, with details on the adopted joint tensor sparse coding model outlined below.

### 9.2.2 Problem Modeling

Unlike the conventional approach of flattening data into vectors, we opt to represent the seismic data as a tensor $\mathcal{X} \in \mathbb{R}^{N_1 \times N_2 \times N_3}$ in order to preserve its inherent three-dimensional structure. This approach enables us to train dictionaries in the tensor space. The mathematical formulation of the tensor dictionary learning model is as follows:

$$\begin{aligned} \min_{\mathcal{D},\mathcal{Z}} \tfrac{1}{2}\|\mathcal{X} - \mathcal{D} * \mathcal{Z}\|_F^2 + \lambda'\|\mathcal{Z}\|_1, \\ \text{s.t. } \|\mathcal{D}(:, j, :)\|_F^2 \leq 1, j \in [r], \end{aligned} \tag{9.4}$$

where $\mathcal{D} \in \mathbb{R}^{N_1 \times r \times N_3}$ represents the tensor dictionary, where each lateral slice $\mathcal{D}(:, j, :)$ corresponds to a basis element. The tensor coefficients are represented by the matrix $\mathcal{Z} \in \mathbb{R}^{r \times N_2 \times N_3}$. The parameter $\lambda'$ controls the trade-off between the approximation error and the sparsity of the tensor coefficients, while $r$ denotes the number of atoms in the dictionary.

The underground layers often exhibit stable depositional patterns, leading to consistent and laterally continuous seismic responses within the same seismic field [13]. As a result, both the low-granularity seismic data $\mathcal{Y}_l$ and high-granularity seismic data $\mathcal{X}_h$ from the same field tend to be sparse and approximate. Leveraging this property, we propose a joint training strategy to learn high-granularity dictionary $\mathcal{D}_h$ and low-granularity dictionary $\mathcal{D}_l$, by simultaneously training on the low-granularity data $\mathcal{Y}_l$ and high-granularity data $\mathcal{X}_h$. These dictionaries are constrained to share the same sparse representation $\mathcal{Z}$, ensuring that data from both granularities are reconstructed under a unified sparse representation. This leads to the following joint dictionary learning model:

$$\mathcal{D}_h = \underset{\mathcal{D}_h, \mathcal{Z}}{\operatorname{argmin}} \|\mathcal{X}_h - \mathcal{D}_h * \mathcal{Z}\|_F^2 + \lambda'\|\mathcal{Z}\|_1, \tag{9.5}$$

$$\mathcal{D}_l = \underset{\mathcal{D}_l, \mathcal{Z}}{\operatorname{argmin}} \|\mathcal{Y}_l - \mathcal{D}_l * \mathcal{Z}\|_F^2 + \lambda'\|\mathcal{Z}\|_1, \tag{9.6}$$

where $\mathcal{D}_h$ represents the high-granularity dictionary, and $\mathcal{D}_l$ represents the low-granularity dictionary. The low-granularity and high-granularity seismic data, $\mathcal{Y}_l$ and $\mathcal{X}_h$, are generated through the following steps:

By combining Eqs. (9.5) and (9.6), we enforce a consistent sparse representation for the training pairs, resulting in the following expression:

$$\min_{\mathcal{D}_h,\mathcal{D}_l,\mathcal{Z}} \frac{1}{P}\|\mathcal{X}_h - \mathcal{D}_h * \mathcal{Z}\|_F^2 + \frac{1}{Q}\|\mathcal{Y}_l - \mathcal{D}_l * \mathcal{Z}\|_F^2 + \lambda'\left(\frac{1}{P}+\frac{1}{Q}\right)\|\mathcal{Z}\|_1, \tag{9.7}$$

where $P$ and $Q$ denote the sizes of the first dimension in $\mathcal{X}_h$ and $\mathcal{Y}_l$, respectively. The factors $1/P$ and $1/Q$ are used to balance the contributions of the two cost terms from Eqs. (9.5) and (9.6).

Equation (9.7) can be converted to the following form:

$$\min_{\mathcal{D},\mathcal{Z}} \|\mathcal{X} - \mathcal{D} * \mathcal{Z}\|_F^2 + \lambda\|\mathcal{Z}\|_1 \tag{9.8}$$

where

$$\mathcal{X} = \begin{bmatrix} \frac{1}{\sqrt{P}}\mathcal{X}_h \\ \frac{1}{\sqrt{Q}}\mathcal{Y}_l \end{bmatrix}, \mathcal{D} = \begin{bmatrix} \frac{1}{\sqrt{P}}\mathcal{D}_h \\ \frac{1}{\sqrt{Q}}\mathcal{D}_l \end{bmatrix}, \lambda = \lambda'\left(\frac{1}{P}+\frac{1}{Q}\right) \tag{9.9}$$

To efficiently solve this optimization problem, we alternately update $\mathcal{D}$ and $\mathcal{Z}$ within the tensor domain. As a result, Eq. (9.8) is decomposed into two main subproblems: (i) computing the tensor sparse coefficients, which is addressed using the tensor-based iterative shrinkage thresholding algorithm (TFISTA) that leverages the tensor product; (ii) updating the tensor dictionary, which is optimized using the Lagrange dual method. The detailed solution process will be discussed in the subsequent sections.

## 9.3 Joint Tensor Dictionary Learning (JTDL)

To solve problem (9.8), we employ an alternating optimization approach, where one variable is fixed at each step while optimizing the other variable, thereby iteratively updating the tensor dictionary $\mathcal{D}$ and the coefficients $\mathcal{Z}$.

### *9.3.1 Tensor Coefficients Learning*

To solve for the tensor sparse coefficients, we fix the tensor dictionary $\mathcal{D} \in \mathbb{R}^{N_1\times r\times N_3}$ and apply the tensor-based iterative shrinkage thresholding algorithm [12], which efficiently computes the sparse representation $\mathcal{Z} \in \mathbb{R}^{r\times N_2\times N_3}$ for the seismic data $\mathcal{X} \in \mathbb{R}^{N_1\times N_2\times N_3}$. Consequently, problem (9.8) is reformulated as follows:

$$\min_{\mathcal{Z}\in\mathbb{R}^{r\times N_2\times N_3}} \frac{1}{2}\|\mathcal{X}-\mathcal{D}*\mathcal{Z}\|_F^2+\lambda\|\mathcal{Z}\|_1 \tag{9.10}$$

To facilitate computation, (9.10) can be expressed in matrix-product form using Lemma 1:

$$\min_{\underline{\mathcal{Z}}\in\mathbb{R}^{rN_2\times N_3}} \frac{1}{2}\left\|\underline{\mathcal{X}}-\underline{\mathcal{D}^c}\underline{\mathcal{Z}}\right\|_F^2+\lambda\|\underline{\mathcal{Z}}\|_1. \tag{9.11}$$

We employ an enhanced iterative shrinkage thresholding algorithm (ISTA-T), which is based on tensor product techniques [12], to directly solve (9.10). To achieve this, we first reformulate (9.10) as follows:

$$\min_{\mathcal{Z}} f(\mathcal{Z})+\lambda g(\mathcal{Z}) \tag{9.12}$$

In this approach, the objective function $f(\mathcal{Z})$ is defined as $\frac{1}{2}\|\mathcal{X}-\mathcal{D}*\mathcal{Z}\|_F^2$, which has a Lipschitz constant $L=\left\|\underline{(\mathcal{D}^{\dagger}*\mathcal{D})^c}\right\|_F^2$ [14]. Specifically, the constant $L$ can be expressed as $L=\sum_{k=1}^{N_3}\left\|\widetilde{\mathcal{D}}^{(k)^H}\widetilde{\mathcal{D}}^{(k)}\right\|_F^2$. The sparsity constraint term $\|\mathcal{Z}\|_1$ is represented as $\|\mathcal{Z}\|_1$. The variable $\mathcal{Z}$ is solved iteratively, and Eq. (9.12) is reformulated as a linearized function around the current estimate $\mathcal{Z}_p$, incorporating proximal regularization and nonsmooth regularization terms. Thus, at the $(p+1)$-th iteration, the update for $\mathcal{Z}_{p+1}$ is performed as follows:

$$\mathcal{Z}_{p+1}=\underset{\mathcal{Z}}{\operatorname{argmin}} f\left(\mathcal{Z}_p\right)+\left\langle\nabla f\left(\mathcal{Z}_p\right),\mathcal{Z}-\mathcal{Z}_p\right\rangle+\tfrac{L}{2}\left\|\mathcal{Z}-\mathcal{Z}_p\right\|_F^2+\lambda g(\mathcal{Z}), \tag{9.13}$$

where $\nabla f(\mathcal{Z})$ represents the gradient defined in the context of tensor space. Consequently, (9.13) can be expressed as:

$$\mathcal{Z}_{p+1}=\underset{\mathcal{Z}}{\operatorname{argmin}}\tfrac{1}{2}\left\|\mathcal{Z}-\left(\mathcal{Z}_p-\tfrac{1}{L}\nabla f\left(\mathcal{Z}_p\right)\right)\right\|_F^2+\tfrac{\lambda}{L}\|\mathcal{Z}\|_1, \tag{9.14}$$

Finally, the solution to (9.14) can be obtained by employing the technique of the proximal operator.

$$\mathcal{Z}_{p+1}=\operatorname{Prox}_{\beta}\left(\mathcal{Z}_p-\frac{1}{L}\nabla f\left(\mathcal{Z}_p\right)\right), \tag{9.15}$$

where $\operatorname{Prox}_{\beta}(\cdot)$ represents the soft-thresholding operator [15]

$$\operatorname{Prox}_{\tau}(\cdot)\rightarrow\operatorname{sign}(\cdot)\max(|\cdot|-\tau,0). \tag{9.16}$$

To enhance the efficiency of the ISTA-T, an extrapolation operator was employed [16].

### 9.3.2 Tensor Dictionary Learning

The learning process of the dictionary $\mathcal{D}$, with $\mathcal{Z}$ held constant, can be described by the following optimization problem:

$$\begin{aligned} &\min_{\mathcal{D}\in\mathbb{R}^{N_1\times r\times N_3}} \frac{1}{2}\|\mathcal{X}-\mathcal{D}*\mathcal{Z}\|_F^2\,, \\ &\text{s.t. } \|\mathcal{D}(:,j,:)\|_F^2 \le 1,\ j\in[r]\,, \end{aligned} \tag{9.17}$$

Due to the interaction between atoms through the $*$ operation, we apply DCT to transform Eq. (9.17) into the frequency domain, decomposing it into $k$ nearly independent problems, which are coupled only through the norm constraint, as shown below:

$$\begin{aligned} &\min_{\widetilde{\mathcal{D}}^{(k)},k\in[N_3]} \sum_{k=1}^{N_3} \left\|\widetilde{\mathcal{X}}^{(k)} - \widetilde{\mathcal{D}}^{(k)}\widetilde{\mathcal{Z}}^{(k)}\right\|_F^2 \\ &\text{s.t. } \sum_{k=1}^{N_3} \left\|\widetilde{\mathcal{D}}^{(k)}(:,j)\right\|_F^2 \le N_3,\,[j]\in r. \end{aligned} \tag{9.18}$$

Next, we apply the Lagrange dual [17] approach to solve Eq. (9.18) in the frequency domain. To begin, we reformulate it into the following Lagrangian form:

$$\begin{aligned} \mathcal{L}(\widetilde{\mathcal{D}},\Lambda) = &\textstyle\sum_{k=1}^{N_3} \left\|\widetilde{\mathcal{X}}^{(k)} - \widetilde{\mathcal{D}}^{(k)}\widetilde{\mathcal{Z}}^{(k)}\right\|_F^2 + \\ &\textstyle\sum_{j=1}^{r}\lambda_j\left(\sum_{k=1}^{N_3}\left\|\widetilde{\mathcal{D}}^{(k)}(:,j)\right\|_F^2 - N_3\right), \end{aligned} \tag{9.19}$$

where $\lambda_j \ge 0,\ j\in[r]$ are the dual variables, and $\Lambda = \text{diag}(\lambda)$.

Next, by analytically minimizing over $\widetilde{\mathcal{D}}$, we derive the optimal expression for $\widetilde{\mathcal{D}}$

$$\widetilde{\mathcal{D}}^{(k)} = \left(\widetilde{\mathcal{X}}^{(k)}\widetilde{\mathcal{Z}}^{(k)^H}\right)\left(\widetilde{\mathcal{Z}}^{(k)}\widetilde{\mathcal{Z}}^{(k)^H} + \Lambda\right)^{-1},\, k\in[N_3]\,. \tag{9.20}$$

By substituting Eq. (9.20) into the Lagrangian function $\mathcal{L}(\widehat{\mathcal{D}},\Lambda)$, we derive the corresponding Lagrange dual function $\mathcal{D}(\Lambda)$:

$$\mathcal{D}(\Lambda) = -\sum_{k=1}^{N_3}\text{Tr}\left(\widetilde{\mathcal{D}}^{(k)^H}\widetilde{\mathcal{X}}^{(k)}\widetilde{\mathcal{Z}}^{(k)^H}\right) - N_3\sum_{j=1}^{r}\lambda_j \tag{9.21}$$

The problem is solved using Newton's method. Once the dual variables are obtained, the dictionary can be recovered using Eq. (9.21).

## 9.4 Field Examples of JTDL

### *9.4.1 Experimental Setup*

To reconstruct the low-granularity seismic data, we first collected a three-dimensional high-granularity seismic dataset $\mathcal{X} \in \mathbb{R}^{100\times20\times100}$ from the same seismic field. By applying half-sampling, we obtained the low-granularity seismic data $\mathcal{Y} \in \mathbb{R}^{50\times20\times50}$. Next, we utilized a feature extraction operator $f_r$ to extract gradient and curvature features along three directions from the low-granularity data $\mathcal{Y}$, where $r = 1, 2, \cdots, 6$. By convolving the feature operators $f_r$ with $\mathcal{Y}$ along the six directions, we generated the low-granularity training dataset $\mathcal{Y}_l$. In the subsequent steps, both the high-granularity and low-granularity datasets were divided into small cubes of size $\mathbb{R}^{3\times3\times3}$ for processing. Specifically, the high-granularity dataset was constructed as $\mathcal{X}_h \in \mathbb{R}^{9\times10000\times81}$, and the low-granularity dataset as $\mathcal{Y}_l \in \mathbb{R}^{54\times10000\times81}$. A detailed illustration of the data construction process is provided in Fig. 9.2.

Next, we constructed a high-granularity dictionary $\mathcal{D}_h \in \mathbb{R}^{9\times512\times81}$ using the high-granularity seismic data $\mathcal{X}_h$, and a low-granularity dictionary $\mathcal{D}_l \in \mathbb{R}^{54\times512\times81}$ using the low-granularity seismic data $\mathcal{Y}_l$. The low-granularity data were generated by extracting six features, as illustrated in Figs. 9.3 and 9.4. Although the structures

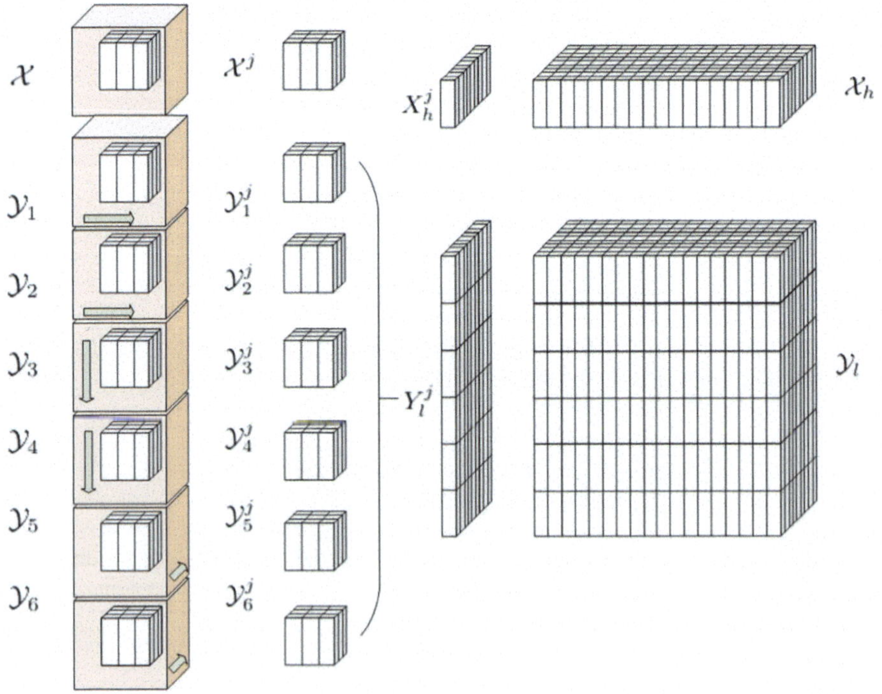

**Fig. 9.2** Creating training pairs for low- and high-granularity seismic data

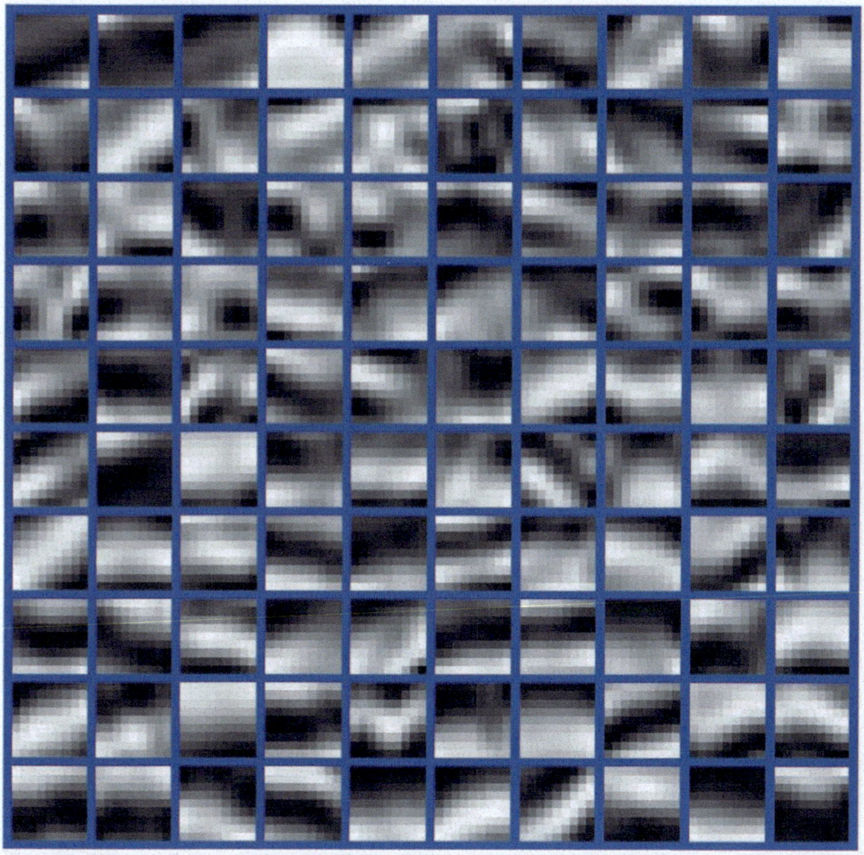

**Fig. 9.3** Dictionaries learned from low-granularity

of the two dictionaries are similar, the high-granularity dictionary captures finer and more precise features, reflecting the differences in data granularity.

To validate the effectiveness of the proposed seismic trace interpolation algorithm, we conducted experiments using data from a real-world seismic field and presented the interpolation results. Figure 9.3 shows the low-granularity seismic data, which were obtained by sampling the original high-granularity seismic data in both spatial and temporal dimensions, as shown in Fig. 9.4. Figure 9.5c presents the interpolation results generated using our method. In contrast, the original high-granularity seismic data were generated using a GAN [18] generator. The generated high-granularity data successfully reveals a weak reflector, which is crucial for seismic interpretation. Figure 9.5e shows the reconstructed single trace waveforms generated by both our method and the GAN [18], highlighting the reconstruction errors in the high-frequency components for both approaches.

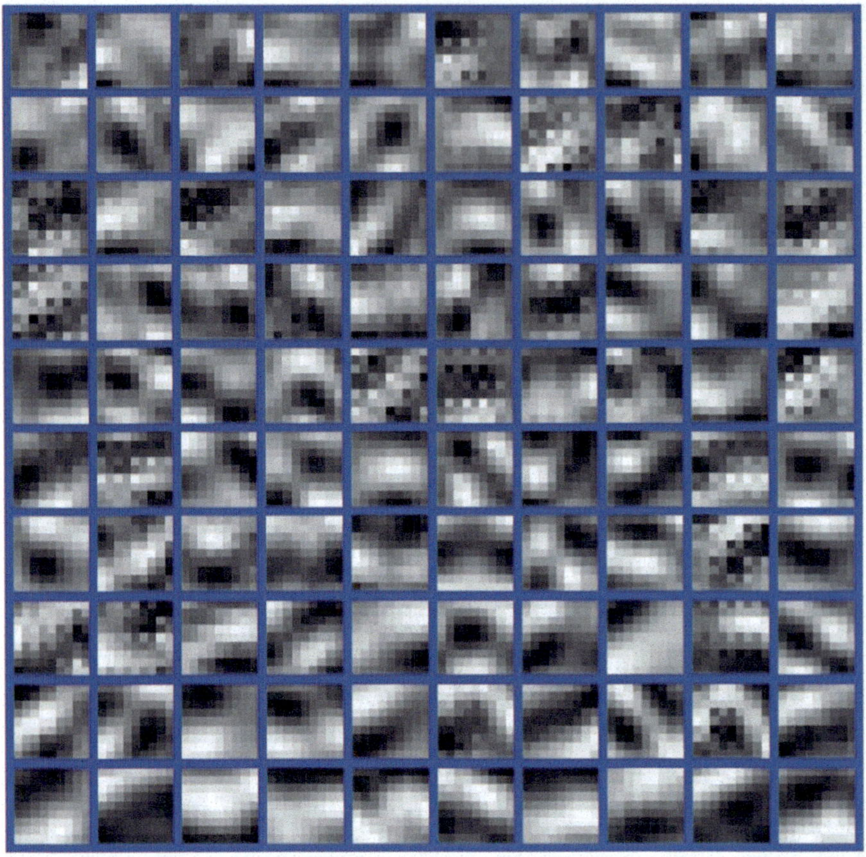

**Fig. 9.4** Dictionaries learned from high-granularity

## 9.5 Summary

This chapter presents an innovative tensor joint sparse coding approach and applies it to the task of 3D seismic trace interpolation. By leveraging tensor products, the method extends two-dimensional dictionary learning to three-dimensional space, thereby capturing the complex spatial information inherent in seismic data more effectively. The optimization process alternates between updating the tensor sparse coefficients and the tensor dictionary. Specifically, the ISTA-T algorithm is employed to compute the tensor sparse coefficients, while the tensor dictionary is updated using the Lagrange dual method. Experimental validation on real-field seismic data demonstrates that the proposed method significantly enhances the reconstruction quality of seismic traces and effectively improves the resolution of seismic data, highlighting its potential and advantages in seismic data processing.

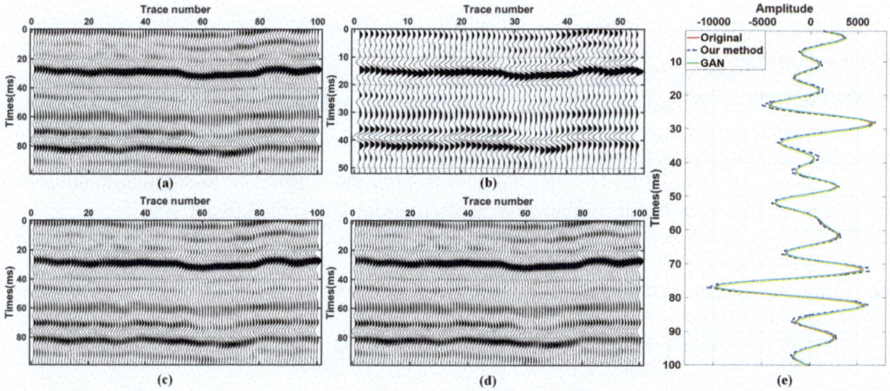


**Fig. 9.5** **a** Original high-resolution seismic data. **b** Low-resolution seismic data. **c** High-resolution seismic data reconstructed using the proposed method. **d** High-resolution seismic data generated by Generative Adversarial Networks (GAN). **e** Single trace seismic data reconstructed using the proposed method and GAN

## References


1. Yong-Gang Zhang, Yun Wang, and Jun-Jie Yin. Single point high density seismic data processing analysis and initial evaluation. *Shiyou Diqiu Wuli Kantan(Oil Geophysical Prospecting)*, 45(2):201–207, 2010
2. F Xiao, J Yang, B Liang, M Zhang, R Li, F Li, H Xiao, X Lei, Q Liu, and T Heesom. High-density 3d point receiver seismic acquisition and processing–a case study from the sichuan basin, china. *first break*, 32(1), 2014
3. R Soubaras, Rs Dowle, and R Sablon. Broadseis: Enhancing interpretation and inversion with broadband marine seismic. *CSEG Recorder*, 37(7):40–46, 2012
4. Thomas Rebert, Ronan Sablon, Nicolas Vidal, Pierre Charrier, and Robert Soubaras. Improving pre-salt imaging with variable-depth streamer data. In *SEG Technical Program Expanded Abstracts 2012*, pages 1–5. Society of Exploration Geophysicists, 2012
5. Yandong Wang, Jianhua Wang, Xiaoliu Wang, Wenbo Sun, and Jinmiao Zhang. Broadband processing key technology research and application on slant streamer. In *International Geophysical Conference, Beijing, China, 24-27 April 2018*, pages 135–138. Society of Exploration Geophysicists and Chinese Petroleum Society, 2018
6. James A Cadzow and XingKang Li. Blind deconvolution. *Digital signal processing*, 5(1):3–20, 1995
7. Ke Zhang, YD Li, GZ Liu, and YL Jia. Multiresolution seismic signal deconvolution. *CHINESE JOURNAL OF GEOPHYSICS-CHINESE EDITION*, 42(4):529–535, 1999
8. Stephen D Billings, Brian R Minty, and Garry N Newsam. Deconvolution and spatial resolution of airborne gamma-ray surveys. *Geophysics*, 68(4):1257–1266, 2003
9. Liu Cewen, Sun Mengyao, Dai Nanxun, Wei Wu, Wei Yanwen, Guo Mingjie, Haohuan Fu (2022) Deep learning-based point-spread function deconvolution for migration image deblurring. Geophysics 87(4):S249–S265
10. Li Yong, Zhang Gulan, Duan Jing, He Chengjie, Hao Du, Luo Fan, Zhan Yizong, Wang Jia (2020) Extended stable factor method for the inverse q-filter. Geophysics 85(3):T155–T163
11. Shang XM, Diao R, Feng YP, Zhao CX (2014) The application of spectral modeling method to high resolution processing of seismic data. Geophysical & Geochemical Exploration (in Chinese) 38(1):75–80

12. Fei Jiang, Xiao-Yang Liu, Hongtao Lu, and Ruimin Shen. Efficient multi-dimensional tensor sparse coding using t-linear combination. In *Proceedings of the AAAI Conference on Artificial Intelligence*, volume 32, 2018
13. She Bin, Wang Yaojun, Liang Jiandong, Liu Zhining, Song Chengyun, Guangmin Hu (2018) A data-driven amplitude variation with offset inversion method via learned dictionaries and sparse representation. Geophysics 83(6):R725–R748
14. Fei Jiang, Xiao-Yang Liu, Hongtao Lu, and Ruimin Shen. Efficient two-dimensional sparse coding using tensor-linear combination. arXiv preprint arXiv:1703.09690, 2017
15. Neal Parikh, Stephen Boyd, et al. Proximal algorithms. *Foundations and trends® in Optimization*, 1(3):127–239, 2014
16. Yangyang Xu and Wotao Yin. A fast patch-dictionary method for whole image recovery. arXiv preprint arXiv:1408.3740, 2014
17. Honglak Lee, Alexis Battle, Rajat Raina, and Andrew Ng. Efficient sparse coding algorithms. *Advances in neural information processing systems*, 19, 2006
18. Christian Ledig, Lucas Theis, Ferenc Huszár, Jose Caballero, Andrew Cunningham, Alejandro Acosta, Andrew Aitken, Alykhan Tejani, Johannes Totz, Zehan Wang, et al. Photo-realistic single image super-resolution using a generative adversarial network. In *Proceedings of the IEEE conference on computer vision and pattern recognition*, pages 4681–4690, 2017

# Chapter 10
# Conclusion and Future Research Directions

## 10.1 Concluding Remarks

In this book, we introduce the fundamentals of tensor computation and its applications in modern seismic data processing. There are various tensor computation methods from various backgrounds, with the core ones covered in this book including tensor-tensor product (t-product), tensor completion, tensor deep learning, and tensor dictionary learning. These techniques collectively enable seismic data processing methods to avoid vectorization or matricization high-dimensional seismic data. Through tensor computation, the seismic data processing tasks discussed in this book can be efficiently computed in tensor form, preventing potential damage to the high-dimensional geological structures that may occur in classical seismic data analysis. Therefore, this tensor-based approach yields better results than traditional matrix-based processing methods.

After introducing the basic concepts of tensor computation, we proceed to present some newly proposed modern seismic data processing methods tailored for tensor computation. Specifically, these methods include seismic data interpolation, seismic footprint removal, seismic noise attenuation, seismic erratic noise suppression, and seismic data super-resolution. For seismic data interpolation, we present a framework based on structured low-tubal-rank tensor completion. This approach integrates the advantages of both the Hankel structure and the low-tubal-rank constraint, aiming to improve the overall performance of the interpolation process. For seismic noise attenuation, we propose a new framework combining a SURE-based tensor convolutional neural network (TCNN) for denoising with an order-$p$ tNN model utilizing the framelet transform. This framework, referred to as FPTNN, is designed to tackle the difficulties in revealing the complete structural details embedded in both 3-D and multi-dimensional (M-D) seismic images. For suppressing seismic erratic noise, we introduce a robust tensor deep learning (RTDL) approach for unsupervised 3-D seismic noise reduction, which combines tensor sparse representation (SR) with a tensor neural network (tNN). For eliminating seismic footprints, we present a novel tensor

 
F. Qian et al., *Tensor Computation for Seismic Data Processing*,
Earth Systems Data and Models 6, https://doi.org/10.1007/978-3-031-78900-7_10

model tailored for 3-D random and footprint noise suppression, employing unidirectional total variation regularized low-rank tensor approximation (UTV-LRTA). Additionally, for enhancing seismic data resolution, we propose an innovative method to generate high-resolution 3-D seismic data from low-resolution inputs by using tensor sparse coding, which involves the joint training of both high-resolution and low-resolution dictionaries.

## 10.2 Potential Future Works

Tensor computation will be one of the significant trends in modern seismic data processing. Based on the supporting techniques and relevant application researches introduced in this book, it is evident that this trend is already underway. However, this book primarily focuses on the design of denoising and interpolation algorithms in modern seismic data processing. With the exciting prospects of tensor computation, there are several potential future directions and practical applications. These include, but are not limited to:

- *Application in more processing scenarios*: Although tensor computation has been successfully applied to seismic denoising and interpolation tasks in this book, its potential extends far beyond. Tensor computation can be expanded to more processing steps in seismic data processing, such as stacking, migration, deconvolution, and beyond. Even in other processing steps involving high-dimensional data manipulation, theoretically, tensor computation can achieve superior results compared to traditional matrix computation because it does not require the destruction of the original high-dimensional image structure during processing.
- *Employing more advanced DL models*: Tensor deep learning, when practically implemented, is decomposed into traditional matrix deep learning for solving. In this book, without loss of generality, we adopt relatively typical matrix deep learning networks, such as deep convolutional networks and deep convolutional autoencoder networks. Indeed, using more cutting-edge deep learning models undoubtedly enhances the effectiveness of tensor deep learning in handling high-dimensional seismic data, such as diffusion networks, self-attention mechanism networks, large models, and others.
- *Exploring more machine learning approaches*: Currently, most of the tensor computation applications to seismic data processing mainly involve supervised and unsupervised learning approaches. With the rapid development of machine learning theory, combined with the sample situation in seismic data processing, we can further extend tensor computation in seismic data processing to weakly supervised learning, transfer learning, self-supervised learning, and other learning approaches, thereby further improving the effectiveness and applicability of tensor-based seismic processing methods.

- *Developing more tensor computation models suitable for seismic data*: In this book, the tensor computation models primarily used in seismic data processing are based on the tSVD model, which can be extended to other tensor computation models, such as the TKD model, CPD model, and TT model. Additionally, seismic data exhibits higher-dimensional characteristics compared to ordinary data, ranging from 3D to 7D. Therefore, it is necessary to develop tensor computation models tailored to the practicalities of seismic data.
- *Expanding model-data-driven framework to more tensor computation models*: Although we have employed a model-data-driven tensor computation method for erratic noise suppression in Chap. 8, we can apply this mixed-driven framework to more tensor-based seismic data processing tasks. This is because model-data-driven methods can leverage both strictly valid model assumptions and data-driven approaches to avoid modeling errors where model assumptions are not strictly met while maintaining model interpretability.

# Appendix A
# Tensor Computation Software

There is an urgent need to explore and investigate tensor computation software packages in processing high-dimensional modern seismic data involving tensor decompositions, which is convenient for readers to develop and customize their own tensor-based seismic data analysis methods. With the flourishing development of tensor computation software, various feature-rich software packages have emerged as indispensable tools to drive applied research in tensor computation. Therefore, drawing inspiration from [1]'s work, it is necessary to conduct a review to select the most suitable software package for readers' usage, with detailed summaries in Table A.1.

In Table A.1, we observed that Python, MATLAB, and C/C++ are the primary languages for developing tensor computation software packages, and each language offers a different way of processing tensor data. For example, TensorLy [2] and tntorch [3] are developed for Python, TensorToolbox [4], Tensorlab [5] are developed for MATLAB, while Uni10 [6] and SPLATT [7] are developed using C/C++. This means that readers can choose different programming languages based on the convenience of these software packages, which helps readers accomplish their seismic data processing tasks through their usage habits.

These tensor computation software packages have been developed to focus on different application domains and tasks. However, no dedicated software package exists to carry out tensor continuation for seismic data processing, which might be one of our future directions. Fortunately, most existing packages can perform the requested tensor operations and decompositions of the seismic data processing tasks discussed in this book. Moreover, these software packages offer readers numerous tensor decompositions not covered in this book, providing possibilities for future task extensions. For instance, tntorch [3] and hottbox [8] can handle tensor regression, classification, and statistical analysis problems. Notably, no single software package integrates all required tensor operations on one platform. For this reason, we will turn to different platforms to achieve their diverse goals and refer to the corresponding user manuals.



F. Qian et al., *Tensor Computation for Seismic Data Processing*,
Earth Systems Data and Models 6, https://doi.org/10.1007/978-3-031-78900-7

**Table A.1** The landscape of software package for tensor computation

| Package name | Language | Package homepage | Description |
|---|---|---|---|
| TensorLy | Python | https://github.com/tensorly/tensorly | Designed to simplify tensor learning, offering straightforward tensor decomposition, learning, and algebra |
| tntorch | Python | https://tntorch.readthedocs.io/en/latest/ | Facilitates tensor modeling and learning with seamless support for various frameworks, including the tensor train (TT) model, CANDECOMP/PARAFAC (CP), the Tucker model, and others |
| TensorD | Python | https://github.com/litpuvn/tensorD | Includes fundamental decomposition techniques like Tucker and CP decompositions, along with newer methods such as Pairwise Interaction Tensor Decomposition |
| Scikit-tensor | Python | https://github.com/mnick/scikit-tensor | Multilinear algebra and tensor factorizations, supporting basic operations such as folding/unfolding and tensor-matrix/vector products, as well as factorizations including CP, Tucker, RESCAL, DEDICOM, and INDSCAL |
| hottbox | Python | https://github.com/hottbox/hottbox | Offers a suite of tools for multi-dimensional operations, tensor decompositions, and sophisticated algorithms designed for multi-linear classification and data fusion. These include methods such as the Support Tensor Machine (STM) and Tensor Ensemble Learning (TEL) |
| TT toolbox | MATLAB and Python | https://github.com/oseledets/TT-Toolbox | Performing basic operations on high-dimensional tensors using the efficient Tensor Train (TT) format |
| Tensor toolbox | MATLAB | https://www.tensortoolbox.org/ | Basic tensor operations and decompositions (such as CP and Tucker) for dense, sparse, and structured tensors |
| Tensorlab | MATLAB | https://www.tensorlab.net/ | Supplies a variety of tools for tensor operations, (coupled) tensor factorizations, and advanced optimization |
| TTeMPS | MATLAB | https://www.epfl.ch/labs/anchp/index-html/software/ttemps/ | Created for exploring algorithms in the Tensor Train (TT) and Matrix Product States (MPS) formats |
| TDALAB | MATLAB | https://github.com/andrewssobral/TDALAB | Delivers core data structures and methods for managing and processing tensor data |
| HT toolbox | MATLAB | https://www.epfl.ch/labs/anchp/index-html/software/htucker/ | Facilitates efficient tensor storage and manipulation in hierarchical Tucker format, while also providing tools for developing advanced algorithms |
| n-way toolbox | MATLAB | https://github.com/andrewssobral/nway | Offers methods for fitting multi-way PARAFAC models, multi-way PLS regression models, and multi-way Tucker models |
| Uni10 | C++ | https://gitlab.com/uni10/uni10 | Created for the advancement of tensor network algorithms |
| mptensor | C++ | https://github.com/smorita/mptensor | Offers standard operations such as tensor contraction and t-SVD |
| iTensor | C++ | https://itensor.org/ | Designing and executing cutting-edge tensor network computations |
| cuTensor | C++ | https://github.com/NVIDIA/CUDALibrarySamples | Fast implementation of operations related to t-SVD |
| SPLATT | C and OpenMP | https://github.com/ShadenSmith/splatt | Sparse tensor factorization: optimized and parallelized multiplication of sparse tensors and matrices |

## References

1. Liu Y, Liu J, Long Z, Zhu C (2022) Tensor computation for data analysis. Springer
2. Kossaifi J, Panagakis Y, Anandkumar A, Pantic M (2016) Tensorly: tensor learning in python. arXiv preprint arXiv:1610.09555
3. Usvyatsov M, Ballester-Ripoll R, Schindler K (2022) tntorch: Tensor network learning with PyTorch. J Mach Learn Res 23(208):1–6
4. Kolda TG, Bader BW (2006) Matlab tensor toolbox. Technical report, Sandia National Laboratories (SNL), Albuquerque, NM, and Livermore, CA
5. Vervliet N, Debals O, De Lathauwer L (2016) Tensorlab 3.0-numerical optimization strategies for large-scale constrained and coupled matrix/tensor factorization. In: 2016 50th Asilomar conference on signals, systems and computers. IEEE, pp 1733–1738
6. Kao Y-J, Hsieh Y-D, Chen P (2015) Uni10: an open-source library for tensor network algorithms. J Phys: Conf Ser (1):012040. IOP Publishing
7. Smith S, Ravindran N, Sidiropoulos ND, Karypis G (2015) SPLATT: efficient and parallel sparse tensor-matrix multiplication. In: 2015 IEEE international parallel and distributed processing symposium. IEEE, pp 61–70
8. Kisil I, Calvi GG, Dees BS, Mandic DP (2011) Hottbox: higher order tensor toolbox (2021). arXiv preprint arXiv:2111.15662

# Index





F. Qian et al., *Tensor Computation for Seismic Data Processing*, Earth Systems Data and Models 6, https://doi.org/10.1007/978-3-031-78900-7

The manufacturer's authorised representative in the EU is Springer Nature Customer Service Centre GmbH, Europaplatz 3, 69115 Heidelberg, Germany. If you have any concerns regarding our products, please contact ProductSafety@springernature.com

Printed and bound by CPI Group (UK) Ltd, Croydon, CR0 4YY

01/10/2026

02225969-0005